Stem Cell Applications in Disease and Health

STEM CELL APPLICATIONS IN DISEASE AND HEALTH

WESLEY B. BURNSIDES
AND
RAPHAEL H. ELLSLEY
EDITORS

Nova Science Publishers, Inc.
New York

For permission to use material from this book please contact us:
Telephone 631-231-7269; Fax 631-231-8175
Web Site: http://www.novapublishers.com

Library of Congress Cataloging-in-Publication Data

Stem cell applications in disease and ahealth / Wesley B. Burnsides and Raphael H. Ellsley(editors)
p. cm.
Includes bibliographical references and index.
ISBN 978-1-60456-255-2(hardcover)
1.Stem cells.2.Stem cells—Transplantion.I.Burnsides,Wesley B.II.Ellsley,Raphael H.
[DNLM: 1.Stem cells2. Stem cell Transplantation. QU 325 S82356 2008]
QH588.S83S7386 2008
616'.02774—dc22 2007052512

Published by Nova Science Publishers, Inc. ≃ New York

Contents

Preface

Stem cells are the foundation cells for every organ, tissue and cell in the body. They are like a blank microchip that can ultimately be programmed to perform any number of specialized tasks. Stem cells are undifferentiated, "blank" cells that do not yet have a specific function. Under proper conditions, stem cells begin to develop into specialized tissues and organs. Additionally, stem cells are self-sustaining and can replicate themselves for long periods of time. These unique characteristics make stem cells very promising for supplying cells to treat debilitating diseases like Alzheimer's disease, cancer, Parkinson's disease, type-1 diabetes, spinal cord injury, stroke, burns, heart disease, osteoarthritis and rheumatoid arthritis. This book gathers important leading-edge research in the field.

Expert Commentary - ES cells are expected to be a strong tool for regenerative medicine, although several problems should be resolved before their practical application. Among them, tumorigenesity of transplanted ES cells are one of the biggest problems for recipients. This is due to the immortality of ES cells even under standard culture conditions. Most somatic cells have limited proliferative potential under standard culture conditions and suffer permanent cell cycle arrest, called cellular senescence, after serial culture passages, while ES cells are only one exception as primary cells. One big question that arises is why ES cells are immortal as primary cells without genetic alteration. The author and associates recently reported that the metabolic shift to enhanced glycolysis would be critical to protect primary cells from the senescence effect of culture stress. Interestingly, ES cells show a similar metabolic property to that of immortalised cells, which would support their proliferation in vitro. These findings indicate that metabolic shift can be a good target to modulate the immortality of ES cells.

Short Communication A - This study sought to determine whether the THLE-5b immortalized non-tumorigenic human liver epithelial cell line possesses liver progenitor-like properties using cell-type specific markers and flow cytometry analysis following Hoechst 33342 staining. Under basic culture conditions, THLE-5b cells showed undifferentiated phenotypes. When grown as cell aggregates, THLE-5b cells exhibited not only up-regulation of hepatocyte-specific markers, but also the appearance of an hepatocyte-like ultrastructure and increased ammonia metabolic activity suggestive of hepatocytic maturation. Under these conditions, transcriptional factors such as hepatocyte nuclear factor (HNF)-1 and HNF-4α were also up-regulated or induced. In Matrigel culture of THLE-5b cells, expression of γ-glutamyl transpeptidase was up-regulated, suggesting a potential for biliary cell differentiation. THLE-5b cells also exhibited hepatocyte-specific functions in severe combined immunodeficiency mice. Conversely, side population (SP) cells have been shown to be highly enriched stem/progenitor cells. Flow cytometry analysis, following Hoechst 33342 staining, demonstrated that THLE-5b cells contain SP cells. In addition, these SP cells expressed a universal marker for stem cell ABCG2/BCRP1 mRNA. These results indicate that THLE-5b cells possess liver progenitor-like properties and imply that bipotent progenitor cell populations similar to THLE-5b cells may exist in the adult human liver.

Short Communication B - Mesenchymal Stem Cells (MSCs) are considered one of the leading adult stem cell candidates for future applications in regenerative medicine. The lead is partly accounted for by ease of expansion and unique immune properties. MSCs mediate immune-suppression and appear capable of being implanted across allogeneic barriers. Additionally, MSCs have been reported to express Major Histocompatibility Complex II (MHC-II) and act as antigen presenting cells. MSCs can also differentiate and transdifferentiate into cells of different tissues. Reduced ethical concerns underscore their potential for medical application in cell therapy. Studies in several laboratories have reported on the transdifferentiation of MSCs to functional neurons. The neurons have been characterized at the molecular, cellular, and functional levels. These studies show promise for neural regeneration and also bring an understanding of the interaction between the stem cells and regions of tissue injuries. However, one significant concern regarding successful development of these neurons is the immune-enhancing effects of MSCs. For example, MSC-induced neurons could probably re-express MHC-II and induce allorejection. This review discusses the immune properties of MSCs and their relevance during neurogenesis.

Chapter I - The rapid development of microarray technology has led to important progress in many biomedical disciplines. In just over a decade, the technology has evolved from first small-scale prints to commercial and custom platforms of truly genomic scale. Though numerous types of microarray platforms exist, cDNA and oligonucleotide-based DNA microarrays are considered being the platforms of choice for the range of applications. However, the success of every study depends upon a careful selection of a platform appropriate for a designed experiment. In this short review, various technological and analytical aspects of microarray technology are discussed; some advantages and drawbacks of individual platform types are compared.

Chapter II - Transplantation of stem and progenitor cells is approved for diseases of the central and peripheral nervous systems, cardiovascular pathology, endocrine diseases, retina and optic nerve pathology. However, results of similar investigations are usually indefinite and unconvincing. The negative consequences are held back in many cases. Existing problems of cell therapy (strict patient selection, host immune response, dispersion of transplanted cells in organism, their fast utilization, and the short-term effect of treatment) could be solved or minimized by means of using carriers – i.e., incubators (scaffolds) for cellular material. They allow questions of creation, growing and implantation of artificial organs, and prolongation of therapeutic effects of transplanted cells to be raised. For all this, physical-chemical stressors are permitted analogously to regulate activity of both pathogenic microbes and mammalian stem cells.

This chapter is devoted to: 1) the modulating effect of artificial surfaces, electrical impulses, visible light, some microelements, nano-scale particles and directed fluid flow under parameters approximate to physiological, on the vitality of microbial and mammalian colony forming cells; 2) certain applications of manipulations with microbial and mammalian cells in chronic disease treatments.

Chapter III - The mammalian central nervous system (CNS), as a non-renewable tissue, can be seriously affected by injuries and neurodegenerative diseases. Research carried out during the last decades revealed that neural stem cells persist within the adult brain, providing a source of neuronal and glial cell precursors throughout life. Yet, in mammals, including humans, adult neurogenesis occurs within restricted areas of the brain, giving rise prevalently to neuronal precursors that integrate inside two regions: the hippocampus and the olfactory bulb. Newly-born neurons of the olfactory bulb are generated in the forebrain subventricular zone (SVZ), now considered as the major stem cell reservoir in the CNS. In the SVZ, the neural stem cells have been identified as a subpopulation of astrocytes, originating during the postnatal period, from embryonic radial glia. This counterintuitive finding overlaps with the recent discovery that embryonic radial glia can themselves act as stem cells, capable of producing both neurons and glia during development. In the adult brain, stem cells reside in specialized 'niches' endowed with molecular and cellular signals capable of regulating their symmetric/asymmetric division, as well as the early specification and fate of their progeny. Although radial glia generally disappear early postnatally by transformation into parenchymal astrocytes, it has recently been demonstrated that some radial glial cells persist within the adult forebrain SVZ, hidden inside a dense astrocytic meshwork called 'glial tubes'. Some of these astrocytes occasionally divide to give rise to transit amplifying cells that actively proliferate, thus generating a rapidly expanding neuronal progeny. In parallel with restricted stem cell niches, widespread neural precursor cells in the CNS parenchyma retain the capability/potentiality to divide and possibly stemness properties. Despite true advancements in neural stem cell biology, the very mode, time, and pattern of cell division, as well as the mechanisms involved in its regulation within the stem cell niche and elsewhere are at present under investigation. Recent data indicate that both radial glia during development and astrocytes in the adult neurogenic niches play a dual role, participating both in the genesis of new cells (stem cell function) and in their regulation (paracrine/modulatory function). Also, the reactive behaviour of stem/progenitor cells in different CNS lesion paradigms has started to be elucidated. A further understanding of neural stem cell origin, and of mechanisms allowing the retention of stemness properties, could be fundamental in designing new strategies for brain repair, aimed at modulating in situ the endogenous sources of neuronal and glial progenitor cells. Nevertheless, since brain neurogenic niches show differences in different species and remarkably change through development and postnatally in order to adapt their function to the mature nervous tissue, a thorough knowledge of transformation from radial glia to neurogenic astrocytes across vertebrate and mammalian species is crucial in unraveling neural stem cell biology and regulation within the brain.

Chapter IV – The discovery of neural stem cells in the central nervous system of adult mammals in 1992 raised great expectations for regenerative therapy after lesions or degenerative diseases. A tremendous accumulation of in vitro data about the dynamics of neural stem cells occurred in the last decade. However, these results do not yet allow successful cell therapy, neither in animal models nor in humans. The major issue is to identify the tissular factors that commit stem cell progeny to a given fate, which is still unresolved. It is well established indeed that neural stem cell function is primarily determined by extrinsic factors from the microenvironment. But the nature of these regulators mostly remains to unravel.

The authors review current knowledge of in vivo competent signals of the micro-environnement, i.e. both matricial and diffusible ones. Also reviewed are the published responses of adult neurogenesis systems to in vivo experimental models of neurogenic proliferation. These in vivo models are likely to provide original cues for identification of extracellular regulators, especially the post-lesional ones which elicit massive and synchronous reactions of the stem cells. Informative models to date are axotomies, apoptosis-mediated lesions, and moderate ischemic damages, but also physiological contexts like stress and various hormonal challenges. Emerging mediators of in vivo recruitment of intrinsic stem cells are some growth factors and cytokines, inflammatory signals, astrocytic and microglial messengers.

The current state-of-the-art in heterotopic graft of lineage-labelled stem cells point to the expected requirement for combination with pharmacological treatment using competent chemical signals. The source of primary stem cells for grafting is a critical issue: embryonic versus adult-derived, neural versus non-neural (hemetopoietic) stem cells, human versus animal donors.

Altogether, many pertinent tracks are being opened to cell therapy progress, that all deserve basic research and animal model studies in the immediate future.

Chapter V - Cancer stem cell population within the solid tumor with indefinite proliferation potential drives the growth and metastasis of cancer. Under the model of cancer stem cells, it is suggested that present therapy strategy may not be efficient at targeting cancer stem cells and consequently, could not really cure these patients. Cancer stem cells exhibit low rate of division and proliferation in their niche that help them to withstand chemotherapy and radiation. Epithelial cancers are believed to originate from transformation of tissue stem cells. Bone marrow-derived cells, which are frequently recruited to sites of tissue injury and inflammation, might also represent a potential source of malignancy in the gastrointestinal tract. Pancreatic cancer is one of most common causes of cancer-related death. Pancreatic cancer stem cells have been characterized recently through serial transplantation of human pancreatic cancer cells. The phenotype of pancreatic cancer stem cells has been defined as $CD24^{+}CD44^{+}CD326(ESA)^{+}$ cells within the cancer. More recently, other cancer stem cells in the gastrointestinal tract, such as colon cancer stem cells and liver cancer stem cells, have been also characterized in their phenotype. These advances clearly will lead to a new strategy in cancer treatment and control in the gastrointestinal tract. In this chapter, the authors review the current status and progress of normal stem cells and cancer stem cell research in the gastrointestinal tract and discusses the implications of these studies in new cancer treatment strategy.

Chapter VI – The ability of self-renewal division of germline stem cells (GSCs) is responsible for continuous production of functional gametes in a gonad throughout an animal's lifetime. The developmental dynamics of germline cells are well studied using several model animals. Germline cells are specified by maternal inherited determinants (preformation) in the fly or by inductive signals (epigenesis) in the mouse during early embryonic development, respectively. After migrating into the presumptive gonadal regions, primordial germ cells (PGCs) acquire stem cell status. GSCs are located at the apical tip of the gonads by associating with neighboring somatic cells to form a 'niche' in the fly. Growth factors produced by niche cells or surrounding tissues are believed to function for the maintenance and division of GSCs. In addition to the biological interest in understanding the cellular and molecular mechanisms of the stem cell system, understanding of the mechanisms of stem cell behavior is crucial to therapeutic usage of these cells, since malfunction of these processes would result in infertility in humans. Recent advances in the establishment of stable cell lines of GSCs would provide new opportunities to study GSCs and their niche cells in detail. Furthermore, these in vitro systems have enabled genetic modification and cryoreservation.

Chapter VII - The cornea forms the clear front of the eye and its clarity is critical for normal visual perception. The corneal surface is covered by an epithelium which is continuous with that of the surrounding conjunctiva. The corneal epithelium is renewed by stem cells located in the transition zone between the cornea and conjunctiva known as the limbus. A loss of these so-called limbal stem cells results in the disease of limbal stem cell deficiency, most commonly caused by chemical burns to the eye. This disease is characterised by failure of corneal epithelial maintenance and encroachment of the corneal surface by the phenotypically different conjunctival epithelium and its associated blood vessels. Both these characteristics result in a painful and blinding disease. In this chapter, the theory of limbal stem cell maintenance of the corneal epithelium, limbal stem cell deficiency and its causes, the diagnosis of limbal stem cell deficiency and its various management options (including the use of ex vivo expanded limbal epithelium) will be discussed. In particular, the role played by the basic science of limbal stem cell biology in the management of limbal stem cell deficiency will be highlighted.

Chapter VIII - Heart failure is one of the most important cardiovascular health problems throughout the world, with high mortality, and there is a need to develop more effective therapeutic strategies to replace specialized treatment such as mechanical circulatory support or cardiac transplantation.

Cell therapy is currently attracting growing interest as a potential new means of improving the prognosis of patients with heart failure. Autologous skeletal myoblasts have been the first to be tested in clinical trials; however, cardiovascular researchers have recently explored many other cell types, including bone marrow cells, endothelial progenitor cells, mesenchymal stem cells, embryonic stem cells, and resident cardiac stem cells. While recent experimental studies and early-phase clinical trials seem to support the concept that cell therapy may enhance cardiac repair, many challenges remain before achieving this goal. Further studies should focus on finding the optimal donor cells and cell numbers for transplantation, the mechanism by which engrafted cells improve cardiac function, controlling the survival, proliferation and differentiation of transplanted cells, and the development of more efficient cell delivery techniques.

Chapter IX - Parkinson's disease (PD) is a progressive neurodegenerative disorder characterized by rigidity, bradykinesia, postural instability, and resting tremor. The major symptoms are related to the progressive loss of dopaminergic neurons in the substantia nigra pars compacta (SNc). Although approximately 95% of PD cases are sporadic, the recent discovery of specific mutations in genes (PARK) that cause familial PD has led to a new approach in the study of the disease. Identification of PARK genes has allowed study of the pathogenesis of PD at the molecular level, and led to a new approach in the study of the disease. Numerous recent studies suggest that PARK gene products, such as α-synuclein, Parkin, DJ-1, PINK1, ATP13A2, UCH-L1, or LRRK2, interact with each other and that their loss of function or gain of toxic functions play an important role in the pathogenesis of both familial and sporadic PD. Nonetheless, the cause and pathogenesis of PD remains unknown; mitochondrial dysfunction, oxidative damage, endoplasmic reticulum stress, failure of the ubiquitin-proteasome system, environmental factors, and genetic predisposition might all be involved. To investigate the molecular pathogenesis and the novel treatment of the disease, toxin-induced PD animal models and genetic animal models that mimic familial PD have contributed. A mouse model, a Drosophila model, a cell model, and a stem cell-derived neuron model carrying the disease-causing PARK gene with genetic modification have been generated. Using PD models, recent experimental studies have demonstrated that embryonic stem cells, neuronal stem cells, or bone marrow cells can be used as donor cells for transplantation to reverse symptoms. Moreover, neurogenesis in the SNc has been reported. Recently, molecular mechanisms for dopaminergic neuronal death in PD have been clarified and many advances in stem cell-based therapeutic approaches have been made. Nonetheless, PD involves multiple neuronal systems and the pathology is widespread: not only the SNc, but also the vagus, locus coeruleus, raphe, thalamus, amygdala, olfactory nuclei, and cerebral cortex are all involved. Progression of the disease is associated with pathological staging based on the topographical extent of the brain lesions. This review discusses molecular genetics and pathogenesis, experimental disease models, and the basis for the successful stem cell-based therapeutic approaches for PD.

Chapter X - Recently, stem cell technology has been developed and be exploited for clinical use. Among stem cells, ES cells have a high application potential as a source to produce therapeutic cells in vitro because of their high growth capacity and differentiation competence. Endoderm cells compose respiratory, digestive and urological organs such as thyroid, lung, liver, pancreas, intestine, prostrate, and bladder. In-vitro manufactured hepatocytes and insulin-producing cells via endoderm cells were expected for use in the treatment of liver diseases and diabetes mellitus. There are many protocols to produce ES cells-derived endoderm. The basic technologies can be categorized as 1. embryoid body, 2. transducing key gene, 3. coculture with stromal cells, 4. chemically defined medium, and 5. linage-tracing with specific markers. However, many methods are focused on an examination of the final differentiation step only. There are problems such as low reproducibility, low differentiation efficiency and lower function. The strategy of the author and associates is to monitor intermediate stages of in-vivo endoderm development. Increasing of intermediate stage cells growth and differentiation efficiency can improve the therapeutic cell differentiation efficiency. A higher function is expected using a differentiation system according to in vivo development. The author and associates have developed the endoderm-specific induction and monitoring system to identify the Goosecoid+ Sox17+ definitive endoderm stem cells with cell surface markers CXCR4 and E-cadherin. Moreover, purified definitive endoderm stem cells can differentiate into various visceral organ cells in each secondary induction assay.

Chapter XI - Hematopoietic stem cells (HSC) are a rare population of self renewing cells within the bone marrow, which constantly reproduce all hematopoietic lineages. The homing and migration capacities of HSC are of fundamental importance for successful transplantation during treatment of hematological diseases. Transplantation of HSC collected from peripheral blood has become more and more significant, requiring directed homing to the bone marrow after intravenous infusion. The multi-step process of HSC homing, involving adhesion, transendothelial migration and migration across the bone marrow stroma, depends on active migration along a gradient of the chemokine SDF-1α. The potential of HSC to undergo self-renewing cell divisions has stimulated investigations of ex vivo stem cell expansion, which would be valuable to improve the quality of stem cell transplants by increasing the number of stem cells. Hematopoietic stem and progenitor cells (HSPC) can be expanded in vitro by supplementation with hematopoietic cytokines, for instance, Flt3-Ligand and stem cell factor. However, analyses of migratory activity using a collagen matrix migration assay recently revealed a dramatic impact of various hematopoietic cytokines on the SDF-1α-directed migration of these ex vivo expanded HSPC.

Although most hematopoietic progenitors reside in the bone marrow, a small fraction of these motile cells constantly circulate via the blood vessel system during steady state homeostasis. In stress situations, like bleeding, organ injury and inflammation, the number of circulating HSPC increases dramatically. During inflammation and tissue damage, SDF-1α as well as other chemokines are released, attracting migrating HSPC to sites of injury, which thereupon participate in tissue repair. Due to their plasticity and pluripotency as well as their capacity to fuse with other cells, HSPC develop into hepatocytes, muscle and neuronal cell types. Thus, the question arises regarding what will be the outcome of these events of trans-differentiation and fusion will be, if a hematopoietic stem cell enters chronically inflamed or even tumor tissue. This directed migration and homing of HSPC to their target tissues underlies the influence of start and stop signals, exerted, apart from CXCR4/SDF-1α interaction, by various factors. Recent observations indicate a modulating effect of the chemokine IL-8 and even neurotransmitters on the migratory activity of hematopoietic stem cells. Further elucidation of these mechanisms may facilitate the selective manipulation of HSPC homing to their target tissues.

Thus, it appears that understanding the mechanisms influencing and regulating the migratory processes of HSPC, comprising homing, retention, release and mobilization, may allow the improvement of their clinical applications.

Chapter XII - There is considerable interest worldwide in being able to expand stem cells ex vivo in order to develop therapies for blood cancers and to repair tissues such as the heart and central nervous system. Such efforts rely on having a sound conceptual framework of the nature of stem cells and how they adopt terminal developmental pathways. Lessons learned from studies of blood cell development in the bone marrow, in particular how self-renewing hematopoietic stem cells give rise to lineage committed progenitors, have laid important foundations for our understanding of how stem cells generate differentiated progeny. However, several models have been proposed to describe hematopoiesis. Many models consider development of blood cell lineages to be an ordered process and a longstanding view has been that there is an early and irrevocable choice made between progenitors developing towards either the lymphoid or myeloid families of cells. Recent studies have identified new subsets of progenitor cells, each possessing particular combinations of lineage potentials – for example, Early Progenitors with Lymphoid and Myeloid potential (EPLM) and a myeloid/lymphoid progenitor lacking erythroid and megakaryocyte lineage potential. These and other progenitors having both lymphoid and myeloid potential would seem to negate a strict lympho/myeloid dichotomy. The notion that progenitors possessing certain combinations of differentiation options indicate which fates are developmentally closely related led the authors to propose some time ago a Sequential Determination model whereby stem cells undergo stepwise progression through binary options all the way from megakaryocytes to T lymphocytes. In this article, the authors review recent developments in the field of hematopoiesis with particular emphasis on how they can be incorporated into the Sequential Determination model.

In adult mammals, hematopoietic stem cells (HSCs) in the bone marrow generate neutrophils, basophils, eosinophils, macrophages, at least two types of dendritic cells, several types of B and T lymphocytes, natural killer cells, megakaryocytes/platelets, erythrocytes and maybe others. Except for T lymphocytes, all these cell types are generated in large numbers throughout life. As to unravelling how stem cells in general can give rise to a number of terminal differentiated progeny, Don Metcalf wrote: "If paradise exists for an experimental biologist, then hematopoiesis has surely got to be it".

Chapter XIII - The use of embryonic stem (ES) cells in regenerative medicine is one of the most important promises of this decade. The knowledge about culture medium, isolation protocols and pluripotency-related genes is growing, but their therapeutic use is still restricted by a risk of teratoma formation. To test the hypothesis that some undifferentiated cells remain after the differentiation culture process, the authors have electroporated mouse ES cells with a construct that comprise the mTert promoter coupled to green fluorescent protein (GFP). When ES cell colonies are in culture, they express the green fluorescent protein as a consequence of their telomerase activity. After LIF removal a natural differentiation process into embryoid bodies (EB) occurs, so the GFP signal gradually diminishes due to this differentiation. However, after two weeks of incubation, small groups of fluorescent cells remaining in the differentiated monolayer were observed. After isolation and expansion of these groups under ES cell culture conditions, ES cell morphology was recovered. Molecular analysis showed that after several passages in ES medium, pluripotency quality characteristics of these cells are completely recovered. According to the authors' results, specific screening of differentiated cells before their therapeutic use should be done to avoid the risk of teratoma formation stimulated by an adequate environment.

In: Stem Cell Applications in Disease and Health
Editors: W. Burnsides and R. Ellsley

ISBN: 978-1-60456-255-2

Expert Commentary

Commentary on Stem Cell Applications in Diseases: The Metabolic Property of ES Cells, the Key to Escape from Culture Stress

Hiroshi Kondoh*

Department of Geriatric Medicine, Graduate School of Medicine
Kyoto University, 54 Kawahara-cho, Shogoin, Sakyo-ku,
Kyoto 606-8507, Japan

Abstract

ES cells are expected to be a strong tool for regenerative medicine, although several problems should be resolved before their practical application. Among them, tumorigenesity of transplanted ES cells are one of the biggest problems for recipients. This is due to the immortality of ES cells even under standard culture conditions. Most somatic cells have limited proliferative potential under standard culture conditions and suffer permanent cell cycle arrest, called cellular senescence, after serial culture passages, while ES cells are only one exception as primary cells. One big question that arises is why ES cells are immortal as primary cells without genetic alteration. We recently reported that the metabolic shift to enhanced glycolysis would be critical to protect primary cells from senescence effect of culture stress. Interestingly, ES cells show a similar metabolic property to that of immortalised cells, which would support their proliferation in vitro. These findings indicate that metabolic shift can be a good target to modulate the immortality of ES cells.

* E-mail address: hkondoh@kuhp.kyoto-u.ac.jp, Tel 0081(0) 75-751-3465, Fax 0081(0) 75-771-9784(Corresponding author)

1. Immortality of ES Cells: Barrier against Their Clinical Application

There is considerable hope that ES cells are promising donor sources for treating several human diseases. They are now facing several critical problems to be overcome: immune rejection after transplantation, ethical issues regarding the usage of human embryos, complicated culture process, and so on. One possible resolution for these concerns is to establish pluripotent stem cells directly from patients' somatic cells. In recent works, several groups reported that murine stem-like cells could be induced from mouse fibroblasts after retrovirus-mediated introduction of four transcription factors (Oct3/4, Sox2, c-Myc, and Klf4) [1, 2]. Although these induced pluripotent stem cells, called iPS cells, can be good tools for cell transplantation therapy, they still need to be better characterized and improved. One big problem of iPS cells is their observed tumourigenesity in recipients, indicating that genetic manipulation of oncogenes or tumour suppressor genes to drive stem cells are not favourable for these purposes [3].

Indeed, teratoma formation or tumorigenesity derived from transplanted cells are one of the biggest problems also in the clinical application of ES cells. This is mainly due to the immortality of both murine and human ES cells under standard tissue culture conditions, unlike other primary cells [4]. Stem cells are characterized by their immortality and indefinite self-renewal potential while retaining their ability to differentiate. This ability of ES cells to self-renew indefinitely is lost upon differentiation, suggesting that the immortality of ES cells is strictly regulated by unknown intrinsic mechanisms.

Thus, differentiation of ES cells and modulating their immortality are becoming central concerns for ES cell research. The first question that arises is how ES cells maintain their immortality without any tumorigenic genetic alteration. ES cells are genome-protected as primary cells, while immortalization requires several genetic alterations in any other primary cell. To understand why ES cells are immortal, it would be useful to explore and compare the immortalization of other primary cells with that of ES cells.

2. Resistance to 'Culture Shock' Observed in ES Cells

Since Hayflick's discovery [5], it is widely known that most primary cells have limited replicative potential in vitro. Upon reaching replicative exhaustion, cells suffer permanent cell cycle arrest, called replicative senescence. This phenomenon is mainly due to telomere erosion, which can be bypassed by ectopic expression of telomerase. Recent works suggest that telomere-independent senescence can be also induced by several environmental stressors; cell culture condition, oxidative damage, oncogenic activation and so on. Among others, murine primary cells are quite sensitive to culture stress and thought to be a good example of culture stress-induced senescence for the following reasons [6]:

1). Normal aboratory mice have exceptionally long telomeres compared with those of humans: 60kb vs 12kb, respectively [7].

2). Mouse embryo fibroblasts (MEFs) with intact telomerase activity generally have a very short lifespan of only 10-20 population doublings under normal culture conditions [8].
3). Lifespan of MEFs is expanded by inactivation of two major tumor suppressor axes; Arf/p53 and Ink4/Rb pathway.
4). The modification of tissue culture conditions such as the use of serum-free medium [9] or culture in low oxygen conditions [10] can avoid this premature senescence in MEFs.
5). Finally, through senescence-bypassing screening, it was found that metabolic shift into enhanced glycolysis is enough to immortalise MEFs [11].

The above-described senescence response observed in MEFs has been attributed to 'culture shock' [12] and is not observed in ES cells, which argues for additional mechanisms besides telomere regulation involved in ES cell immortality. In other words, ES cells are only one exception among primary cells that can bypass telomere–independent senescence in vitro by unknown mechanisms. There are possible mechanistic clues to explain immortality of ES cells as primary cells.

For example, co-expression of four defined transcriptional factors (Oct3/4, Sox2, c-Myc, and Klf4) can transform MEFs into murine stem-like cells. The other works also indicated that tumor suppressor pathways are blunted in ES cells [13], which might explain their immortality. But these genetic manipulations in ES cells are tumorigenic and not favourable for the purpose of their clinical application.

3. Modulation of ES Cell Immortality by Metabolic Shift

It would be most ideal in the clinical application of ES cells if we could modulate their immortality without any tumorigenic genetic alteration. One possible approach is to target the metabolic property of primary cells to modulate their lifespan.

We recently found the common metabolic profile shared between ES cells and immortalized primary cells: enhanced glycolysis with reduced mitochondrial respiration [14]. First, in a senescence bypassing screening in MEFs using a retroviral cDNA library, we isolated the glycolytic enzyme phosphoglycerate mutase (PGM) and glucophosphate isomerase (GPI) as immortalizing clones [11]. Ectopic expression of PGM or GPI increases glycolytic flux and extends the life span of primary MEFs. Conversely, knockdown of PGM or GPI via specific siRNA induces premature senescence. Secondly, we also found that murine ES cells display increased glycolysis, which is essential to support their proliferation [14]. When glycolysis was inhibited by adding a low dose of 2-deoxyglucose (2DG), MEFs were not affected, and most ES cells died. Once differentiated by the removal of cytokine LIF from culture medium, mouse ES cells stopped proliferation and their glycolytic flux reduced to half. These findings suggest that enhanced glycolysis could play a very important role in the proliferative potential of ES cells, not only in other primary cells. Finally, we found enhaced glycolysis can protect primary cells and ES cells from the senescence effect of oxidative damage under standard culture condition [11, 14]. It may be partly because primary cells showing enhanced glycolysis suffer less oxidative damage with less mitochondrial consumption of oxygen in culture.

Thus, if we can modulate metabolic property, especially glycolysis in ES cells, it would be much easier and more practical to apply them as a donor source. First, the maintenance of enhanced glycolysis will help to propagate sufficient populations of ES cells in culture. Then, after their differentiation into target tissue, the mild inhibition of glycolysis will terminate residual undifferentiated ES cells. This protocol will render recipients free from risky tumorigenesis of transplanted ES cells. Future plans should aim for the clarification of the detailed regulatory mechanism of glycolysis and identification of regulatory chemicals against glycolysis for these purposes.

References

[1] Takahashi, K., and S. Yamanaka. (2006). Induction of pluripotent stem cells from mouse embryonic and adult fibroblast cultures by defined factors. *Cell* 126:663-76.

[2] Wernig, M., A. Meissner, R. Foreman, T. Brambrink, M. Ku, K. Hochedlinger, B. E. Bernstein, and R. Jaenisch. (2007). In vitro reprogramming of fibroblasts into a pluripotent ES-cell-like state. *Nature* 448:318-24.

[3] Okita, K., T. Ichisaka, and S. Yamanaka. (2007). Generation of germline-competent induced pluripotent stem cells. *Nature* 448:313-7.

[4] Kondoh, H., and J. Gil. (2006). Protection from the senescence effects of oxidative stress in ES cells. *Stem Cell Therapy*:175-190.

[5] Hayflick, L., and P. S. Moorhead. (1961). The serial cultivation of human diploid cell strains. *Experimental Cell Research* 25:585-621.

[6] Kondoh, H., M. E. Lleonart, J. Gil, D. Beach, and G. Peters. (2005). Glycolysis and cellular immortalization. *Drug Discovery Today* 2:263-267.

[7] Serrano, M., and M. A. Blasco. (2001). Putting the stress on senescence. *Current Opinion in Cell Biology* 13:748-53.

[8] Wright, W. E., and J. W. Shay. (2002). Historical claims and current interpretations of replicative aging. *Nat Biotechnol* 20:682-8.

[9] Loo, D. T., J. I. Fuquay, C. L. Rawson, and D. W. Barnes. (1987). Extended culture of mouse embryo cells without senescence: inhibition by serum. *Science* 236:200-2.

[10] Parrinello, S., E. Samper, A. Krtolica, J. Goldstein, S. Melov, and J. Campisi. (2003). Oxygen sensitivity severely limits the replicative lifespan of murine fibroblasts. *Nat Cell Biol.* 5:741-7.

[11] Kondoh, H., M. E. Lleonart, J. Gil, J. Wang, P. Degan, G. Peters, D. Martinez, A. Carnero, and D. Beach. (2005). Glycolytic enzymes can modulate cellular life span. *Cancer Res* 65:177-85.

[12] Sherr, C. J., and R. A. DePinho. (2000). Cellular senescence: mitotic clock or culture shock? *Cell* 102:407-10.

[13] Miura, T., M. P. Mattson, and M. S. Rao. (2004). Cellular lifespan and senescence signaling in embryonic stem cells. *Aging Cell* 3:333-43.

[14] Kondoh, H., M. E. Lleonart, Y. Nakashima, M. Yokode, M. Tanaka, D. Bernard, J. Gil, and D. Beach. (2007). A high glycolytic flux supports the proliferative potential of murine embryonic stem cells. *Antioxid Redox Signal* 9:293-9.

In: Stem Cell Applications in Disease and Health
Editors: W. Burnsides and R. Ellsley

ISBN: 978-1-60456-255-2

Short Communication A

Evidence Supporting the Existence of Liver Progenitor Cells in Human Livers

Takayoshi Tokiwa[1,*], Taisuke Yamazaki[1] and Takashi Tsukiyama[2]
[1]Department of Liver Cell Biology, Kohno Clinical Medicine Research Institute
Kitashinagawa 1-28-15, Shinagawa
Tokyo 140-0001, Japan
[2]Kitashinagawa Hospital, Kohno Clinical Medicine Research Institute
Tokyo, Japan

Abstract

This study sought to determine whether the THLE-5b immortalized non-tumorigenic human liver epithelial cell line possesses liver progenitor-like properties using cell-type specific markers and flow cytometry analysis following Hoechst 33342 staining. Under basic culture conditions, THLE-5b cells showed undifferentiated phenotypes. When grown as cell aggregates, THLE-5b cells exhibited not only up-regulation of hepatocyte-specific markers, but also the appearance of a hepatocyte-like ultrastructure and increased ammonia metabolic activity suggestive of hepatocytic maturation. Under these conditions, transcriptional factors such as hepatocyte nuclear factor (HNF)-1 and HNF-4α were also up-regulated or induced. In Matrigel culture of THLE-5b cells, expression of γ-glutamyl transpeptidase was up-regulated, suggesting a potential for biliary cell differentiation. THLE-5b cells also exhibited hepatocyte-specific functions in severe combined immunodeficiency mice. Conversely, side population (SP) cells have been shown to be highly enriched stem/progenitor cells. Flow cytometry analysis, following Hoechst 33342 staining, demonstrated that THLE-5b cells contain SP cells. In addition, these SP cells expressed a universal marker for stem cell ABCG2/BCRP1 mRNA. These results indicate that THLE-5b cells possess liver progenitor-like properties and imply that

[*] E-mail address: t.tokiwa@kcmi.or.jp. Tel & Fax: 81-3-3474-1833 (Corresponding author)

bipotent progenitor cell populations similar to THLE-5b cells may exist in the adult human liver.

Key Words: immortalized human cell lines, non-tumorigenic, liver progenitor cells, SP cells

Introduction

Identifying and characterizing liver stem/progenitor cells offers major therapeutic opportunities in fields as diverse as cell transplantation, cell therapy, and liver reconstruction. True adult liver stem/progenitor cells have not yet been characterized, but are generally accepted to be present in adult rats and other animals [1-3]. However, whether stem/progenitor cells exist in the human liver has been controversial [4,5]. This controversy may be partly resolved with the use of human liver progenitor cell lines. We have previously developed the SV40 T antigen-immortalized, non-tumorigenic liver epithelial cell line, THLE-5b, from an adult human liver [6] and found that cells were morphologically and functionally immature, suggesting a potential to differentiate into a hepatocyte phenotype [7]. Cell-type specific markers are often used to identify liver stem/progenitor cells. Using these markers, we demonstrated that a cell line isolated from scattered cell clusters in primary cultures of adult porcine livers is bipotential, able to differentiate into hepatocytes or bile-duct cells [1,3].

Conversely, side-population (SP) cells have been shown to be highly enriched stem cells and are identified by flow cytometry on the basis of a unique ability to efflux Hoechst 33342 dye [8]. Hepatic oval cells exhibit the SP phenotype [9], and Hussain et al. [5] reported that SP cells derived from human livers possess the potential of hepatic stem cells. The present study examined whether THLE-5b cells possess liver progenitor-like properties using cell-type specific markers and flow cytometry analysis following Hoechst 33342 staining, to obtain evidence supporting the existence of liver progenitor cells in the human liver.

Materials and Methods

Cells

THLE-5b cells were CK18-positive, but did not show typical epithelial morphology (Fig. 1A). THLE-5b cells were plated in soft agar (10^5 cells/dish) to determine whether the cells are anchorage-dependent, and failed to grow after 4 weeks [unpublished data].

For tumorigenity testing, 5 million cells were injected subcutaneously into Balb/c nu/nu mice, and no tumors were found after 6 months [6]. THLE-5b cells were certified as being the designated type, free of mycoplasma contamination, and have been used from young stock. Human hepatoma cell lines, HuH-7 and HepG2, were obtained from the Japanese Cancer Research Resource Bank and used as controls. Cultures were grown in culture medium RPMI 1640 (Invitrogen Corp., Carlsbad, CA) supplemented with 10% fetal calf serum (FCS, Invitrogen), and antibiotics in a humidified atmosphere of 5% CO_2 and 95% air at 37°C.

Aggregate Formation

After reaching confluence, THLE-5b cells were trypsinized and the obtained cell suspensions were filtered through a 150-μm nylon filter. Approximately 5 x 10^5 cells/ flask were then inoculated into 25-ml Meyer's flasks containing 3 ml of culture medium DMEM/F12 (Invitrogen). Meyer's flasks were precoated with 2.5% poly(2-hydroxyethyl methacrylate) (poly-HEMA, Sigma Chemical Co., St. Louis, Mo.) in 95% ethanol as described previously [4]. Cell aggregate formation was carried out under the following two conditions:

i. static culture on poly-HEMA substrate (aggregate culture-1); and
ii. rotation culture, in which the Meyer's flasks were rotated at 70 rpm on an orbital shaker in a chamber, on poly-HEMA substrate (aggregate culture-2).

The culture was maintained in a humidified atmosphere of 5% CO_2 and 95% air at 37°C. Cell aggregates formed were removed after 24 hours from the flasks and used for the experiments.

Immunocytochemistry

Cells in monolayers were fixed in acetone/methanol (1:1) and cell mass was fixed in a 10% buffered formalin solution. The cell mass was then dehydrated, cleared in xylene, and embedded in paraffin. Sections were prepared for immunocytochemistry [4]. We used antibodies against human albumin, rabbit human α1-antitrypsin, human CK19, human CK18, and human CK14.

RNA Extraction, Reverse Transcription and Polymerase Chain Reaction (RT-PCR)

Total RNAs were isolated as described previously [4]. The cDNAs were synthesized by first strand cDNA kit (ReverTraAce-α, Toyobo Co., LTD, Osaka, Japan). PCR amplification was performed with a 5 μl aliquot of reverse transcriptase reaction product [except to amplify the CK19 (1 μl of 1st round product)], in a total volume of 50 μl and in the presence of 0.2 mM dNTP, 0.25 μM primers and 1.25 U Taq DNA polymerase (MBI Fermentas) and using a 25 to 40 cycle program (CK19: 20 cycles for 1st round, 15-17 cycles for 2nd round) consisting of denaturation at 94 °C for 1 min, annealing 55°C for 1 min, and extension at 72°C for 2 min. The PCR primer sequences specific for albumin, α-fetoprotein (AFP), transferrin, α1-antitrypsin, CYP1A1, CYP1A2, CYP3A4, tryptophane 2,3-dioxygenase (TO), E-cadherin, γ-glutamyl transpeptidase (GGT), Thy-1, c-kit, CK19, HNF-1, HNF-4α, HNF-6, and glyceraldehydes-3-phosphate dehydrogenase (G3PDH) were described previously [4]. The PCR primers for the ATP-binding cassette transporter ABCG2/BCRP1 are as follows: 5'-agttccatggcactggccata-3'(sense), 5'-tcaggtaggcaattgtgagg-3' (antisense).

Cell Transplantation

THLE-5b cells were attached to microcarrier beads (Cytodex 3^{TM}, Amersham Pharmacia) by incubating 10^7 cells per 1 ml of swollen beads overnight at 37°C. These cells were injected intraperitoneally into severe combined immunodeficiency mice in the Balb/c background (SCID) as described elsewhere [10]. Twenty one days after transplantation, the mass of cells adjacent to beads was submitted to the immunocytochemistry test as mentioned above.

SP Cell Sorting by Flow Cytometry

After reaching semi-confluence, THLE-5b cells were trypsinized and resuspended at 1×10^6/ml in pre-warmed RPMI 1640 medium supplemented with 2 % FCS and 25 mM *N*-2-hydroxyethylpiperazine-*N'*-2-ethane-sulfonic acid (Hepes). The cells were filtered through a 40 μm cell strainer, and then incubated with 10 μg/ml Hoechst 33342 dye (Sigma) for 2 h at 37° C in a water bath with gently agitation every 15 min. To identify SP, the cells were pre-incubated with 50 to 200 μM verapamil (Sigma), ABC transporter inhibitor, for 20 min at 37° C before adding Hoechst 33342 dye. After incubation, the cells were centrifuged and suspended in cold-Hanks' balanced saline solution (HBSS) supplemented with 2 % FCS and 10 mM Hepes containing 2 μg/ml propidiumiodide for dead cells discrimination. These cells were kept on ice until flow cytometry analysis is performed. The viable cells were analyzed for Hoechst 33342 dye exclusion by EPICS ARTLA (Beckman Coulter). Hoechst 33342 dye was excited with the UV laser at 350 nm, and its fluorescence emission was measured with two wave lengths using a 450 DF10 and a 675LP optical filter for detection of Hoechst blue and red, respectively. The SP cells were visualized with tight gating based on the verapamil control.

Results

Phenotypic properties of THLE-5b cells under basic culture conditions THLE-5b cells expressed CK18 (Fig. 1A), CK19 (Fig. 1D) and CK14 (data not shown), but did not express albumin and α1-antitrypsin (Figs. 1B, 1C). In addition, the cells also expressed oval cell/hematopoietic markers Thy-1 and c-kit (data not shown).

Culture as Cell Aggregates

1. Formation of Aggregates

THLE-5b cells formed cell aggregates within 1-2 days of inoculation. The aggregation pattern in aggregate culture-1 differed substantially from that in aggregate culture-2 (data not shown). In aggregate culture-1, smaller aggregates were formed than those formed in aggregate culture-2. THLE-5b cells formed compact aggregates in aggregate culture-2.

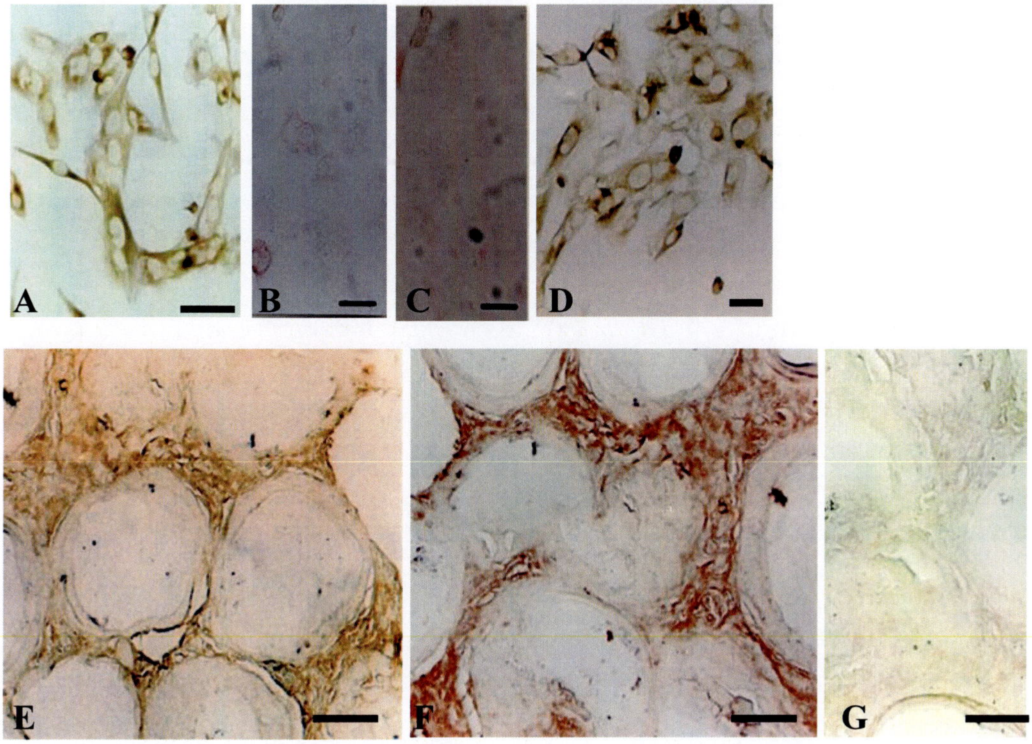

Figure 1. Immunocytochemistry of THLE-5b cells. A-D) Cells under basic culture conditions. E-G) Cell mass adjacent to microcarrier beads in SCID mice after 3 weeks. **A)** CK18; **B, E)** albumin; **C, F)** α1-antitrypsin; **D,G)** CK19. Cells are not counterstained. Scale bar, 25 μm.

2. Gene *Expression*

By RT-PCR analysis, expression of CYP3A4, TO and E-cadherin mRNA of THLE-5b cells in aggregate culture-2 was up-regulated or induced, whereas that of TO and E-cadherin mRNA was slightly up-regulated in aggregate culture-1. The level of expression of these genes was considerably higher in aggregate culture-2 than in aggregate culture-1 (Fig. 2A).

Expression of albumin and transferrin mRNA was induced in aggregate culture-1 as well as in aggregate culture-2, although that of α1-antitrypsin and CYP1A1 mRNA was up-regulated only in aggregate culture-2. No expression of AFP and CYP1A2 mRNA was found in either of the culture conditions examined (data not shown). These results indicate that hepatocytic differentiation of THLE-5b cells with a few exceptions is promoted by cell aggregation (Fig. 2B). In aggregate culture-2, expression of GGT mRNA was down-regulated, although the level of CK19 mRNA was only slightly down-regulated (Fig. 2C). On the other hand, under these conditions, expression of HNF-1, HNF-4α and HNF-6 mRNA was up-regulated or induced in aggregate culture-1 and/or aggregate culture-2 (Fig. 2D). These data are summarized in Table 1.

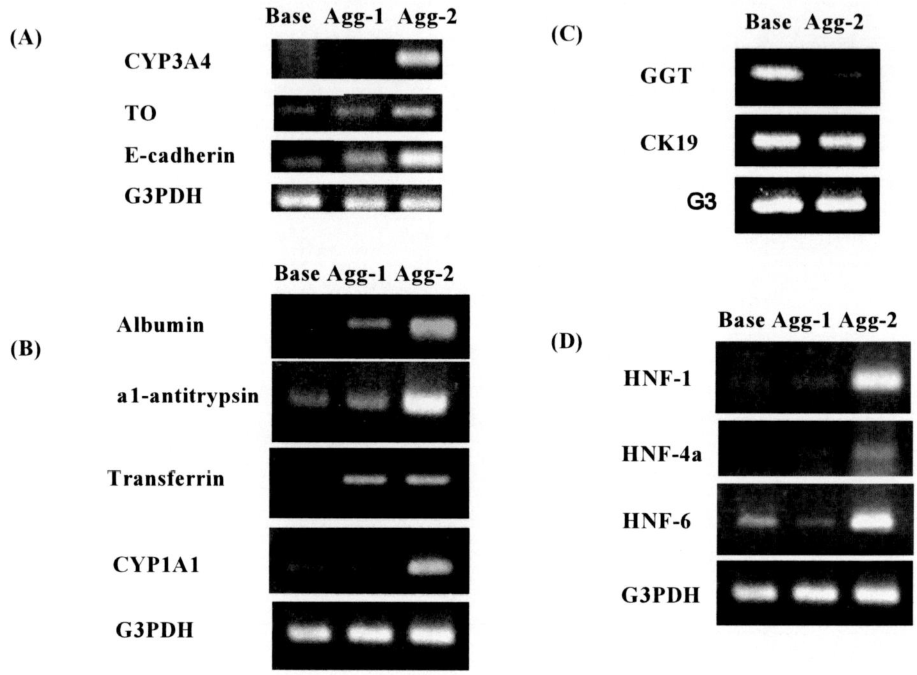

Figure 2. RT-PCR analysis of THLE-5b cells. **A)** Expression of liver-specific genes. **B)** Expression of liver-specific genes differing from A). **C)** Expression of biliary cell markers. **D)** Expression of liver-enriched transcription factors. Basic, cells under basic culture conditions; Agg-1, cell aggregates in aggregate culture-1; Agg-2, cell aggregates in aggregate culture-2. G3PDH, internal loading control.[Reprinted from Cell Biology International, Vol 30, pp 992-998 (2006) with permission from Elsevier Ltd.]

Table 1. THLE-5b Cells Are Bipotential* NE, not examined; – , no expression; + , expressed; ++ , strongly expressed

Markrs	Basic culture condition	Aggregate culture-1	Aggregate culture-2
Hepatocytic CYP3A4	–	–	++
TO	–	+	++
E-Cad	–	+	++
Alb	–	+	++
AAT	+	+	++
TF	–	+	+
CYP1A1	–	–	+
AFP	–	–	–
HNF-1	–	–	++
HNF-4	–	–	+
Bile ductular HNF-6	+	+	++
GGT	++	NE	–
CK19	++	NE	++

*Determined by RT-PCR. Abbrebiations: TO, tryptophane 2,3-dioxygenase; E-cad, E-cadherin; Alb, albumin; AAT, α1-antitrypsin; TF, transferrin; AFP, α-fetoprotein; GGT, γ–glutamyl transpeptidase

Cell Transplantation

THLE-5b cells transplanted in the peritoneal cavity of SCID mice were well localized after 3 weeks of experiments. Transplanted THLE-5b cells expressed albumin and α1-antitrypsin, but did not express CK19 (Figs. 1E, 1F, 1G).

Flow Cytometry

First, we measured the number of SP cells in HuH-7 and HepG2 cells as controls. HuH-7 and HepG2 contained 0. 2% and 0% SP cells, respectively (data not shown), which is similar to previously reported values [11].

Next, we measured the number of SP cells in the THLE-5b cells under the same conditions as used for the HuH-7 cells. We found that THLE-5b cells contained 1.73 ± 0.81% (n=4) SP cells. These SP cells were essentially eliminated by treatment with verapamil (Fig. 3A). ABCG2/BCRP1 has been proposed as a universal marker for stem cells. We found that these SP cells (THLE-5b SP cells) strongly express ABCG2/BCRP1 mRNA (Fig. 3B).In order to examine tumorigenicty of SP cells, cells were grown in culture plates containing RPMI 1640 supplemented with 10% fetal calf serum and injected subcutaneously into 3 Balb/c nu/nu mice at 5 million SP cells/ mouse. No tumor formation occurred 6 months after inoculation (data not shown).

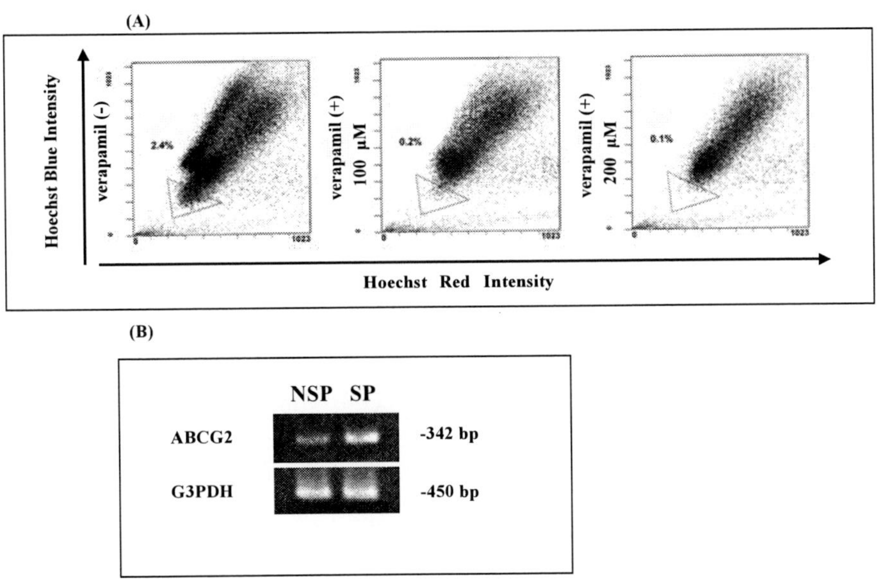

Figure 3. Hoechst 33342 staining for SP cell analysis in THLE-5b cells. **A)** SP cells were detected when THLE-5b cells were incubated in RPMI-1640. SP cells, which disappear under treatment with verapamil, are outlined and shown as a percentage of the total cell population. **B)** Expression of ABCG2/BCRP1 mRNA in THLE-5b SP cells. NSP, non-SP; G3PDH, internal loading control.

Discussion

THLE-5b cells grown as a monolayer expressed CK14, Thy-1 and c-kit, which are rarely detected in mature hepatocytes [12]. CK14 are found in liver progenitor cells or oval cells and may be a marker for liver progenitor cells or oval cells that are capable of differentiating into hepatocytes [13]. Thy-1 is detected only in oval cells [14], which in turn are identified by co-expression of stem cell and hepatic lineage markers [4]. THLE-5b cells expressed hepatocytic and bile duct epithelial cell markers under basic culture conditions. These data suggest that the undifferentiated phenotype of THLE-5b cells is reminiscent of liver progenitor cells.

Up-regulation or induction of liver-specific genes was observed in THLE-5b cells when the cells were grown as aggregates. Significantly, induction of hepatocytic markers by aggregation coincided with down-regulation of liver progenitor cell marker CK14 [4] and bile duct/oval cell marker GGT. Furthermore, ultrastructural study showed more differentiated structures, including the formation of bile canaliculi- and desomosome-like structures, which are thought to be characteristic of mature hepatocytes, when THLE-5b cells were grown as aggregates [4]. Under these conditions, increased ammonia-metabolic activity was also apparent [4]. These results indicate that THLE-5b cells within aggregates differentiated as hepatocyte-like cells. In addition, THLE-5b cells expressed albumin and α1-antitrypsin, but lost expression of CK19 when transplanted intraperitoneally. These results suggest that THLE-5b cells remain functional in vivo and could provide novel tools for use in cell transplantation.

Culture in Matrigel is shown to favor bile duct cell differentiation [4]. Bile duct/oval cell marker GGT was up-regulated in Matrigel culture of THLE-5b cells [4]. However, cells did not form bile duct-like structures and the bile duct/oval cell marker CK19 was slightly down-regulated [4]. This demonstrates that THLE-5b cells have differentiated into a biliary phenotype, but only partially, a finding that may reflect either that induction to differentiate in Matrigel is not as specific as suggested previously [15], or may reflect a limited differentiation potential inherent in these cells. Taken together, these data indicate that THLE-5b cells possess liver progenitor-like properties that can be induced to differentiate into either hepatocytes or bile duct cells.

Meanwhile, RT-PCR analysis of THLE-5b cells indicated that of the 8 hepatocyte-specific genes tested, 4 (including E-cadherin) and 7 were up-regulated or induced in aggregate culture-1 and -2, respectively. These results indicate that cells in aggregate culture-1 may display similar phenotypes to those in aggregate culture-2 and suggest that cell aggregation may enrich progenitor cells or a subset of stem cells, SP cells, in THLE-5b cells. SP cells have been isolated using flow cytometry and the DNA-binding dye Hoechst 33342 in mammalian tissues and cell lines [8]. Flow cytometry and Hoechst 33342 dye analysis of THLE-5b cells indicated the inclusion of SP cells, which express a universal marker for stem cells [16], ABCG2/BCRP1 mRNA.

Use of immortalized cell lines offers advantages over primary cultured cells in allowing extremely reproducible studies. THLE-5b cells are an SV40 large T antigen-immortalized cell line. THLE-5b cells are negative for several tumor parameters, including anchorage-independent growth on soft agar and tumor formation in nude mice, suggesting that cells are only weakly transformed.

The present study found that THLE-5b cells possess liver progenitor-like properties using cell-type specific markers and flow cytometry analysis following Hoechst 33342 staining. The results imply that bipotent progenitor cell populations similar to THLE-5b cells may exist in adult human livers.

Acknowledgments

We thank Dr. Harris (NCI, Bethesda, MD) for providing us with THLE-5b cells.

Conclusion

An immortalized non-tumorigenic human liver epithelial cell line possesses liver progenitor-like properties. These results have fundamental implications on the existence of liver progenitor cells in human livers.

References

[1] Kano, J., Noguchi, M., Kodama, M., Tokiwa, T. (2000). The in vitro differentiating capacity of nonparenchymal epithelial cells derived from adult porcine livers. *American Journal of Pathology,* 156, 2033-2043.

[2] Tokiwa, T., Yamazaki, T., Kano, J., Noguchi, M., Enosawa, S., Kohno, M. (2002).Colony-forming capacity of porcine liver epithelial cells in culture.*In Vitro Cellular Developmental Biology* 38, 440-442.

[3] Tokiwa, T., Yamazaki, T., Ono, M., Enosawa, S., Tsukiyama, T. (2008).Cloning and characterization of liver progenitor cells from the scattered cellclusters in primary culture of porcine livers. *Cell Transplantation* (in press).

[4] Tokiwa, T., Yamazaki, T., Xin, W., Sugae, N., Noguchi, M., Enosawa, S.,Tsukiyama, T. (2006). Differentiation potential of an immortalized non-tumorigenichuman liver epithelial cell line as liver progenitor cells. *Cell Biology International* 30, 992-998.

[5] Hussain, S., Strom, S.C., Kirby, M.R., Burns, S., Langemeijer, S., Ueda, T., Hsieh, M., Tisdale, J.F. (2005). Side population cells derived from adult human liver generate hepatocyte-like cells in vitro. *Digestive Diseases and Sciences* 50, 1755-1763.

[6] Lechner, J.F., Smoot, D.T., Pffeifer, A.M., Tokiwa, T., Harris, C.C. (1990).A non-tumorigenic human liver epithelial cell culture model for chemical and biological carcinogenesis investigations. In J.S. Rhim, & A. Drifschlo (Eds.),Neoplastic transformation in human cell system. Mechanism of *carcinogenesis*. (pp. 307-321). NewYork, Humana Press.

[7] Tokiwa, T., Kano, J., Noguchi, M., Kodama, M., Tateishi, T. (1998).Restoration of differentiated functions in multicellular aggregates of a human liver epithelial cell line. *Materials Science & Engineering C* 6, 249-252.

[8] Goodell, M.A., Brose, K., Paradis, G., Conner, A.S., Mulligan, R.C. (1996). Isolation and functional properties of murine hematopoietic stem cells that are replicating in vivo. *Journal of Experimental Medicine* 186, 1797-1806.

[9] Shimano, K., Satake, M., Okaya, A., Kitanaka, J., Kitanaka, N., Takemura, M., Sakagami, M., Tearada, N., Tsujimura, T. (2003). Hepatic oval cells have the side population phenotype defined by expression of ATP-binding cassette transporter ABCG2/BCRP1. *American Journal of Pathology* 163, 3-9.

[10] Malhi, H., Irani, A.N., Gagandeep, S., Gupta, S. (2002). Isolation of human progenitor liver epithelial cells with extensive replication capacity and differenitiation into mature hepatocytes. *Journal of Cell Science* 115, 2679-2688.

[11] Chiba, T., Kita, K., Zheng, Y.Z., Yokosuka, O., Iwama, A., Nakauchi, H.,Taniguchi, H. (2006). Side population purified from hepatocellular carcinoma cellsharbors cancer stem cell-like properties. *Hepatology* 44, 240-251.

[12] Baumann, U., Crosby, H., Ramani, P., Kelly, D.A., Strain, A.J. (1999). Expression of the stem cell factor c-kit in normal and diseased pediatric liver: identification of a human hepatic progenitor cell? *Hepatology 30*, 112-117. Florino, A.S. Diehl, A.M., Lin, H.Z., Lemischka, I.R., Reid, L.M. (1998).

[13] Maturation-dependent gene expression in a conditionally transformed liver progenitor cell line. *In Vitro Cellular Developmental Biology* 34, 247-258. Petersen, B.E., Goff, J.P., Greenberger, J.S., Michalopoulos, G.K. (1998). Hepatic oval cells express the hematopoietic stem cell marker Thy-1 in the rat. *Hepatology* 27, 433-445.

[14] Strick-Marchant, H., Weiss, M.C. (2002). Inducible differentiation and morphogenesis of bipotent liver cell lines from wild-type mouse embryo.*Hepatology 36,* 794-804.

[15] Zhou, S., Schuetz, J.D., Bunting, K.D., Colapietro, A.M., Sampath, J., Morris, J.J., Lagutina, I., Grosveld, G.C., Osawa, M., Nakauchi, H., Sorrentino,

[16] B.P. (2001). The ABC transporter Bcrp1/ABCG2 is expressed in a wide variety of stem cells and is a molecular determinant of the side-population phenotype. *Nature Medcine* 7, 1028-1034.

In: Stem Cell Applications in Disease and Health
Editors: W. Burnsides and R. Ellsley
ISBN: 978-1-60456-255-2

Short Communication B

Immune Threat to Neurogenesis from Human Adult Mesenchymal Stem Cells

Peter Navarro, Marianne D. Castillo and Pranela Rameshwar*
University of Medicine and Dentistry of New Jersey – New Jersey Medical School
Department of Medicine – Hematology/Oncology

Abstract

Mesenchymal Stem Cells (MSCs) are considered one of the leading adult stem cell candidates for future applications in regenerative medicine. The lead is partly accounted for by ease of expansion and unique immune properties. MSCs mediate immune-suppression and appear capable of being implanted across allogeneic barriers. Additionally, MSCs have been reported to express Major Histocompatibility Complex II (MHC-II) and act as antigen presenting cells. MSCs can also differentiate and transdifferentiate into cells of different tissues. Reduced ethical concerns underscore their potential for medical application in cell therapy. Studies in several laboratories have reported on the transdifferentiation of MSCs to functional neurons. The neurons have been characterized at the molecular, cellular, and functional levels. These studies show promise for neural regeneration and also bring an understanding of the interaction between the stem cells and regions of tissue injuries. However, one significant concern regarding successful development of these neurons is the immune-enhancing effects of MSCs. For example, MSC-induced neurons could probably re-express MHC-II and induce allorejection. This review discusses the immune properties of MSCs and their relevance during neurogenesis.

Keywords: MSC: Mesenchymal Stem Cells; CTL: cytotoxic T cells; APC: Antigen presenting cells; MHC: Major Histocompatibility Complex; IFNγ: Interferon gamma; NK: Natural killer cells

* E-mail address: rameshwa@umdnj.edu, Tel. (973) 972 0625; Fax (973) 972 8854.

Grant Support

This work was supported by a grant from the F.M. Kirby Foundation.

Introduction

Efforts to regenerate nervous tissue have come up short for many years until the pluripotent property of stem cells was discovered. Stem cells, both embryonic and adult, are capable of transdifferentiation, providing a promising solution to the problem of neurogenesis. Human adult mesenchymal stem cells [MSC] in particular, are considered an effective and uncontroversial alternative to their embryonic counterparts. MSCs are found in the bone marrow and normally differentiate into stromal cells that provide support for the growth of hematopoietic stem cells. MSCs constitute a small and heterogeneous population of cells that have the ability to differentiate into osteoblasts, chondrocytes and adipocytes and also transdifferentiate into neurons, even years after their initial preservation [Xiang et al, 2007]. Their pluripotency is already being investigated in the treatment of: cystic fibrosis, osteogenesis imperfecta and graft versus host disease (Leblanc et al., 2005; Min et al., 2007; Wang et al., 2005).

Under the appropriate setting, MSCs have been shown to mediate immunosuppression by downregulating the effects of B-, T- and natural killer cells. As veto cells, MSCs are able to downregulate the immune response in a setting of graft vs. host response (Potian et al, 2003). The veto function is clinically relevant for bone marrow and organ transplantation when MSCs can be added as third party cells to attenuate rejection and thereby facilitate engraftment. Interestingly, despite their veto functions, MSCs can elicit cytotoxic T-cell (CTL) responses indicating their ability to allow for viral clearance (Kang et al., 2005). Based on the varied properties of MSCs, one wonders why they reside in bone marrow. We deduce that they have a critical role in bone marrow protection from inflammatory responses and invading infectious agents, perhaps acting in the capacity of a `gate-keeper' role (Razonable et al., 2003; Kang et al., 2005).

MSCs also possess immunoenhancing effects. As compared to reports on the immune suppressive properties, there are relatively less investigations on the immune stimulatory properties of these stem cells. MSCs mediate antigen presenting properties, and also act as phagocytes (Chan et al., 2006; Stagg 2007). In addition, these stem cells have been shown to affect B-cell and NK cell functions (Rasmusson et al., 2007; Spaggiari et al., 2007). These properties of MSCs raise important questions about their ability to be used in regenerative therapy of the nervous system. Thus, during the discussion of potential neurogenesis by MSCs, the material will include information on the immune characteristics of MSCs, and argue their potential for interfering with neurogenesis.

Transdifferentiation into Neurons

MSCs constitute a small number of the total cells in the BM microenvironment. Implicitly, MSCs are obtained from bone marrow, but they can also be acquired from many other tissues such as brain, spleen, liver, kidney and lung (Meirelles et al., 2006). MSCs can differentiate into adipocytes, chondrocytes and, for the focus of this review, transdifferentiate into neurons. This review refers to transdifferentiation as the ability of mesodermal MSCs to cross germ layer to form ectodermal neurons.

Transdifferentiation to neurons was demonstrated by *in vitro* studies with various induction protocols (Bertani et al., 2005). The ability of MSCs to form neurons was demonstrated by morphological methods, histochemical for neuronal markers and functional by electrophysiology. The latter include synaptic transmission (Cho et al., 2005). These *in vitro* findings are consistent with the *in vivo* models to use MSCs in the treatment of neurodegenerative diseases. An advantage of selecting MSCs for neural regeneration is the ability to deliver them from autologous source, if taken from bone marrow. The major advantage for autologous use is to prevent rejection (Harvey and Chop, 2003; Chopp et al., 2007).

It is important to include that MSCs have been shown to generate dopaminergic progenitors *in vitro,* underscoring their potential for Parkinson's disease, and traumatic brain injury (Trzaska et al., 2007). The other type of neurons generated from MSCs might also be used in the treatment of spinal cord injury and traumatic brain injury (Greco and Rameshwar 2007; Cho et al., 2005; Trzaska et al., 2007). Additionally, MSCs appear to mature via a neural progenitor type cells, suggesting that they might be able to form oligodendrocytes (Greco et al., 2007). These in vitro studies are consistent with the neuroprotective and neuroregenerative effect of MSCs in transplantation experiments within the brain (Rivera et al., 2006). MSCs have also been shown to promote proliferation, migration, and differentiation of neural stem cells into mature neural cells within the hippocampus of mice (Munoz et al., 2005). Taken together, MSCs seem to provide an avenue of research towards novel treatments of degenerative disorders within the nervous system.

Potential Allo-Rejection of MSC-Derived Neurons

A subset of MSCs expresses the Major Histocompatiblity Complex II (MHC-II). This provides them with the ability to function as antigen presenting cells (APCs) in the presence of low levels of Interferon–γ (IFNγ) (Chan et al, 2007). Contradictory to this, MSCs are also able to function as immune suppressors by down-regulating the expression of MHC-II under high levels of IFNγ (Chan et al., 2007; Romieu-Mourez et al, 2007). Due to this dual property of IFNγ in bone marrow, we have determined these cells as physiologically important with the designation of 'bone marrow gatekeeper cells'.

The anatomic location of MSCs supports of the gate-keeper function of these stem cells. In human, MSCs are mostly found surrounding the vasculature and trabeculae of bone marrow (Bianco et al., 2001). This anatomic location provides the MSCs with the ability to monitor invading organisms and cells entering and exiting the BM. Relatively low levels of IFNγ, which impart immune-enhancement by mediating APC functions, and high IFNγ levels, which mediate immune suppressor functions, provide the MSC with the role of preventing exacerbated inflammation in bone marrow. This is relevant for bone marrow homeostasis since inflammation could cause dysfunctional hematopoiesis.

The question is whether MHC-II can be re-expressed on neurons if MSCs are applied in an allogeneic host. If this occurs, the new neurons can potentially elicit an allogeneic response with graft vs. host response, and also begin to function as APC cells. Liu et al demonstrated enhanced expressions of both MHC-I and MHC-II during the development of MSCs to neurons (Liu et al., 2007). Interestingly, despite the expression of MHC molecules, the MSC-derived neurons failed to elicit significant allogeneic response. This was partly explained by deficiency of co-stimulatory molecules and inadequate MHC signaling by the neurons (Liu et al., 2007). Although this report is encouraging with regards to stem cell therapy with `off the shelf' MSCs, this finding is significant and requires experimental studies to determine the effects of long-term transplantation. In a clinical setting, if a patient encounters a viral infection, MSC derived neurons may revert to their immune enhancing properties and present antigen to recipient T-cells, which may cause a detrimental immune response. Additionally, it is difficult to predict how the MSC-derived neurons will react within an inflammatory microenvironment, such as in acute sites of tissue injuries.

Immune Effects of MSCs on Natural Killer (NK) Cells

Two different groups have explored the effects of MSCs on NK cells and both have reported similar findings (Spaggairi et al., 2006; Sotiropoulou et al., 2006). Freshly isolated, inactivated NK cells will not be able to function as killer cells in the presence of MSCs. However, MSCs are unable to suppress the killer function of IL-2 activated NK cells. The latter can be prevented if IFNγ is administered to the MSCs, as it will promote MHC class I upregulation and thereby spare them from NK lysis (Sotiropoulou et al., 2006).

In another study, Spaggiaro et al explored the mechanism by which NK activating surface receptors present on MSCs can inhibit the proliferation of NK cells (Spaggiaro et al., 2007). The authors showed that MSCs co-cultured with NK cells and IL-2 resulted in diminished cytotoxicity as compared to NK cells activated with IL-2 alone. Since the studies showed decreases in cytokine production by NK cells, this could be partly accounted by MSC-mediated dysfunction of NK activity. Also, the inhibition of NK activity as well as cytokine production was found to be mediated by the synergistic effects of soluble indoleamine 2,3-dioxygenase and prostaglandin E2.

A common mechanism by immune cells is the role of microenvironmental influence on the fate of the MSCs. IL-2 and IFNγ, which are representative microenvironmental mediators, could be involved in determining the behavior of immune cells regardless of MSCs. Both cytokines could affect the functions of B-cells and NK activities, which might depend on the levels of the cytokines and/or the synergistic effect of cytokines. Since MSCs have been shown to mediate suppression on the major immune cells, T-, B- and NK cells, these findings will need to be carefully reviewed as MSCs, and other types of stem cells are transplanted to patients.

Other Immune Effects of MSCs

Reports on the effects on B-cells by MSCs are conflicting. Recent studies have reported on the inhibition of murine B-cell functions via contact with MSCs, or with soluble factors produced from MSCs (Augello et al., 2007). In studies with human cells, B-cell-MSC interactions showed diminished B-cell proliferation via cell cycle arrest and decreased chemotaxis via downregulated chemokine production (Corcione et al., 2006). The B-cells showed a decrease in IgG production at 1:1 MSC:B-cell ratio. In contrast to these studies, other report showed evidence of increased IgG production by enriched human B cells cultured with MSCs (Rasmusson et al., 2007). The production of IgG depended on the level of B-cell stimulation with various stimulators. IgG production was increased by low levels of stimulators and diminished with increases in antigen challenges. These are significant findings since translation of MSCs to patients will need careful monitoring during infection.

The difference between the two conflicting papers lies in the amount of MSCs that were cultured with the B cells. Corcione et al. cultured MSCs and B cells at 1:1 ratios, the levels of which are not seen in the bone marrow microenvironment where the frequency of MSCs is low as compared to T- and B-cells. However, Rasmusson et al. studied 1:10 ratios of MSCs to B cells, which might be more representative of bone marrow cellularity. At this ratio the authors observed increased levels of immunoglobulin production. The data further supported MSCs-mediated expression of IL-6 (Majumdar et al., 1998). This relationship with IL-6 is important since IL-6 is necessary for immunoglobulin production as well as B-cell differentiation.

The regulatory effect of MSCs on B-cells depends on the amount of antigen molecules that the lymphocytes encounter. Since MSCs are found endogenously, the question is why they have varied effects on B-cell functions. Perhaps a role for MSC is to prevent the B cells from over-expressing immunoglobulin when the antigen threat is high. On the other hand, the MSCs could be increasing Ig production by B cells in order to amplify the immune response to low levels of antigen and rapid clearance. The complexity of MSCs on B-cell functions underscores the need to ensure that MSC transplant patients are not immunocompromised.

Conclusion

MSCs are not considered as immune privileged stem cells. They can, under certain conditions, present antigen, stimulate antibody secretion and are susceptible to lysis by natural killer cells. Since their application to translational science requires differentiating and transdifferentiating to various tissue-specific cells, further exploration into their immune behavior is required. Lacking is adequate information on the immune-enhancing functions of MSCs, as compared to the large body of literature on the immune suppressive properties. Regarding MSC-derived neurons, there is a possibility that the neurons might revert to immune- enhancing cells as its stem cell property. It is unclear how the MSC-derived neurons may act as stimulators to immune cells, including the antibody-producing B-cells and natural killer cells. While MSCs may be able to downregulate the immune response in some instances, it is unclear if similar behavior would be true after transdifferentiated to neurons. It is understandable to explain how the bone marrow microenvironment, which has a wealth of cytokines and other soluble mediators, could mediate both stimulatory and suppressive immune responses by MSCs. However, this information might not be extrapolated to regions of the brain to predict the behavior of MSC-derived neurons. Further research with *in vivo* models is needed for a better understanding of microenvironmental effects on MSC-neurons, which will be beneficial to patients. Recent studies implicating MSCs in promoting tumorigenesis underscores how the stem cells behave out of its microenvironment to promote pathogenesis (Miura et al., 2006; Matushansky et al., 2007). Thus, at this time, MSCs cannot be implanted in pathogenic sites, unless controlled experimental analyses are investigated.

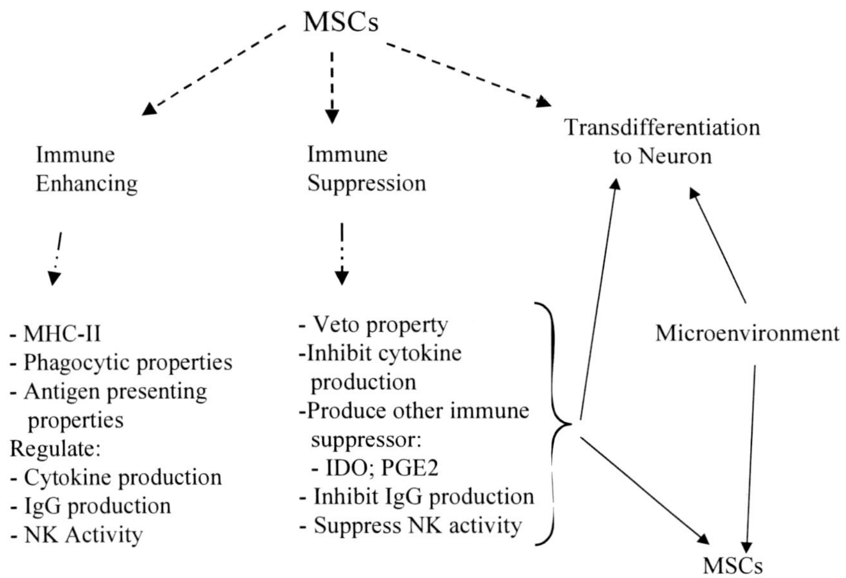

Figure 1. Shown are MSCs and the dual immune functions. These stem cells can also generate functional neurons. The varied immune properties of MSCs as well as other microenvironmental factors could interact with MSCs or the generated neurons at sites of tissue injuries.

Figure 1 describes the mechanism by which both the immune-enhancing and –inhibitory effects of MSCs interact with other microenvironmental factors to affect MSCs placed in vivo, or the transdifferentiated neurons. The question is how MSCs or the neurons that they form function in a changing milieu of factors that are likely to be present during tissue injury. In summary, MSCs, which are obtained easily, and with minimum ethical issues, show potential for future application in medicine.

References

Augello, A; Tasso, R; Negrini, SM; Amateis, A; Indiveri, F; Cancedda, R; Pennesi, G. Cell therapy using allogeneic bone marrow mesenchymal stem cells prevents tissue damage in collagen-induced arthritis. *Arthritis Rheum*, 2007, 56:1175-86.

Bertani, N; Malatesta, P; Volpi, G; Sonego, P; Perris, R. Neurogenic potential of human mesenchymal stem cells revisited: analysis by immunostaining, time-lapse video and microarray. *J Cell Sci*, 2005, 118:3925-36.

Bianco, P; Riminucci, M; Gronthos, S; Robey, PG. Bone marrow stromal stem cells: Nature, biology, and potential applications. *Stem Cells*, 2001;19:180.

Chan, JL; Tang, KC; Patel, AP; Bonilla, LM; Pierobon, N; Ponzio, NM; Rameshwar, P. Antigen-presenting property of mesenchymal stem cells occurs during a narrow window at low levels of interferon-gamma. *Blood*, 2006;107:4817-4824.

Cho, KJ; Trzaska, KA; Greco, SJ; McArdle, J; Wang, FS; Ye, JH; Rameshwar, P. Neurons derived from human mesenchymal stem cells show synaptic transmission and can be induced to produce the neurotransmitter substance P by interleukin-1 alpha. *Stem Cells*, 2005, 23:383-91.

Chopp, M; Li, Y; Zhang, J. Plasticity and remodeling of brain. *J Neurol Sci*, 2007.

Corcione, A; Benvenuto, F; Ferretti, E; Giunti, D; Cappiello, V; Cazzanti, F; Risso, M; Gualandi, F; Mancardi, GL; Pistoia, V; Uccelli, A. Human mesenchymal stem cells modulate B-cell functions. *Blood*, 2006, 107:367-372.

Greco, SJ; Rameshwar, P. Enhancing effect of IL-1 on neurogenesis from adult human mesenchymal stem cells: Implication for inflammatory mediators in regenerative medicine. *J Immunol,* 2007, 179:3342-3350.

Greco, SJ; Zhou, C; Ye, JH; Rameshwar, P. An interdisciplinary approach and characterization of neuronal cells transdifferentiated from human mesenchymal stem cells. *Stem Cells Dev,* 2007, (In press).

Harvey, RL; Chopp, M. The therapeutic effects of cellular therapy for functional recovery after brain injury. *Phys Med Rehabil Clin N Am*, 2003, 14:S143-151.

Kang, HS; Habib, M; Chan, J; Abavana, C; Potian, JA; Ponzio, NM; Rameshwar, P. A paradoxical role for IFN-gamma in the immune properties of mesenchymal stem cells during viral challenge. *Exp Hematol*, 2005, 33:796-803.

Le Blanc, K; Gotherstrom, C; Ringden, O; Hassan, M; McMahon, R; Horwitz, E; Anneren, G; Axelsson, O; Nunn, J; Ewald, U; Norden-Lindeberg, S; Jansson, M; Dalton, A; Astrom, E; Westgren, M. Fetal mesenchymal stem-cell engraftment in bone after in utero transplantation in a patient with severe osteogenesis imperfecta. *Transplantation*, 2005, 79:1607-1614.

Liu, K; Castillo, DM; Murthy, RG; Patel, N; Rameshwar, P. Tachykinins and Hematopoiesis. *Clinica Chimica Acta* 2007, 385:28-34.

Majumdar, MK; Thiede, MA; Mosca, JD; Moorman, M; Gerson, SL. Phenotypic and functional comparison of cultures of marrow-derived mesenchymal stem cells (MSCs) and stromal cells. *J Cell Physiol*, 1998, 176:57-66.

Matushansky, I; Hernando, E; Socci, ND; Mills, JE; Matos, TA; Edgar, MA; Singer, S; Maki, RG; Cordon-Cardo, C. Derivation of sarcomas from mesenchymal stem cells via inactivation of the Wnt pathway. *J Clin Invest*, 2007 [In Press]

Meirelles, L. S.; Chagastelles, PC; Nardi, NB. Mesenchymal stem cells reside in virtually all post-natal organs and tissues. *J Cell Sci*, 2006, 119:2204-13.

Min, CK; Kim, BG; Park, G; Cho, B; Oh, IH. IL-10-transduced bone marrow mesenchymal stem cells can attenuate the severity of acute graft-versus-host disease after experimental allogeneic stem cell transplantation. *Bone Marrow Transplant*, 2007, 39:637-45.

Miura, M; Miura, Y; Padilla-Nash, HM; Molinolo, AA; Fu, B; Patel, V; Seo, BM; Sonoyama, W; Zheng, JJ; Baker, CC; Chen, W; Ried, T; Shi, S. Accumulated chromosomal instability in murine bone marrow mesenchymal stem cells leads to malignant transformation. *Stem Cells*, 2006, 24:1095-1103.

Munoz, JR; Stoutenger, BR; Robinson, AP; Spees, JL; Prockop, DJ. Human stem/progenitor cells from bone marrow promote neurogenesis of endogenous neural stem cells in the hippocampus of mice. *Proc Natl Acad Sci U S A*, 2005, 102:18171-18176.

Potian, JA; Aviv, H; Ponzio, NM; Harrison, JS; Rameshwar, P: Veto-like activity of mesenchymal stem cells: functional discrimination between cellular responses to alloantigens and recall antigens. *J Immunol,* 2003, 171:3426-3434.

Rasmusson, I; Le Blanc, K; Sundberg, B; Ringden, O. Mesenchymal stem cells stimulate antibody secretion in human B cells. *Scand J Immunol*, 2007, 65:336-343.

Razonable, RR; Paya, CV. Herpesvirus infections in transplant recipients: current challenges in the clinical management of cytomegalovirus and Epstein-Barr virus infections. *Herpes,* 2003;10:60-65.

Rivera, FJ; Couillard-Despres, S; Pedre, X; Ploetz, S; Caioni, M; Lois, C; Bogdahn, U; Aigner, L. Mesenchymal stem cells instruct oligodendrogenic fate decision on adult neural stem cells. *Stem Cells*, 2006, 24:2209-2219.

Romieu-Mourez, R; Francois, M; Boivin, MN; Stagg, J; Galipeau, J. Regulation of MHC class II expression and antigen processing in murine and human mesenchymal stromal cells by IFN-gamma, TGF-beta, and cell density. *J Immunol*, 2007, 179:1549-1558.

Sotiropoulou, PA; Perez, SA; Gritzapis, AD; Baxevanis, CN; Papamichail, M. Interactions between human mesenchymal stem cells and natural killer cells. *Stem Cells*, 2006, 24 74-85.

Spaggiari, GM; Capobianco, A; Abdelrazik, H; Becchetti, F; Mingari, MC; Moretta, L. Mesenchymal stem cells inhibit natural killer cell proliferation, cytotoxicity and cytokine production: role of indoleamine 2,3-dioxygenase and prostaglandin E2. *Blood*, 2007 [In Press]

Spaggiari, GM; Capobianco, A; Becchetti, S; Mingari, MC; Moretta, L. Mesenchymal stem cell-natural killer cell interactions: evidence that activated NK cells are capable of killing MSCs, whereas MSCs can inhibit IL-2-induced NK-cell proliferation. *Blood*, 2006, 107:1484-1490.

Stagg, J. Immune regulation by mesenchymal stem cells: two sides to the coin 40. *Tissue Antigens*, 2007, 69:1-9.

Trzaska, KA; Kuzhikandathil, EV; Rameshwar, P. Specification of a Dopaminergic Phenotype from Adult Human Mesenchymal Stem Cells. *Stem Cells*, 2007 [In Press]

Wang, G; Bunnell, BA; Painter, RG; Quiniones, BC; Tom, S; Lanson, NA Jr; Spees, JL; Bertucci, D; Peister, A; Weiss, DJ; Valentine, VG; Prockop, DJ; Kolls, JK. Adult stem cells from bone marrow stroma differentiate into airway epithelial cells: potential therapy for cystic fibrosis. *Proc Natl Acad Sci U S A*, 2005, 102:186-191.

Xiang, Y; Zheng, Q; Jia, BB; Huang, GP; Xu, YL; Wang, JF; Pan, ZJ. Ex vivo expansion and pluripotential differentiation of cryopreserved human bone marrow mesenchymal stem cells. *J Zhejiang Univ Sci B*, 2007, 8:136.

In: Stem Cell Applications in Disease and Health
Editors: W. Burnsides and R. Ellsley
ISBN: 978-1-60456-255-2

Chapter I

Selecting a Proper Platform for Microarray Experiment

Sergey V. Anisimov*
V.A. Almazov Federal Center for Heart, Blood and Endocrinology,
Saint-Petersburg, 197341, Russia and
Cureline, Inc., San Francisco, CA 94080, USA

Abstract

The rapid development of microarray technology has led to important progress in many biomedical disciplines. In just over a decade, the technology has evolved from first small-scale prints to commercial and custom platforms of truly genomic scale. Though numerous types of microarray platforms exist, cDNA and oligonucleotide-based DNA microarrays are considered being the platforms of choice for the range of applications. However, the success of every study depends upon a careful selection of a platform appropriate for a designed experiment. In this short review, various technological and analytical aspects of microarray technology are discussed; some advantages and drawbacks of individual platform types are compared.

Keywords: DNA microarray, platform selection

The rapid development of microarray technology has led to important progress in many biomedical disciplines, facilitating unraveling basic mechanisms underlying many important biological processes, gene discovery, drug discovery, disease diagnostics, etc. [1,2]. An ability to identify novel prognostic and diagnostic markers, potential therapeutic targets and molecular markers of toxicogenomic response in fast and reliable manner by microarray

* E-mail address: sergey-anisimov@cureline.com. Tel: +7 (812) 347 5546. FAX: +7 (812) 596 3896. Correspondence: Department of Hematology Research, V.A.Almazov Federal Center for Heart, Blood and Endocrinology Akkuratova Str., 2, 197341, Saint-Petersburg, Russia

means has undoubtedly transformed modern biology. This point could be vividly illustrated by a simple query to the PubMed database (a service of the U.S. National Library of Medicine and the National Institutes of Health), which lists 9 original microarray technology-related articles and 1 review in a whole period of 1993-1997 (the earliest reports by [3-5], 106 relevant papers in 1998-1999 and over 25000 papers since (Fig. (**1**)). This expansion is in parallel with a volume of microarray platforms designed and utilized, as further illustrated with a number of that deposited in Gene Expression Omnibus (GEO, National Center for Biotechnology Information (NCBI); http://www.ncbi.nlm.nih.gov/geo/) database: data from over 3000 individual microarray platforms are present in this database alone (as of early 2008).

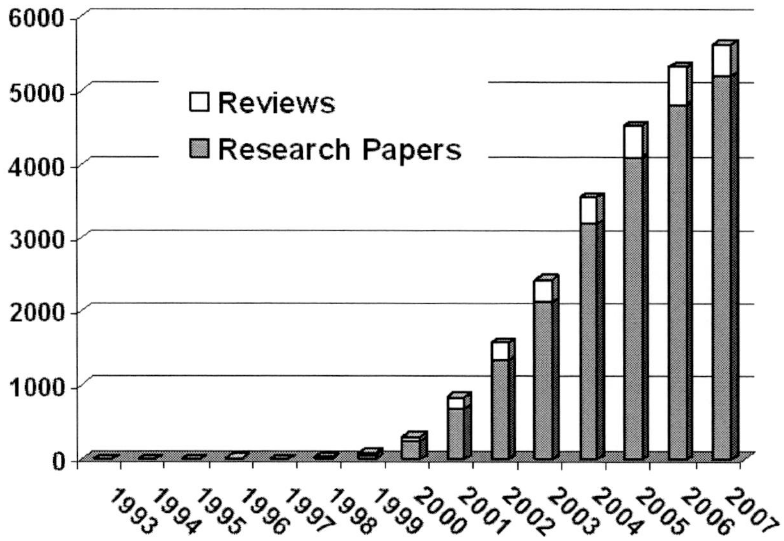

Figure 1. Microarray technology-relevant entries in PubMed: 1993-2007. Entries are sorted as original articles (nearly 22000 in total) and reviews (over 2500). Taken together, over the period of 1993-2007, >24000 published papers have relied on microarray technology. This expansion is in parallel with a volume of microarray platforms designed and utilized.

Although no established classification exists for microarray subtypes, it could be provisionally classified based on the object pursued (e.g., microarray-based expression analysis, mutation/single nucleotide polymorphism (SNP) analysis, comparative genomic hybridization), substrate type (e.g., whole tissue, DNA, RNA, antibody, protein and peptide microarrays), design (flat (2D) or volumetric (3D) microarrays), detection method (microelectronic, fluorescent and radioisotope detection-based microarrays), etc.. It was hard if possible to predict in the early years of microarray era what principle type of the microarray platform will be most suitable. Numerous types of the microarray platforms now considered 'unorthodox' were therefore tested, with a various degree of success. Even though some of those types (e.g., radio transponder-based, volumetric (3D) microarrays and some others) are still used in the limited niches of their specific applications, it was DNA microarray-based expression analysis that have gradually took over the largest share of the microarray technology-related applications.

DNA microarrays consist of two major categories, namely cDNA microarrays and oligonucleotide microarrays. The platforms belonging to the first group (printed on glass slides or on nylon filter membranes) rely on collections of bacterial clones containing inserts typically 500-2,000 nucleotides long, representing either fully sequenced genes of known function or partially sequenced cDNA derived from expressed sequence tags (ESTs). The latter corresponds to the messenger RNAs of genes representing either a certain degree of similarity to know genes, or to fully unknown, novel genes. Oligonucleotide microarrays rely on the synthesized molecules printed over the solid surfaces. Generally, microarray-bound oligonucleotide probes (matching targeted genes) could be 20-35 nucleotides (short oligonucleotide-based DNA microarrays), 35-45 nucleotides (intermediate) or 65-75 (long) nucleotides long. Each manufacturer seeks finding a better balance of economical (e.g., oligonucleotide synthesis expenses) and technical (e.g., hybridization-related parameters, including oligonucleotide design parameters like guanine/cytosine content, secondary structure, melting temperature, etc.) features when selecting the properties of oligonucleotides for the corresponding platforms. Furthermore, similar features affect a decision to place a large number of short oligonucleotide probe sets for each gene on the same platform, or rather use a single long oligonucleotide per gene. Thus, a wide assay of oligonucleotide-based DNA microarray platforms is available, and a scientist can select one most utterly satisfying the needs of the planned research experiment [6]. While longer oligonucleotides generally promise higher hybridization specificity and affinity (thus being more suitable for studying low-abundance genes), one may rather prefer using a higher number of shorter oligonucleotides spotted over the same microarray platform. Similarly, two major intrinsic factors affect the precision of gene expression data generated when utilizing cDNA microarray platforms. First, cDNA clones may have moderately different insert sizes, thus the signal intensities might to certain extent vary as a function of insert length. Second, cDNA clones can be derived from the alternative regions within the target gene, again affecting a quality of quantitative assessments of the hybridization data. Those factors contrast with the features of oligonucleotide microarrays, for which the probes are designed to be similar in regards to hybridization properties, - and therefore suggest cDNA microarray platforms might be more suitable for comparative studies than for gene expression profiling of individual samples ('snapshot gene expression profiling'). Interestingly, an emerging class of microarrays relying on extremely long oligonucleotides (synthesized molecules up to the thousands nucleotides long) might become a format of choice in the following decade.

Similarly, certain advantages and disadvantages are associated with available labeling techniques, with radioactive (radioisotope) and fluorescent labeling being most common among those. Although highly effective and most sensitive, radioactive labeling is relatively inexpensive; additionally, labeled membranes could be stripped and re-labeled, if necessary. At the same time, applying this type of labeling technique rightfully requires strict control at each step (storage and handling of both the radioactive isotope stock and labeled probes, liquid and solid waste disposal, etc.). All works related to the radioactive isotopes are subject to the governmental and institutional regulations in the context of safety risks, and in many cases require specific certification and licensing, designation of the dedicated work zones, etc. Radioactive labeling is also considered not being suitable for high-density DNA arrays, due to the risk of high radioactive signal 'leaking' from the neighboring spots, causing cross-contamination of individual spot-derived signal intensity value records (Fig. (**2**)). At the same time, fluorescent labeling allows performing simultaneous two-color hybridizations (e.g., for the experiment vs. control) and its application is not associated with safety risks of the scale mentioned above. In comparison to radioactive isotopes routinely applied for microarray probe labeling, fluorescent dyes (including commonly used Cy3 and Cy5 cyanine dyes) have longer shelf life. An excess of the labeled probe could be frozen for re-hybridization, which could take place many weeks following the labeling procedure itself. At the same time, those dyes are highly light-sensitive, which requires a good practice of laboratory procedures performed. Moreover, cyanine dyes are oxidation-sensitive (Cy5 in particular).

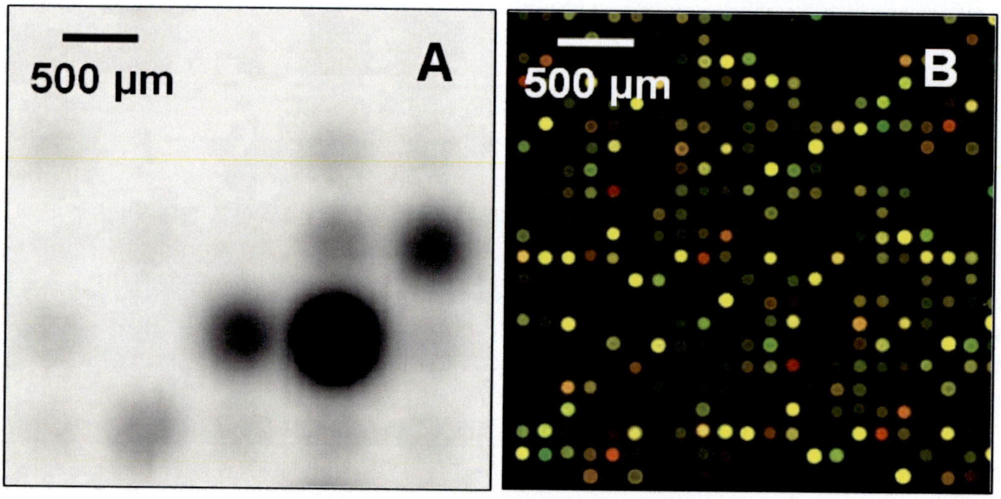

Figure 2. (A), Raw signal generated by cDNA microarray spotted over nylon membrane by GMS417 arrayer (Genetic Microsystems, USA). Distance between spot centers, 665 μm; spot diameter, 300 μm. (B) Raw signal generated by oligonucleotide microarray spotted over glass slides by MicroGrid II 600R arrayer (Harvard BioRobotics, USA). Distance between spot centers, 140 μm; average spot diameter, 90–110 μm. Scale bars, 500 μm (0.5 mm).

Over the course of just over a decade, numerous technological advances have led to the remarkable evolution of DNA microarrays [7-9]. While in the early years of microarray platform development those have carried hundreds of spots of DNA material [7,10-12], after just a few years large-scale DNA microarray platforms could carry some tens of thousands of these [13-16]. By the moment, a number of available DNA microarray platforms carry over hundred thousand spots of DNA material, thus reaching truly genomic scale. Still, specialized microarray platforms of small-to-intermediate scale are most valuable in many fields of study, allowing a concentrated (and thus less expensive) analysis of a limited number of pre-selected gene (or protein, etc.) targets: e.g., belonging to one or a few biological pathways [17]. Notably, a progress in microarray platform manufacturing (exemplified by the possibility to print a substrate material over the platform in the precise manner and with even higher density), closely follow the development of detection tool (high-resolution, high-throughput scanners). Clearly, individual properties of the materials utilized can affect platform selection: for example, cDNA microarrays printed over nylon membranes have significantly longer shelf life than oligonucleotide microarrays printed over the glass slides, and generally do not require controlled storage conditions.

Importantly, many areas of research could not directly benefit from the progress of microarray technology itself, due to the amount of biological material essential for analysis. It was therefore essential to refine techniques allowing highly accurate amplification of the starting material, for downstream microarray application [18,19]. Importantly, such techniques should match an existing range of available microarray platform types [20-23]. In its extreme, RNA material purified from a single cell is now enough to perform large-scale microarray analysis [24,25]. It is generally believed, however, that any amplification performed on pre-hybridization stem introduces a degree of bias to the quantitative accuracy of microarray data generated. Novel technological developments (e.g., laser capture microdissection (LCM)) allow reliable and high-throughput selection of the material for the downstream analysis. The researcher must therefore balance between a desire to have cell population as pure and uniform as possible, and a requirement to have a larger volume of starting material, as a mean to decrease amplification-introduced bias. Numerous further measures aim improving quantitative accuracy of microarray-generated data. Among those, repeated hybridizations of microarray platforms with independently purified and labeled probes (so-called 'biological replicates') and/or selfsame probes ('technical replicates') are utilized extensively [26-31] (see also [32]). As already mentioned above, multiplication of the number of spots carrying the same cDNA clone/oligonucleotide within the same platform could serve as an alternative, or rather an addition to this approach [17,33]. A quantity of the spots carrying equal DNA material (being duplicated, triplicated, quadruplicated, etc.) is then limited by the technical properties of the particular microarray platform. This combined approach effectively prevents data loss by individual small-scale hybridization artifacts, in particular. It also allows more accurate estimation of the variability present in an experiment, increases the probability of detecting genes undergoing significant fold changes in expression, and substantially decreased the probability of observing fold changes due to chance rather than true differential expression [34]. 'Pooling' of the samples assessed by microarray technology represents an alternative to purifying those by LCM or any other mean. Though this approach is criticized by many reviewers, it can effectively eliminate a need for amplification of starting material and furthermore reduce the effect of biological variation in the individual samples [35-37] (see also extensive reviews in [38,39]).

As described in this short review, many DNA microarray platforms are currently available commercially or via custom design, providing a multiplicity of choices. Due to the diversity of available DNA microarray platforms, no single established protocol or reagent could be applied to every DNA microarray platform. Moreover, technological means employed for data acquisition and analysis (i.e. scanner model, analytical software available, etc.) vary in different laboratories, which obviously also affects platform selection (Fig. (**3**)). Therefore, variations and adaptations of basic protocols are most common on every stage of microarray experiments (namely, sample labeling, pre-hybridization, hybridization and post-hybridization washes) [40]. Taken together, variations in the performance between individual laboratories, practiced protocols, and analytical algorithms applied to the same array platform are no less meaningful than variations between the platforms [41-43]. At the same time, many of those factors could be successfully compensated for, using appropriate manufacturing, handling and data processing techniques [44].

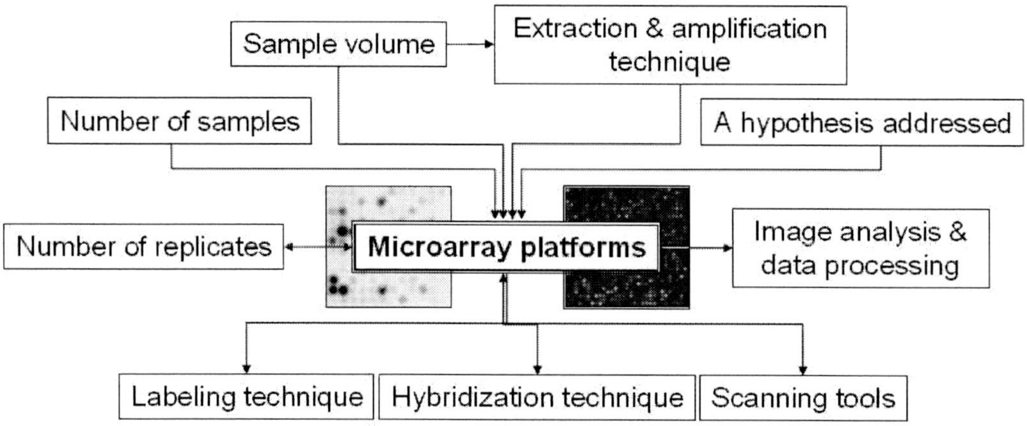

Figure 3. Some factors affecting microarray platform selection. Note that in the majority of cases the availability of the tools (e.g., access to the particular scanner model, license for radioactive isotope studies, etc.) affects the selection.

Finally, it is generally accepted that no standard way of microarray data analysis could address all the diversity of available microarray platforms and individual experiments performed [45]. Analytical algorhythm utilized should therefore be adjusted in each individual case according to the scale of DNA microarray platform, presence and scale of the intrinsic target multiplication (i.e., duplications of the target genes spotted within the same platform), biological replicates, total quantity of samples compared, etc. Numerous options for data analysis could be considered in each individual case [46-50]. Furthermore, blind design of microarray experiment could be a valuable factor [28,51,52]. It should be also stressed that no matter what platform has been selected for a particular microarray application, microarray-derived data should always be validated using an alternative approach (e.g., a conventional RT-PCR, real-time PCR, in-situ hybridization technology, etc.).

References

[1] Pusztai, L. Chips to bedside: incorporation of microarray data into clinical practice. (2006) *Clin. Cancer Res.,* 12(24), 7209-7214.

[2] Trevino, V.; Falciani, F. and Barrera-Saldaña, H.A. DNA microarrays: a powerful genomic tool for biomedical and clinical research. (2007) *Mol. Med.*, 13(9-10), 527-541.

[3] Fodor, S.P.; Rava, R.P.; Huang, X.C.; Pease, A.C.; Holmes, C.P. and Adams, C.L. Multiplexed biochemical assays with biological chips. (1993) *Nature*, 364(6437), 555–556.

[4] Guo, Z.; Guilfoyle, R.A.; Thiel, A.J.; Wang, R. and Smith, L.M. Direct fluorescence analysis of genetic polymorphisms by hybridization with oligonucleotide arrays on glass supports. (1994) *Nucleic Acids Res*., 22(24), 5456-5465.

[5] Pease, A.C.; Solas, D.; Sullivan, E.J.; Cronin, M.T.; Holmes, C.P. and Fodor, SP. Light-generated oligonucleotide arrays for rapid DNA sequence analysis. (1994) *Proc. Natl. Acad. Sci. USA,* 91(11), 5022-5026.

[6] Walker, S.J.; Wang, Y.; Grant, K.A.; Chan, F. and Hellmann, G.M. Long versus short oligonucleotide microarrays for the study of gene expression in nonhuman primates. (2006) *J. Neurosci. Methods*, 152(1-2), 179-189.

[7] Schena, M.; Shalon, D.; Davis, R.W. and Brown, P.O. Quantitative monitoring of gene expression patterns with a complementary DNA microarray. (1995) *Science*, 270(5235), 467-470.

[8] Lipshutz, R.J.; Fodor, S.P.; Gingeras, T.R. and Lockhart, D.J. High density synthetic oligonucleotide arrays. (1999) *Nat. Genet*., 21(1 Suppl), 20-24.

[9] Müller, U.R. Evolution of high-sensitivity microarray technology. (2003) *JALA*, 8(6), 96-100.

[10] Schena, M.; Shalon, D.; Heller, R.; Chai, A.; Brown, P.O. and Davis, R.W. Parallel human genome analysis: microarray-based expression monitoring of 1000 genes. (1996) *Proc. Natl. Acad. Sci. USA*, 93(20), 10614-10619.

[11] DeRisi, J.; Penland, L.; Brown, P.O.; Bittner, M.L.; Meltzer, P.S.; Ray, M.; Chen, Y.; Su, Y.A. and Trent, J.M. Use of a cDNA microarray to analyse gene expression patterns in human cancer. (1996) *Nat. Genet*., 14(4), 457-460.

[12] Shalon, D.; Smith, S.J. and Brown, P.O. A DNA microarray system for analyzing complex DNA samples using two-color fluorescent probe hybridization. (1996) *Genome Res*., 6(7), 639-645.

[13] Kargul, G.J.; Dudekula, D.B.; Qian, Y.; Lim, M.K.; Jaradat, S.A.; Tanaka, T.S.; Carter, M.G. and Ko, M.S. Verification and initial annotation of the NIA mouse 15K cDNA clone set. (2001) *Nat. Genet*., 28(1), 17-18.

[14] Truckenmiller, M.E.; Vawter, M.P.; Cheadle, C.; Coggiano, M.; Donovan, D.M.; Freed, W.J. and Becker, K.G. Gene expression profile in early stage of retinoic acid-induced differentiation of human SH-SY5Y neuroblastoma cells. (2001) *Restor. Neurol, Neurosci*., 18(2-3), 67-80.

[15] VanBuren, V.; Piao, Y.; Dudekula, D.B.; Qian, Y.; Carter, M.G.; Martin, P.R.; Stagg, C.A.; Bassey, U.C.; Aiba, K.; Hamatani, T.; Kargul, G.J.; Luo, A.G.; Kelso, J.; Hide, W. and Ko, M.S. Assembly, verification, and initial annotation of the NIA mouse 7.4K cDNA clone set. (2002) *Genome Res*., 12(12), 1999-2003.

[16] Carter, M.G.; Hamatani, T.; Sharov, A.A.; Carmack, C.E.; Qian, Y.; Aiba, K.; Ko, N.T.; Dudekula, D.B.; Brzoska, P.M.; Hwang, S.S. and Ko, M.S. In situ-synthesized novel microarray optimized for mouse stem cell and early developmental expression profiling. (2003) *Genome Res*., 13(5), 1011-1021.

[17] Anisimov, S.V.; Christophersen, N.S.; Correia, A.S.; Li, J.Y. and Brundin P. "NeuroStem Chip": a novel highly specialized tool to study neural differentiation pathways in human stem cells. (2007) *BMC Genomics*, 8, 46.

[18] Wilson, H.L. and O'Neill, H.C. Preparation of label from small cell numbers for microarray screening. (2004) *OMICS*, 8(3), 190-196.

[19] Viale, A.; Li, J.; Tiesman, J.; Hester, S.; Massimi, A.; Griffin, C.; Grills, G.; Khitrov, G.; Lilley, K.; Knudtson, K.; Ward, B.; Kornacker, K.; Chu, C.Y.; Auer, H. and Brooks, A.I. Big results from small samples: evaluation of amplification protocols for gene expression profiling. (2007) *J. Biomol. Tech.*, 18(3), 150-161.

[20] Nagy, Z.B.; Kelemen, J.Z.; Fehér, L.Z.; Zvara, A.; Juhász, K. and Puskás, L.G. Real-time polymerase chain reaction-based exponential sample amplification for microarray gene expression profiling. (2005) *Anal. Biochem.*, 337(1), 76-83.

[21] Pettersson, E.; Lindskog, M.; Lundeberg, J. and Ahmadian A. Tri-nucleotide threading for parallel amplification of minute amounts of genomic DNA. (2006) *Nucleic Acids Res.*, 34(6), e49.

[22] Laurell, C.; Wirta, V.; Nilsson, P. and Lundeberg J. Comparative analysis of a 3' end tag PCR and a linear RNA amplification approach for microarray analysis. (2007) *J. Biotechnol.*, 127(4), 638-646.

[23] Hu, G.; Yang, Q.; Cui, X.; Yue, G.; Azaro, M.A.; Wang, H.Y. and Li, H. A highly sensitive and specific system for large-scale gene expression profiling. (2008) BMC Genomics, 9, 9.

[24] Brandt, S.; Kloska, S.; Altmann, T. and Kehr J. Using array hybridization to monitor gene expression at the single cell level. (2002) *J. Exp. Bot.*, 53(379), 2315-2323.

[25] Saitou, M.; Yabuta, Y. and Kurimoto, K. Single-cell cDNA high-density oligonucleotide microarray analysis: detection of individual cell types and properties in complex biological processes. (2008) *Reprod. Biomed. Online*. 16(1), 26-40.

[26] Baggerly, K.A.; Coombes, K.R.; Hess, K.R.; Stivers, D.N.; Abruzzo, L.V. and Zhang, W. Identifying differentially expressed genes in cDNA microarray experiments. (2001) *J. Comput. Biol.*, 8(6), 639-659.

[27] Andreadis, J.D.; Mann, T.T.; Russell, A.C.; Stenger, D.A. and Pancrazio, J.J. Identification of differential gene expression profiles in rat cortical cells exposed to the neuroactive agents trimethylolpropane phosphate and bicuculline. (2001) *Biosens. Bioelectron.*, 16(7-8), 593-601.

[28] Anisimov, S.V.; Boheler, K.R.; Khavinson, V.Kh. and Anisimov, V.N. Studies of the effects of Vilon and Epithalon on gene expression in mouse heart using DNA-microarray technology. (2002) *Bull. Exp. Biol. Med.*, 133(3), 293-299.

[29] Anisimov, S.V.; Boheler, K.R. and Anisimov, V.N. Microarray technology in studying the effect of melatonin on gene expression in the mouse heart. (2002) *Dokl. Biol. Sci.*, 383, 90-95.

[30] Anisimov, S.V.; Khavinson, V.Kh. and Anisimov, V.N. Elucidation of the effect of brain cortex tetrapeptide Cortagen on gene expression in mouse heart by microarray. (2004) *Neuro Endocrinol. Lett.*, 25(1-2), 87-93.

[31] Anisimov, S.V.; Khavinson, V.Kh. and Anisimov, V.N. Effect of melatonin and tetrapeptide on gene expression in mouse brain. (2004) *Bull. Exp. Biol. Med.*, 138(5), 504-509.

[32] Wei, C.; Li, J. and Bumgarner, R.E. Sample size for detecting differentially expressed genes in microarray experiments. (2004) *BMC Genomics*, 5(1), 87.

[33] Lee, M.L.; Kuo, F.C.; Whitmore, G.A. and Sklar, J. Importance of replication in microarray gene expression studies: statistical methods and evidence from repetitive cDNA hybridizations. (2000) *Proc. Natl. Acad. Sci. USA,* 97(18), 9834-9839.

[34] Black, M.A. and Doerge, R.W. Calculation of the minimum number of replicate spots required for detection of significant gene expression fold change in microarray experiments. (2002) *Bioinformatics*, 18(12), 1609-1616.

[35] Warrington, J.A.; Nair, A.; Mahadevappa, M. and Tsyganskaya, M. Comparison of human adult and fetal expression and identification of 535 housekeeping/maintenance genes. (2000) *Physiol. Genomics*, 2(3), 143-147.

[36] Walker, D.G.; Lue, L.F. and Beach, T.G. Gene expression profiling of amyloid beta peptide-stimulated human post-mortem brain microglia. (2001) *Neurobiol. Aging*, 22(6), 957-966.

[37] Agrawal, D.; Chen, T.; Irby, R.; Quackenbush, J.; Chambers, A.F.; Szabo, M.; Cantor, A.; Coppola, D. and Yeatman, T.J. Osteopontin identified as lead marker of colon cancer progression, using pooled sample expression profiling. (2002) *J. Natl. Cancer Inst.*, 94(7), 513-521.

[38] Kendziorski, C.M.; Zhang, Y.; Lan, H. and Attie, A.D. The efficiency of pooling mRNA in microarray experiments. (2003) *Biostatistics*, 4(3), 465-477.

[39] Zhang, W.; Carriquiry, A.; Nettleton, D. and Dekkers, J.C. Pooling mRNA in microarray experiments and its effect on power. (2007) *Bioinformatics*, 23(10), 1217-1224.

[40] Yauk, C.L.; Berndt, M.L.; Williams, A. and Douglas, G.R. Comprehensive comparison of six microarray technologies. (2004) *Nucleic Acids Res*., 32(15), e124.

[41] Kuo, W.P.; Jenssen, T.K.; Butte, A.J.; Ohno-Machado, L. and Kohane, I.S. Analysis of matched mRNA measurements from two different microarray technologies. (2002) *Bioinformatics*, 18(3), 405-412.

[42] Irizarry, R.A.; Warren, D.; Spencer, F.; Kim, I.F.; Biswal, S.; Frank, B.C.; Gabrielson, E.; Garcia, J.G.; Geoghegan, J.; Germino, G.; Griffin, C.; Hilmer, S.C.; Hoffman, E.; Jedlicka, A.E.; Kawasaki, E.; Martínez-Murillo, F.; Morsberger, L.; Lee, H.; Petersen, D.; Quackenbush, J.; Scott, A.; Wilson, M.; Yang, Y.; Ye, SQ. and Yu, W. Multiple-laboratory comparison of microarray platforms. (2005) *Nat. Methods*, 2(5), 345-350.

[43] Johnson, D.S.; Li, W.; Gordon, D.B.; Bhattacharjee, A.; Curry, B.; Ghosh, J.; Brizuela, L.; Carroll, J.S.; Brown, M.; Flicek, P.; Koch, C.M.; Dunham, I.; Bieda, M.; Xu, X.; Farnham, P.J.; Kapranov, P.; Nix, D.A.; Gingeras, T.R.; Zhang, X.; Holster, H.; Jiang, N.; Green, R.D.; Song, J.S.; McCuine, S.A.; Anton, E.; Nguyen, L.; Trinklein, N.D.; Ye, Z.; Ching, K.; Hawkins, D.; Ren, B.; Scacheri, P.C.; Rozowsky, J.; Karpikov, A.; Euskirchen, G.; Weissman, S.; Gerstein, M.; Snyder, M.; Yang, A.; Moqtaderi, Z.; Hirsch, H.; Shulha, H.P.; Fu, Y.; Weng, Z.; Struhl, K.; Myers, R.M.; Lieb, J.D. and

Liu, X.S. Systematic evaluation of variability in ChIP-chip experiments using predefined DNA targets. (2008) *Genome Res.*, 18(3), 393-403.

[44] Wilkes, T.; Laux, H. and Foy, C.A. Microarray data quality - review of current developments. (2007) *OMICS*, 11(1), 1-13.

[45] Butte, A. The use and analysis of microarray data. (2002) *Nat. Rev. Drug Discov.*, 1(12), 951-960.

[46] Quackenbush, J. Computational analysis of microarray data. (2001) *Nat. Rev. Genet.*, 2(6), 418-427.

[47] Churchill, G.A. Fundamentals of experimental design for cDNA microarrays. (2002) *Nat. Genet.*, 32(Suppl), 490-495.

[48] Cheadle, C.; Vawter, M.P.; Freed, W.J. and Becker, K.G. Analysis of microarray data using Z score transformation. (2003) *J. Mol. Diagn.*, 5(2), 73-81.

[49] Leung, Y.F. and Cavalieri, D. Fundamentals of cDNA microarray data analysis. (2003) *Trends Genet.*, 19(11), 649-659.

[50] Hackl, H.; Sanchez Cabo, F.S.; Sturn, A.; Wolkenhauer, O. and Trajanoski, Z. Analysis of DNA microarray data. (2004) *Curr. Top. Med. Chem.*, 4(13), 1357-1370.

[51] Cirelli, C.; Shaw, P.J.; Rechtschaffen, A. and Tononi, G. No evidence of brain cell degeneration after long-term sleep deprivation in rats. (1999) *Brain Res.,* 840(1-2), 184-193.

[52] Hakak, Y.; Walker, J.R.; Li, C.; Wong, W.H.; Davis, K.L.; Buxbaum, J.D.; Haroutunian, V. and Fienberg, A.A. Genome-wide expression analysis reveals dysregulation of myelination-related genes in chronic schizophrenia. (2001) *Proc. Natl. Acad. Sci. USA,* 98(8), 4746-4751.

In: Stem Cell Applications in Disease and Health
Editors: W. Burnsides and R. Ellsley
ISBN: 978-1-60456-255-2

Chapter II

Physical-Chemical Manipulations with Microbial and Mammalian Cells: From Experiments to Clinics

Igor Albertovich Khlusov [*], ***Leonid Valentinovich Zagrebin, Sergey Semenovich Shestov and Sergey Aleksandrovich Naumov*** [#]
Limited Company "Center for Informational-Cell Medicine", Moscow;
[#]Limited Company "SibMedAnalyt", Tomsk

Abstract

Transplantation of stem and progenitor cells is approved for diseases of the central and peripheral nervous systems, cardiovascular pathology, endocrine diseases, retina and optic nerve pathology. However, results of similar investigations are usually indefinite and unconvincing. The negative consequences are held back in many cases. Existing problems of cell therapy (strict patient selection, host immune response, dispersion of transplanted cells in organism, their fast utilization, and the short-term effect of treatment) could be solved or minimized by means of using carriers – i.e., incubators (scaffolds) for cellular material. They allow questions of creation, growing and implantation of artificial organs, and prolongation of therapeutic effects of transplanted cells to be raised. For all this, physical-chemical stressors are permitted analogously to regulate activity of both pathogenic microbes and mammalian stem cells.

This chapter is devoted to: 1) the modulating effect of artificial surfaces, electrical impulses, visible light, some microelements, nano-scale particles and directed fluid flow under parameters approximate to physiological, on the vitality of microbial and mammalian colony forming cells; 2) certain applications of manipulations with microbial and mammalian cells in chronic disease treatments.

[*] E-mail address: khl@ultranet.tomsk.ru

Introduction

The ability of adult stem cells to differentiate not only into cells of their existence, but into many other ones, which led to intense development and innovation of cell technologies in clinical medicine, is now described. Cell therapy is developed according to four main areas: 1) transplantation of differentiated, specialized body cells (e.g. pancreatic islets transplantation for correction of pancreatic diabetes), 2) application of stem cells, 3) transplantation of embryonic (fetal) cells and tissues, and 4) application of immortal line of transfected cells [16].

Transplantation of stem and progenitor cells is approved in diseases of central and peripheral nervous systems, cardiovascular pathology, endocrine diseases, retina and optic nerve pathology. However, results of similar investigations are usually indefinite and unconvincing. The negative consequences are held back in many cases. Existing problems of cell therapy (strict patient selection, host immune response, dispersion of transplanted cells in the organism, their fast utilization, and the short-term effect of treatment) could be solved or minimized by means of using carriers – incubators (scaffolds) for cellular material. They allow questions of creation, growing and implantation of artificial organs, and prolongation of therapeutic effects of transplanted cells to be raised.

A variety of physical-chemical characteristics of artificial materials (metals and alloys, ceramics, polymers, composites, and nano-scale particles) and the opportunity to control cell behavior before and after their implantation allow correcting the course of different diseases and results. Throughout the processes, physical-chemical stressors are permitted analogously to regulate activity of both pathogenic microbes and mammalian stem cells.

The aims of this chapter to describe: 1) the modulating effect of artificial surfaces, electrical impulses, visible light, some microelements, nano-scale particles and directed fluid flow under parameters approximate to physiological, on the vitality of microbial and mammalian colony forming cells; and 2) certain applications of manipulations with microbial and mammalian cells in chronic disease treatments.

Experimental Research

1. Hemopoiesis in Bioinert Porous Scaffold in Long-Term Culture

Ex vivo long-term culture of bone marrow cells (LTCBM) according to a model proposed by Dexter [4] is a convenient system to study diverse aspects of hemopoiesis. A principle opportunity of LTCBM use for hemopoietic tissue purging from tumor cells in acute and chronic myelogenous leukemia and acute lymphoblastic leukemia has been established [13]. At the same time, the existing disadnvantages of LTCBM technique and rapid developing technology of cellular system cultivation in porous scaffolds have determined modeling of new artificial organs (bone marrow cells incorporated into porous metal scaffold) and studying its hemopoietic properties in vitro.

Bone marrow cells were washed from femurs of *CBA/CaLac* mice with RPMI-1640 medium with 5 % fetal calf serum (FCS) and were diluted with full cultural medium (without hydrocortisone) down to a necessary concentration of viable nuclear cells (Table 1). A TiNi-alloy brick was placed to hold the myelokaryocytes suspension. TiNi-alloy was transferred into other tubes in 24 hours. Cells seeded in it (Figure 1) were calculated and cellular associations were cultivated at 37° C for 30 days. According to the protocol (Table 1), the cultural medium was completely replaced by a new one. After the first and sixth cultural days the supernatant of mouse splenocytes stimulated by phytohemagglutinin (PHA) as a source of colony stimulating activity (CSA) was added into the medium.

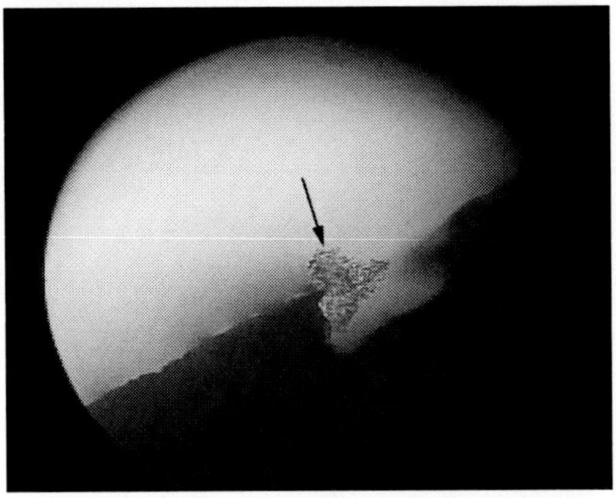

Figure 1. Association of bone marrow cells in porous metal scaffold. Magnification 100.

In another experiment, bone marrow cells in porous TiNi-alloy were grown for six weeks at 37° C and 5 % CO_2 in cultural medium of the following composition: 12.5 % FCS, 12.5 % horse serum, 300 mg/l L-glutamine, 2.5 x 10^{-5} 2-mercaptoethanol, 10^{-6} hydrocortisone and 75 % RPMI-1640 medium. Medium volume (1 ml) was half replaced per each week.

The amount of nucleated and viable cells was counted in the cultural medium, and cellular morphology was estimated. The number of committed precursors of granulomonocytopoiesis (CFU-GM) and granulocytopoiesis (CFU-G) was evaluated in liquid phase by means of the colony formation method in methylcellulose culture under myelokaryocytes cloning for seven days. Supernatants of mouse splenocytes activated by PHA (10 %) with recombinant human CSF-G (6 ng/ml) were used for precursors' growth.

The conducted investigation showed that the cell population in porous TiNi-alloy was directly dependent on initial cellularity in culture (Table 1). At the same time, further production of nucleated myelokaryocytes into the liquid phase of the system (artificial bone marrow - cultural medium) expressed as a percentage is inversely proportional to the number of karyocytes seeded into TiNi-alloy. In total expression, producing variations were 13.5 - 135 %. Apparently, it may be possible to control producing activity of artificial bone marrow in dependence on clinical needs. Throughout the process, the morphology of cells related mainly to granulomonocytopoiesis.

Sources of CSA also regulated a rate of karyocytes production. So, CSA addition increased the number of cells in cultural medium by approximately 1.5-2 times (Table 1, days 6-8). Nuclear cell viability increased an average of 15 % and was 85 %.

Table 1. Rate (υ) of production (10^3 cells/day) and total production (10^3) of cells into liquid medium by mouse myelokaryocytes culture in porous TiNi-alloy in diverse days of cultivation

Initial Cell Count, 10^6	Cell Count in TiNi, 10^6	% Cells in TiNi	days													Total cell production	Total % production
			1 υ of production	1 % production	4 υ	4 %	6 υ	6 %	8 υ	8 %	13 υ	13 %	20 υ	30 υ			
20	6.56	33	440	7	120	1.8	15.8	0.24	22.8	0.35	1.5	0.02	0.36	0.25		889	13.5
15	3.32	22	336	10	74	2.2	21.8	0.65	26.5	0.80	2.4	0.07	1.21	0.45		679.5	20.5
10	1.98	20	210	11	52	2.6	7.8	0.39	19	0.96	2.3	0.12	0.79	0.15		439	22.2
5	1.04	21	115	11	42	4.0	11.3	1.08	21.5	2.07	0.3	0.03	0.5	0.1		312	30
2	0.29	15	48	17	12	4.0	2.8	0.95	6.0	2.06	0.9	0.31	0.36	0.05		108	37.2
1	0. 07	7	40	57	9	10	2.8	3.95	5.5	7.8	1.3	1.86	0.43	0.05		94.5	135

Table 2. Rate (υ) of cell production (10^3/week) into liquid medium by mouse myelokaryocytes culture in porous TiNi-alloy

Initial Cell Count, 10^6	Cell Count in TiNi, 10^6	% Cells in TiNi	weeks											
			1 υ of production	1 % of viability	2 υ	2 %	3 υ	3 %	4 υ	4 %	5 υ	5 %	6 υ	6 %
15	2.03	14	440	55	380	58	280	71	120	88	102	-	54	58
15	10.76	72	1040	81	440	45	260	77	142	90	146	-	98	82
15	7.84	52	640	69	500	44	294	68	164	87	120	-	80	78

An ability of bone marrow in porous TiNi-alloy to produce cells in conditions of long-term (for 6 weeks) tissue culture was examined in the next experiment. It was found under identical initial cellularity in liquid phase that from 14 % to 72 % of karyocytes can seed in a porous bearer in conditions of 3-day cultivation (Table 2). The most significant statistical differences in karyocyte production in cultural medium were noted in the first and sixth week of cultivation (Table 2). During weeks 2-5, the dynamics of indices had stable reductive character little dependent on cellularity of artificial bone marrow in porous TiNi-alloy. The highest cell viability was observed forward to the 4th cultural week. Karyocytes related morphologically to mature granulomonocyte elements mixed with erythrocytes, lymphoid and/or erythroid cells.

An important peculiarity of cellular culture on porous TiNi-alloy is the existence of "clots", groups of cells combined with connective tissue fibers on its surface and, to a less degree, in liquid phase, probably typical of bone marrow structure formation. The content of committed precursors in hemopoietic tissue is known to exceed those in peripheral blood (conception of bone marrow barrier).

Perhaps the organ formation of bone marrow in porous TiNi-alloy is a base of slight deviance from committed granulomonocytic precursors from bearer into liquid phase of growth (table 3). The maximum concentration of granulomonocytic progenitors occurred in liquid medium for the first three weeks forward to the end of the 3d week of observation by the practically identical total number of precursors. There was more than a two-fold decrease in CFU-GM and CFU-G count in medium up to the 4th week of cultivation. This was evidence supporting exhausting hemopoietic properties of artificial bone marrow under conditions of long-term culture (Table 3).

Thus, a principle opportunity to design hybrid implants containing bone marrow on porous scaffold has been established in vitro. Certain hemopoietic properties of bone marrow in porous TiNi-alloy in liquid culture are similar to the character of hemopoietic tissue of the whole organism.

Table 3. The content of granulomonocyte and granulocyte precursor cells in liquid phase of mouse myelokaryocytes culture in porous TiNi-alloy, X ± m

Contents	1 cultural week	2 weeks	3 weeks	4 weeks
on 10^5 cells	1.29 ± 0.42	2.8 ± 0.58	3.25 ± 0.24*	3.0 ± 0.33*
Total count	27.35 ± 8.9	36.96±7.66	27.11 ± 2.0	12.78 ± 1.41

Note:* Differences with indices of 1st week at $p<0.05$ according to Student's t-test.

2. Physical-Chemical Regulation of Hemopoietic and Stromal Stem Cells Pull *In Vitro*

In recent years, some data for life cycle regulation of isolated cell lines have been collected. The existing quantity of pharmaceuticals implies the use of their stimulating or inhibiting effect on multiplication of one cell or another. At the same time, stem element reactions on microenvironment's inorganic factors under the conditions of multicellular systems are less studied and rather contradictory. Physical-chemical factors, in particular irradiation and thermal effect, ultraviolet light, and cytostatic drugs induce, in general, a negative effect on cell growth intensity. From the other side, positive effect of some monochromatic light regimes (laser or light-emitting diode) and alternating magnetic field on cell vital activity are possible. However, exposure doses are extreme, as a rule.

The aim was research of the modulating effects of electrical impulses, visible light and some microelements under parameters approximate to physiological, on the vital activity of marrow progenitor cells.

In vitro cell cultivation was carried out using material of 14 healthy male *CBA/CaLac* mice with a mass of 18-20g. Animals were killed by means of ether-anesthesia.

Autonomous electrostimulator of gastrointestinal tract (AES GIT) with or without sputtering of chromium ions was placed into suspension of myelokaryocytes under sterile conditions. The electrostimulator generate pulse current with a pulse duration of 5-7 msec, a pulse amplitude of 9-15 mA, and 16 pulse packet duration of 320-450 msec. Total energy density of electrical impulses varied in the range of W=0.1-3.6 Joule/cm^3.

In another experiment, myelokaryocytes were cultivated with a photon tablet under the temperature of 37°C for 3 hours (incorporeal irradiation). In one of the experiments analogous light exposure on cells was made by means of a photon tablet, which was outside the cell reservoir (extracorporeal irradiation). Autonomous photostimulator (APS, photonic tablet) represents a mini-device with a supply voltage of 4.5V, current impulse amplitude of 4 mA, impulse duration of 6 msec, 16 impulses in packet, pulse packet period of 3 sec, and wave-length of 655± 20 nm (red region).

Viability of marrow nuclear cells was evaluated, supernatants of myelokaryocytes were collected, and colony stimulating and inhibitory activities were determined by colony formation method. Cloning of granulocytic (CFU-G) and fibroblastic (CFU-F) colony forming units was conducted in semisolid nutrient medium for 6 days in the presence of human recombinant CSF-G (ISN) in concentration of 1 ng/ml. Statistical analysis was done according to Student's t-test, Wilcoxon T-test (P_T) and Mann-Whitney U-test (Pu).

The most important parameters of electromagnetic pulse interaction with biological objects are power density (intensity) I (W/cm^2), energy density W (J/cm^2), wave length, exposure regime and duration [17]. Investigation showed that electropulses can have both positive and negative effect on cell growth in vitro. The direction effect depends on duration of exposure on marrow cell suspension, which determines total energy density (table 4).

The mediated electro pulse effect on the cell cycle of progenitor cells, connected with regulation of myelokaryocyte synthesizes function, is possible under the conditions of multicellular systems, apart from direct effect. It was established that supernatants of marrow cells treated in fluid medium in the range of W=0.2-0.4 J/cm^3 activated granulocytic colony-formation in methylcellulose culture (Figure 2).

Further W increase led to the appearance of soluble factors, which depressed CFU-G growth in the presence of recombinant CSF-G. With respect to the CFU-F, supernatants showed only suppression component (W>0.4 J/cm^3) (Figure 2) under the condition of direct electric stimulation of fibroblast precursor cell proliferation. It is significant that electrolysis products had no effect on CFU-G and CFU-F concentration in cell-free medium under W=2,4 J/cm^3. Precursor output in culture didn't differ from the reference level (point 0) and significantly exceeded corresponding values, obtained by the addition of cell supernatants (on 160 % and 67%, P<0.05).

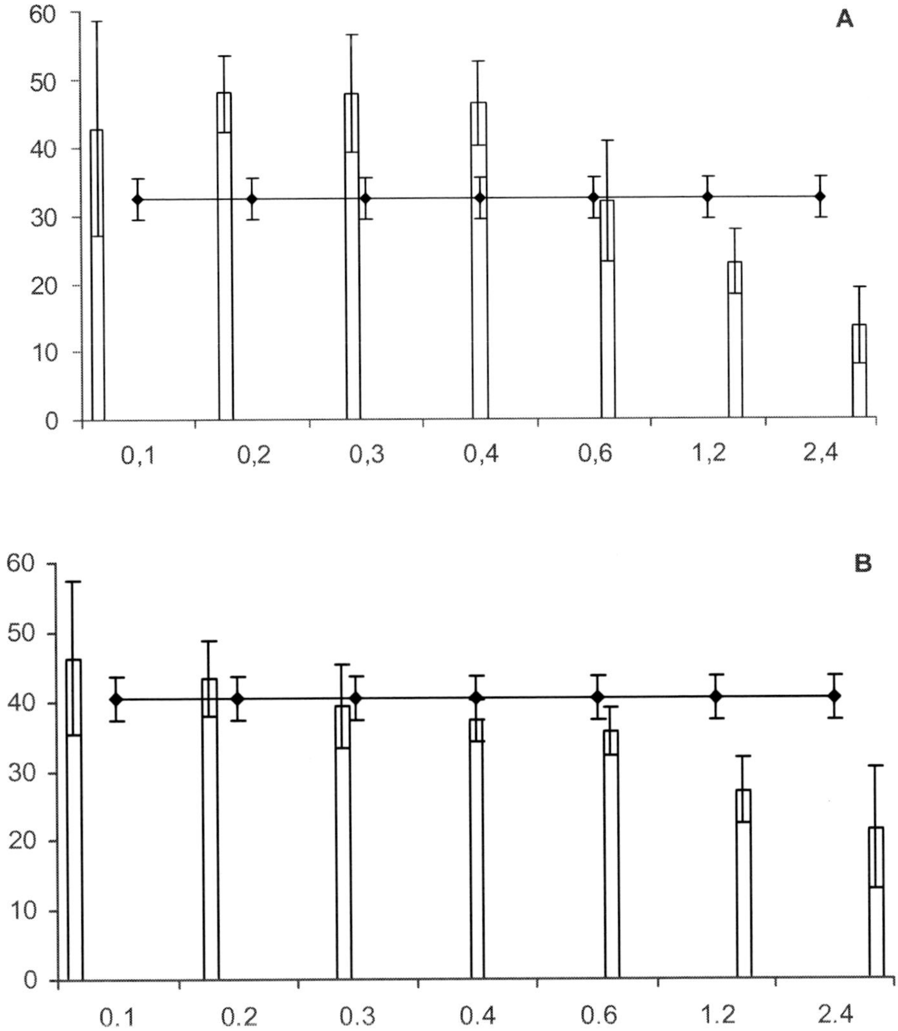

Figure 2. Effect of supernatants of marrow cells treated by electropulses in fluid medium, on CFU-G (A) and CFU-F (B) growth in semisolid culture.

Table 4. Total energy density effect of electropulses on number of CFU-G and CFU-F in culture (% of the control, point 0), X

Energy density, J/cm^3	**0**	**0.1**	**0.2**	**0.4**	**0.6**	**1.2**	**2.4**	**3.6**
CFU-G n=16	100	655*	807*	1010*	706	2116*	100	15*
CFU-F n=16	100	1198*	1928*	1630*	997*	5585*	400	0*

Note: * –reliable differences ($p<0.05$) with control according to Mann-Whitney U-test; n - number of counted wells.

Microelements, in particular chromium, discharge from the electrodes' steel shell during the electrolysis process and can have a known effect on cells. To test the effect of electric exposure ion component, AES GIT with chromium sputtering at the dose, meeting a human daily need, were used. Combined effect of electric field (W=3.6 J/cm^3) and chromium cations on myelokaryocytes inhibited completely the CFU-G and CFU-F growth.

Low-energy, impulse monochromatic light of visible spectrum was also the extreme factor for CFU-G and CFU-F growth. The exposure of red light with a myelokaryocytes suspension inhibited the growth of 80-90% progenitor cells (Table 5) even in presence of CSF-G, which was the powerful apoptosis inhibitor. During the corporeal red impulse light exposure on liquid culture, the energy density was about 1 J/cm^3. The absorbed dose by every myelokaryocyte varied over the range of 0.5 mGy. Naturally, by the removal of the light source from the cell suspension (extracorporeal irradiation), its inhibitory effect on progenitor cells growth and multiplication is leveled (Table 5) as a consequence of photon exposure power decrease, connected, in particular, with intensification of light diffusion and reflection.

Red light photoreception by intracellular oxygen molecules underlies hyper production of oxygen and peroxide active forms, which are mediators of cell death. Our findings are of interest in this respect. According to these data, red impulse light exposure in vitro with energy density of about 1 J/cm^3 was accompanied by significant (P_T) increase of morphological indices of apoptosis and necrosis in all marrow lineages, first of all, in lymphoid cells.

Thus, cell effects of physical-chemical exposures are less specific. Nevertheless, it is possible to induce both positive and negative regulatory effects on the vital activity of hemopoietic and stromal precursors, depending on energy density and irritant concentration. Together with pharmacological and biological medications, it may have a practical importance in preliminary stem cell preparation for transplantation, as they are biologically inert on to a sufficient degree, and require, for example, an induction of proliferation and/or differentiation before use.

Table 5. Progenitor cell concentration (x10^5 nuclear cells) in 6-day culture after the treatment of myelokaryocyte suspension with red light, X

Precursor type	Control (point 0)	Incorporeal irradiation	Extracorporeal irradiation
CFU-G	11.88 n=6	1.25 *n=10	4.38*# n=12
CFU-F	35.83 n=6	8.00 *n=10	21.46*# n=12

Note: (*) – reliable differences with control; (#) – reliable differences with incorporeal irradiation according to Mann-Whitney U-test; n – number of counted wells.

Therefore, the effect on hemopoietic precursors' vital activity can occur not only by means of direct (physical-chemical, biochemical and biophysical) reactions, but also by those mediated through the cell microenvironment, regulating direction and intensity of metabolic intracellular processes. Lymphocytes can be considered, at least, as candidates for microenvironment target cells, which are able to switch the growth program of hemopoietic and stromal marrow precursor cells in multicellular systems depending on irritant strength.

3. Nanoparticles and Magnetic Field Influence on Colony Forming Activity of Unipotent Hemopoietic Precursors *In Vitro*

The biological importance of nanoparticles is reviewed on along the following main lines: 1) the toxicity of the implant's deterioration particles; 2) biocompatibility and specific activity of proper nanoparticles, and 3) properties of nanoparticles as carriers (delivery systems) of drugs and biological molecules.

The active search of biocompatible materials for design of drugs and biological molecule delivery systems is directed on liposomes and solid particles [1]. Here methods of control of a nanoparticle's movement and properties by means of applied magnetic fields are offered. Increasingly more both positive and negative information concerning molecular and cellular effects of nanoparticles is collected.

Functional methods of cell activity determination have a high sensitivity to the nature of the material tested. This work is dedicated to studying metal oxide nanoparticle effects on the colony forming ability of marrow granulocytopoiesis precursors in constant magnetic field.

Metal nano-scale powders were made by method of electrical wire explosion or by mechanochemical synthesis. Electronic microscopy showed (Figure 3) that along with particles of 10-40 nm diameters, their aggregates with size under 100 nm presented in powders. Phase composition of synthesized powders, shown in tables 6-7, is confirmed with X-ray diffraction (XRD) data.

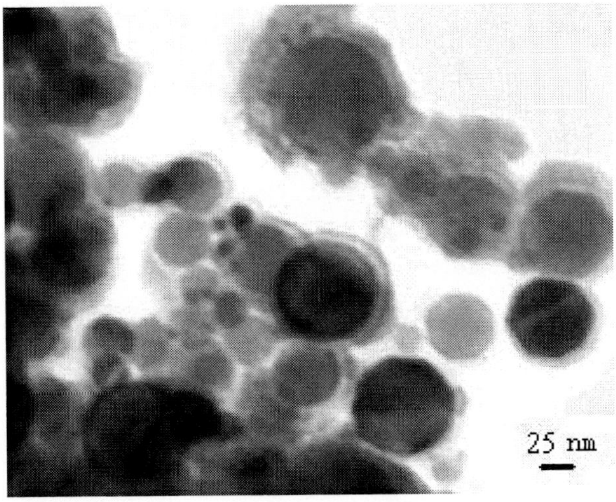

Figure 3. Electronic microscopy of Fe_3O_4 nanosized powder. Magnification 160000.

In vivo experiments were carried out in autumn-winter period using marrow cell material obtained from *CBA/CaLac* mice and *Wistar* rats. Animals were killed by means of ether anesthesia, marrow cells were isolated from femur in concentration of $0.5x10^6$ karyocytes/ml and were cultivated in suspension volume of 4.5ml for 1 hour with powders (3 mg/l of cell culture) of nanoferrimagnetic oxides Fe_2O_3, Fe_3O_4 or $CoFe_2O_4$. A corresponding solvent (sodium chloride isotonic solution, SCIS) volume (0.5 ml) was added into control tubes.

To study sanitary-chemical properties of nanoparticles in the range of 1-10 maximum allowable concentrations (MAC), nanopowders were dissolved in SCIS during 7 days under

37°C according to the orders of ISO 10993-5. The pH of the solution was estimated. A portion of unfractionated cell suspension was used to determine bone marrow colony forming ability. Cytotoxicity of studied nanoparticle specimens was determined in a test with 0.4 % trypan blue before and at 24 hours after their cultivation with myelokaryocytes under the above described conditions.

Long-term cloning (for seven days) was carried out in plastic Petri dishes of 35 mm under 37°C and 5 % CO_2 in concentration of 0.75 x 10^6 of viable marrow karyocytes in 1.5 ml of cultural medium. The nutrient medium (280 mg/l L-glutamine, 40 mg/l of gentamycini sulfas, 15% of embryonic bovine serum, and 85% of medium 199) was used to allow spontaneous colony forming ability of bone marrow cells. Intact myelokaryocyte culture after the addition of SCIS was used as growth control. Half of the cultures were studied for cloning on a magnetic plate with a field intensity of 200 Oersted. Number of colony forming units (CFU) was counted at seven days. Granulocytic precursors (CFU-G) formed colonies of 50 or more nucleated elements, and had granulocytic morphology under azure II - eosin stain. A statistical analysis was done according to Mann-Whitney U-test (Pu).

Aqueous seven-day nanoparticle extracts induced a slight solvent acidulation: in MAC in the range of 0.03-0.12 pH units, and in 10 MAC in the range of 0.17-0.22 pH units. Apparently, nanoparticles can weakly interact with hydroxyl groups, facilitating a hydrogen protons collection in solution. Nevertheless, nanoparticle effect on model biological fluid is within limits, recommended by Russian Standart (+1.0 ед. pH).

Biological experiments showed ambiguous results (Tables 6-7). Nanoparticles had no cytotoxicity in the 24-hour test with trypan blue. No cobalt ferrite effected on CFU-G output in multicellular marrow culture independently of animal species and applied exposures.

At the same time, ferric oxide nanoparticles either did not produce any effect (in rats) or stimulate a functional activity of granulocytic precursors. A constant magnetic field had a modulating (reverse) effect on cells, cultivated with nanoparticles (Tables 6-7).

Table 6. A content of CFU-G (x10^5/ml) in mice cell culture in 7 days of culturing, X

Specimen, n=4	Number of CFU-G in culture	
	Without magnetic field	With magnetic field
Control	3.90	2.75
Fe_3O_4	2.07	5.47* p_1=0,049
$CoFe_2O_4$	4.00	4.20

Note: Here and in table 7, (P_1) – statistic differences with control, (P_2) - with corresponding values without magnetic field.

Thus, functional properties of nanoparticles in multicellular systems, as well as under additional manipulations, are to be studied. They may depend on direct effect, but also be mediated through stem cells' microenvironment factors, the state of which is unsteady in different individuals and under different vital activity. For example, fibroblasts can be target cells of nanoferrimagnetic oxide Fe_3O_4. Magnetite promotes stromal cell adhesion and intercellular contact under magnetic field effect (Figure 4).

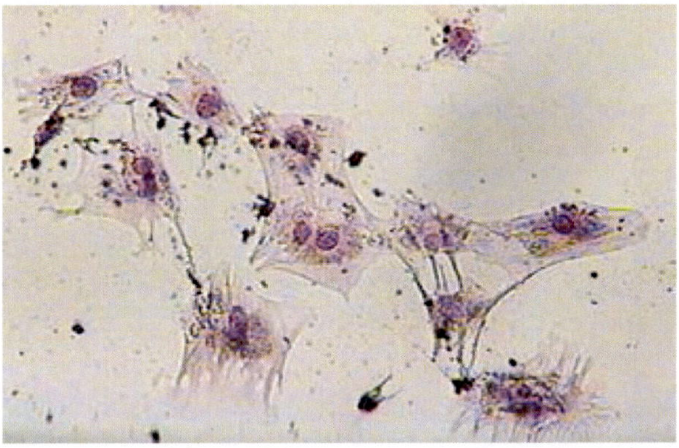

Figure 4. Fibroblasts in bone marrow culture treated by nanosized Fe_3O_4 and constant magnetic field. Hematoxylin–eosin stain. Magnification 700.

Table 7. A content of CFU-G (x10^5 /ml) in rat myelokaryocyte culture at seven days of culturing, X

Specimen,n=4	Number of CFU-G in culture	
	Without magnetic field	With magnetic field
Control	6.22	6.11
Fe_3O_4	9.78*p_1=0,016	6.00
$CoFe_2O_4$	5.33	6.89
Fe_2O_3	9.33*p_1=0,015	4.89* p_2=0,034

4. Blood Cell Differentiation Features in Conditions of Flow-Through Cultivation

An ability of adult stem cells to differentiate not only in cells of their existence, but in many other ones, which led to intense development and innovation of cell technologies in clinic is now described. Transplantation of stem and progenitor cells, isolated from autologous tissues or blood, is approved the diseases of the central and peripheral nervous systems, cardiovascular pathology, endocrine diseases, retina and optic nerve pathology. However, results of similar investigations are usually indefinite and unconvincing. The negative consequences are held back in many cases.

The aim of this work is in vitro study of directed fluid flow effect on the morphofunctional state of stromal precursors circulating in peripheral human blood.

The original device ("bioreactor" in vitro), which allows modeling directed movement of a nutrient medium, was developed. The device is a closed system including a chamber with semipermeable filter. The system was preliminary filled with a nutrient medium consisting of 80% McCoy 5A medium, 20% of embryonic bovine serum and gentamycini sulfas (on the basis of 0.02 ml for 10 ml of medium). Mononuclear cells in the peripheral blood of healthy donors were isolated by means of ficoll-verografin gradient. Isolated cells were brought to the necessary concentration of 3x10^6 cells/ml of a nutrient medium. Cell material was introduced into the chamber and placed on the filter. The chamber was connected with the reservoir through the roller pump, which provided a formation of uniform direct movement of a nutrient medium in a closed system with the constant rate of 2.1-2.4 cm^3/min.

As a control, studied cells were cultivated in steady-state culture on a semipermeable filter, placed in Petri dish with the necessary nutrient medium under the strict observance of temperature regime (37° C), content of CO_2 (5-7%) and 100% humidity level.

Cultivating duration was 24, 48 and 72 hours. On completion of the experiment, cell material was studied by means of cytochemical and histological methods. Examination and photographing of cell material was carried out by means of a morphologist work station, including an IBM PC Pentium personal computer, EPSON digital camera and Karl Zeiss-Yena microscope. Optical density of adherent mononuclear cells was determined in conditional units (c.u.o.d.) by means of the digital image's computer morphometry using PhotoShop 6.0. Statistical analysis was carried out by Mann-Whitney U-test (p_U).

Twenty-four hours after beginning the experiment, mononuclear culture on the filter was presented with strongly adherent cells of round shape with basophilic cytoplasm and large nuclei, which had sometimes a serrate invagination. Nucleus/cytoplasm rate was about 1.

The high cytochemical activity (table 8) of α-naphthylacetateesterase, which was blocked with sodium fluoride, was noted in cells. Alkaline phosphatase was not found in cultivated cells.

Increasing α-naphthylacetateesterase activity as compared with primary index and that of cells cultivated under the standard conditions ($p_U<0.01$), was noted in described cells at 48 hours (Tables 8 and 9). By carrying out the reaction with sodium fluoride, isolated elements, with unsuppressed nonspecific esterase activity and revealed moderate alkaline phosphatase activity, were found on the filter.

In 72 hours, the cell culture on the filter mainly consisted of macrophages with foam cytoplasm, containing vacuoles and lymphocytes. The nucleus was of reniform or round shape with serrate invagination. Nucleus/cytoplasm rate was less than 1. Isolated big cells of irregular, mainly fusiform shape were found among lymphocytes and macrophages. Cytochemically, a higher activity of α-naphthylacetateesterase was noted in mentioned cells in comparison with indexes in 48 hours and with cells grown under the standard (stationary) conditions ($p_U<0.01$; Tables 8 and 9). Nonspecific esterase activity was not suppressed with sodium fluoride. Alkaline phosphatase activity increase as compared with index in 48 hours from the beginning of the experiment ($p_U<0.01$; Table 8) was also found in similar cells. Morphological and cytochemical parameters of the mentioned cells corresponded to active synthesizing fibroblasts.

Table 8. Cytochemical activity (c.u.o.d.) of blood mononuclear cells by the flow-through cultivation in vitro, X, p_U

Cytochemical reaction	Cultivation terms, hours		
	24	48	72
Nonspecific esterase	63.81	69.17p_0	73.39p_1
Fluoride-resistant nonspecific esterase	0	64.93	71.45p_0
Alkaline phosphatase	0	40.17	69.07p_0

Note: Statistically significant differences were noted ($p_0<0.01$) by comparison with primary index; ($p_1<0.01$) by comparison with indices in dynamics of the experiment.

Under standard (stationary) conditions of blood cell cultivation, the cell culture on the filter was presented during each phase of the experiment with cells of beanlike form with beanlike or round nucleus with serrate invagination. Nucleus/cytoplasmatic rate was about or less than 1. Cytochemically, moderate activity of α-naphthylacetateesterase, which was gradually increasing during cultivation ($p_U<0.05$; Table 9), was noted. It is connected with monocyte differentiation into macrophages. Nonspecific esterase activity was inhibited by sodium fluoride. An alkaline phosphatase was not founded in cultivated cells.

Thus, the increase in intracellular enzymatic activity of human peripheral blood mononuclear cells was noted under conditions of directed nutrient medium movement ("bioreactor" in vitro). A monocyte differentiation into macrophages and fast differentiation of circulated stromal precursors into mature collagen synthesized forms are sped up under the modeling effect of microenvironment factors (directed fluid flow, extracellular matrix).

The data allow extending the conceptions about the effect of physical microenvironment factors on the morphofunctional state of young stromal cells and the opportunity to realize their undesirable fybrogenic potential under cell therapy manipulation.

Table 9. In vitro comparison of α-naphthylacetateesterase activity (c.u.o.d.) in blood mononuclear cells under different conditions, X, p_U

Cultivation conditions	Cultivation terms, hours		
	24	48	72
Bioreactor in vitro	63.81 n=25	69.17 p_0 n=25	73.39 p_0 n=25
Steady-state culture	62.38 n=25	63.95 n=25	64.08 n=25

Note: statistically significant differences were noted ($p_0<0.01$) as compared with results under standard conditions; n –number of counted cells.

5. Artificial Surface Effect on Osteogenous Differentiation of Mesenchymal Stromal Marrow Cells *In Situ*

Osteogenous properties of hydroxilapatite and tribasic calcium phosphate, which are the base of mineral bone matrix, are obviously proved per ectopic osteogenesis phenomenon, when bone tissue is generated on the surface of calcium phosphate (CP) materials. Urist was one of the first scientists, who described this phenomenon in 1965.

Multipotential mesenchymal stromal cells are able to differentiate into osteocytes, chondrocytes, fibrocytes, adipocytes, neurocytes, myocytes and stromal cells, which can maintain hemopoiesis. However, the choice of differentiation line under the conditions of multicellular systems is still unclear.

Different specimens of CP materials have different abilities for osteogenesis depending on their physical-chemical properties (crystallinity and porosity degree, solubility, surface roughness, etc.). A key combination of their structure, thickness and dissolution rate for realization of osteogenous potential of mesenchymal stromal cell pull (MSCP) had not been found until now.

In this connection, the complex value of physical-chemical parameters of CP material surface in regulation of stromal precursor's differentiation was both of theoretical and practical interest.

Sixty-two BALB/c male mice were used in the experiment. One implant (diameter of 12 mm, thickness of 1 mm) with applied syngeneic marrow column (average marrow area is 7.5 mm^2), taken from femur under aseptic conditions, was introduced subcutaneously into each animal (49 animals) under ether-anesthesia. Bone marrow was cultivated for the adhesion on the substrate for 45 min in culture medium, containing 95% of RPMI-1640 and 5% of embryonic bovine serum. The marrow was a source of MSCP and growth factors. No generation of tissue plates under separate subcutaneous introduction of scaffolds or marrow fragments was noted in the mice. The implants were released in 45 days. Standard methods of light thin slice microscopy were used for histological analysis of tissue plates.

Titanium discs were used as substrates. Artificial biocompatible (bioinert or calcium phosphate) surfaces were formed on discs by use of hydroxilapatite (HAP) micro- and nanoparticles by the means of anode-spark (microarc) oxidation or schlicker method. Electronic CP surface treatment was done in a vacuum of 10^{-7} tore with constant excitation energy of 100 eV and electron-flux density of 100 nA/cm^2 in a period of 2 sec – 24 hours with the gun of Kimball Physics, Inc. (USA).

Surface morphology of biocompatible coatings, their element and phase compositions were determined by scanning electronic microscopy (SEM), micro X-ray spectroscopic analysis (MRSA), X-ray diffraction (XRD) and atomic-force microscopy (AFM) methods. The coating thickness and roughness were also measured. Statistical analysis was carried out by Mann-Whitney U-test (Pu). Correlation analysis was carried out according to Spearman rank correlation coefficient (r).

At present the major achievements attained are in induction of MSCP differentiation in vitro that demonstrates comprehensive opportunities for cell therapy. Nevertheless, a realization of the necessary potential of stromal precursor cells is more difficult in vivo. In our opinion, scaffold-technology allowing a purposeful creating of specific microenvironment for MSCP is the problem to address in many cases. In particular, CP materials are a compatible carrier for osteogenous cells.

Physical-chemical testing allowed assigning the following groups of biocompatible surfaces conventionally: 1) metal-ceramic (MC); 2) calcium-phosphate dense (CPD) X-ray amorphous and CPD crystal coatings determined as smooth, calcium-deficient (atomic calcium/ phosphorus ratio is less than 1.67), thin, finely-porous (pore diameter less than 50 μm); 3) calcium-phosphate loose (CPL) X-ray amorphous and CPL crystal noted as rough, calcium-deficient, thick (thickness more than 50 μm), finely-porous (pore's diameter less than 50 μm); and 3) calcium-phosphate glass ceramic (CPG) with nonstoichiometric ($Ca/P > 1.67$), thick, macroporous surface (diameter about 200 μm).

Tissue reaction showed that there were no signs of inflammation in any observation group at 45 days after subcutaneous implantation of studied discs with syngeneic marrow. It is necessary to make a note of implant biocompatibility, causing only weak encapsulation, which indicates insignificant reaction of subcutaneous tissue on the implant's introduction. Marrow sites of different size, adipose tissue, loose shapeless connective tissue with ossification elements, and bone tissue were revealed on different artificial surfaces. The presence of osteoblasts indicated an active bone tissue formation (Figure 5).

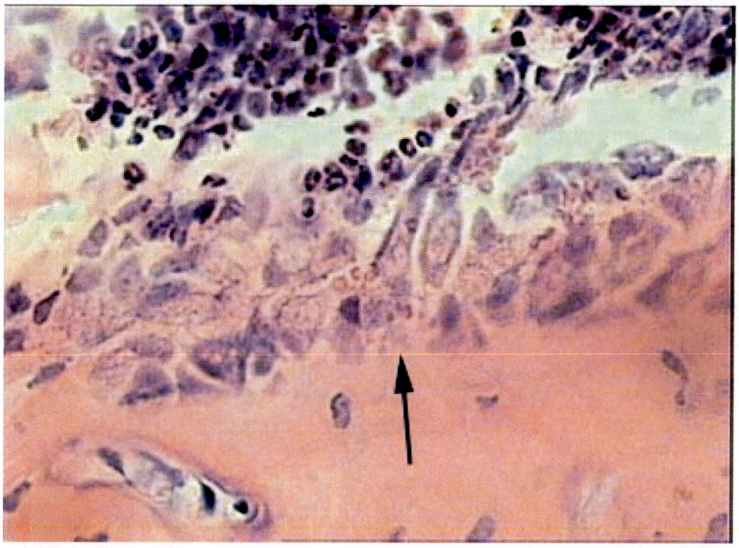

Figure 5. Osteoblasts immersion in bone matrix for forming bone tissue on calcium-phosphate surface. Hematoxylin - eosin stain. Magnification 400.

Thus, biological tests showed that CP surfaces with "primitive" physical-chemical composition, which was not corresponding to the last one in bone, had an active effect on MSCP functional properties. The four of six CP coatings, characterized by marked surface roughness or macroporosity, had an osteoinductive potential. CPL and CPG artificial surfaces had maximum probability (100%) of osteogenesis induction. CPL crystal coatings realized osteogenous potential to a lesser degree (67%).

In this connection, the attempt to clarify statistically the most important question was made: what surface properties are governing for initiation of MSCP adhesion, migration and osteogenous differentiation, leading to metaplasia of marrow and connective tissue into a mature bone? Correlation analysis allowed revealing of differential value of physical and chemical parameters of artificial surfaces for the regulation of stromal cell pool differentiation in situ. Marrow cell adhesion is the initial stage, predetermining their conduction on the artificial surface (the area of tissue plate). Histogenesis success depends on surface roughness.

The optimal surface roughness is 4-5 class according to the Russian standard that allows a differentiating of progenitor stromal cells into connective and adipose tissue under the conditions of a specific microenvironment. Microscopic examination of active hemopoiesis (Figure 5) also implies an additional MSCP differentiation into stromal cells, forming a hemopoiesis inductive microenvironment. Adhesion and movement of stromal cell (including osteogenous precursors) is necessary, but insufficient conditions for their differentiation into bone tissue. Induction of stromal precursor's osteogenous potential (osteoinduction) is determined by ratio of calcium and phosphorus atoms in coating. The presence of hydroxilapatite (atomic ratio Ca/P=1.67) is not the obligate requirement for osteogenesis. In the offered experimental system, MSCP differentiation in osteogenous tissue is blocked only at Ca/P < 0.5. Ectopic osteogenesis courses successfully on "thick" surfaces. Minimal CP layer thickness must be 50-80 μm.

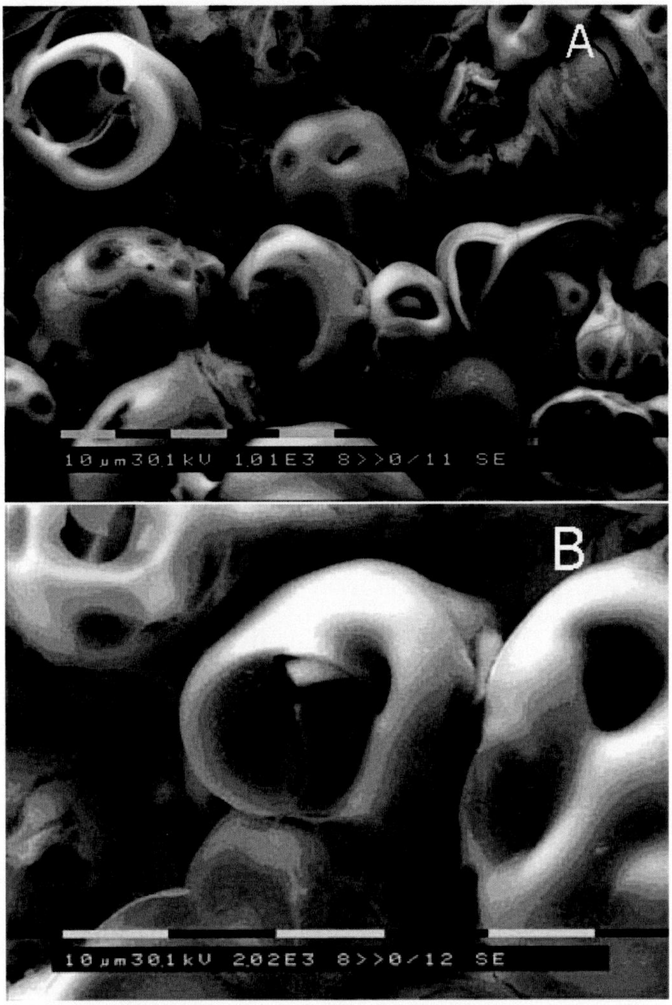

Figure 6. SEM of calcium phosphate coatings designed by microarc oxidation from HAP micro – (A) and nanoparticles (B).

In the case of formation of CP coatings on titanic substrates from nanosized (30-50 nm) HAP particles, relief does not practically differ from the relief by use of HAP microparticles (Figure 6). Nevertheless, the probability of bone plate formation on nanosized CP coatings was only 12.5% (the 1 case of 8). In other cases, loose shapeless connective tissue grew. It was interesting to discover the little-studied fact that additional treatment of nanostructured CP coatings with electron bunch increased by several times marrow cell conduction on the surface. Here, bone plates were formed closed from two sides. The red bone marrow was between these bone plates. An opportunity for bone tissue formation (osteoinduction) has reached 100%.

Stem and committed cells pool is a heterogeneous population that is under the control of distant and local regulatory mechanisms. Structure-functional units (microregions), including special cells, stromal elements, microcirculation vessels and terminal nerve endings are educed in every tissue. Research shows that not only organic, but also well-defined physical and chemical parameters belong to the stromal cell microenvironment. Their local or system changes may have an influence on stromal functional activity.

6. Effect of Human Fetal Liver or Umbilical Cord Cell Implantation on Dynamics of Spontaneous Leukemia

The high frequency (up to 91% forward to 7-10th months) of spontaneous virus-induced thymus lymphoma leading to animals' death has been observed in AKR/JY mice during aging [2]. Lymphoma is considered by different authors to be a variant of haematosarcoma and, on the other hand, as a solid tumor. Hemopoietic cell transplantation is a perspective method of leukemia treatment. In this connection, the effect of diverse human xenografts on the blood system and on tumor progress in AKR/JY mice had an evident interest.

Table 10. Indices of peripheral blood (10^9 / л) in the dynamics of *AKR/JY* mice aging (control) following an implantation of artificial human liver in porous TiNi-alloy (experiment), X±m

Indices	3 months	
	Control	Experiment
Band neutrophils	0.07 ± 0.02	0.22 ± 0.07
Polymorphonucleated neutrophils	1.47 ± 0.15	2.31 ± 0.29 *
Eosinophils	0.2 ± 0.03	0.19 ± 0.03
Monocytes	0.13 ± 0.03	0.23 ± 0.04
Lymphocytes	3.76 ± 0.34	3.86 ± 0.62
Total leukocytes count	5.61 ± 0.31	6.86 ± 0.87
Number of erythrocytes with micronuclears	4.7 ± 0.67	4.17 ± 0.51
Indices	**4 months**	
	Control	**Experiment**
Band neutrophils	0.04 ± 0.03	0.03 ± 0.02
Polymorphonucleated neutrophils	1.34 ± 0.66	1.11 ± 0.22
Eosinophils	0.09 ± 0.03	0.07 ± 0.03
Monocytes	0.23 ± 0.08	0.28 ± 0.10
Lymphocytes	5.12 ± 0.87	6.65 ± 0.74
Total leukocytes count	7.69 ± 0.85	8.14 ± 0.67
Number of erythrocytes with micronuclears	1.69 ± 0.28	1.84 ± 0.16

Table 10. (Continued)

Indices	5 months			
	Control		Experiment	
Band neutrophils	0.01	± 0.01	0.02	± 0.02
Polymorphonucleated neutrophils	1.82	± 0.15	2.07	± 0.56
Eosinophils	0.16	± 0.04	0.15	± 0.05
Monocytes	0,24	± 0.10	0.29	± 0.08
Lymphocytes	7.08	± 1.18	6.17	± 0.85
Total leukocytes count	9.32	± 1.27	8.21	± 0.93
Number of micronuclear erythrocytes	1.82	± 0.32	2.35	± 0.19
Indices	**7 months**			
	Control		**Experiment**	
Band neutrophils	0.01	± 0.01	0.00	± 0.00
Polymorphonucleated neutrophils	1.83	± 0.25	1.84	± 0.41
Eosinophils	0.47	± 0.16	0.27	± 0.21
Monocytes	0.32	± 0.11	0.66	± 0.26
Lymphocytes	7.27	± 0.71	4.49	± 0.39 *
Total leukocytes count	9.92	± 0.93	7.25	± 0.75 *
Number of micronuclear erythrocytes	2.64	± 0.28	2.06	± 0.33

Here, and in table 11-13: *- Significance with control at $p<0.05$ according to Student's t-test.

Human artificial liver was constructed in vitro by means of seeding fetal cells of 11 week gestation in a final concentration of 20 x 10^6 into one plate of porous TiNi-alloy. Xenografts with human umbilical cord cells were designed in vitro by means of seeding mononuclear cells in a final concentration of 3-6x10^6 into one porous polymer carrier. Biodegradable co-polymer of hydroxybutyric and hydroxyvaleric acids (PHB/PHV) was synthesized by *Ralstonia eutropha* B5786 bacteria at the Institute of Biophysics SB RAS (Krasnoyarsk) [18]. The PHB/PHV samples had a weight-average molecular mass of 160 kDa and crystallinity 65%. Their melting points and the temperatures for the onset of decomposition were 168°C and 267°C, respectively.

One of the hybrid implants was implanted intraperitoneally (TiNi-alloy, Figure 7) or subcutaneously (polymer matrix) into 28 mice 2-3 months old under ether-anesthesia. Thirty-two mice were investigated in a control group (without implantation). Blood indices were studied at four days and 4-9 months. An estimation of the state of blood system organs (bone marrow, peripheral blood, thymus, and spleen) was necessary to determine the effectiveness of implants and reasons for animal death in the postimplantation period.

The data show that TiNi-alloy xenograft has no any influence on white blood indices in the dynamics of mice aging (Table 10). The 3rd month was an exception, when postoperative neutrophilia was noted in recipients 4 days after the implantation. Lymphocytes number decreased forward to the 7th month only.

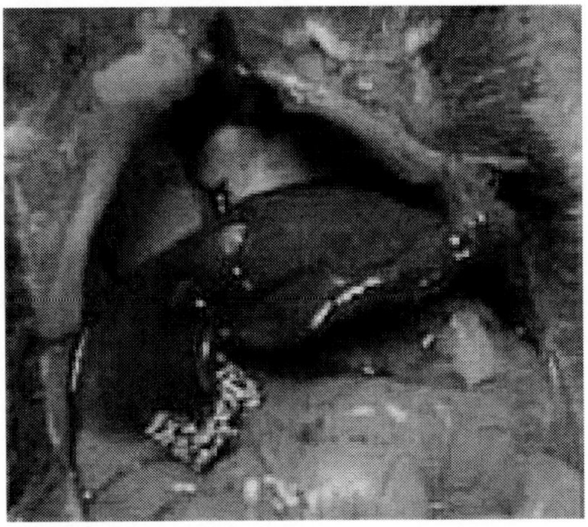

Figure 7. Xenograft connection with liver of recipient.

At the same time, liver xenograft implantation induced a marked stimulating effect on erythron. So, reticulocyte number in blood increased considerably (Table 2), which testified to activation of erythropoiesis processes in recipient's bone marrow.

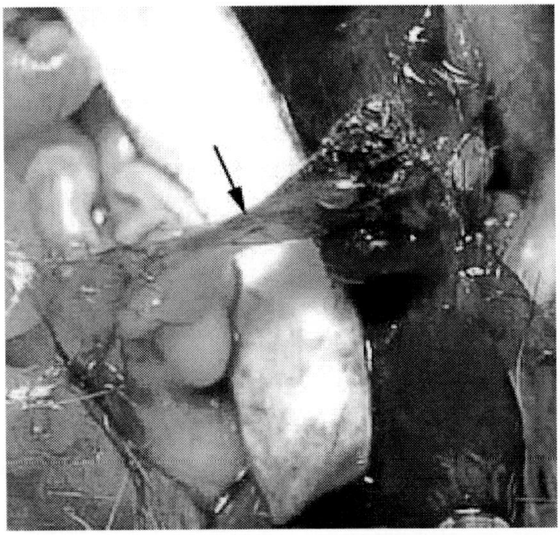

Figure 8. Blood vessels ingrowth into porous structure of xenograft.

On the other hand, content of erythrocytes with fetal hemoglobin in peripheral blood increased 3 times forward to the 4th month (Table 11). Fetal liver is known to be an active hemopoietic organ with high concentration of erythroid progenitor cells. Thus, it may testify active functioning of the xenograft which produces donor erythrocytes with fetal hemoglobin into the blood of the recipient.

Further (5-7th months of mice life), no statistical differences were marked in control and experimental data. It allows concluding that the active work of human fetal liver in mouse organism averages one month. Biocompatibility and functioning of xenograft are confirmed by ingrowth of recipient's blood vessels into the porous structure of TiNi-alloy (Figure 8).

First signs of immunological conflict, accompanied by falling mice body weights, reduction of leukocytes count, splenomegaly and gradual decrease (Table 11) in activity of host erythropoiesis were found two months following the artificial human liver implantation. Amount of animal death from the graft's xenogenecity was 33 % (Table 12).

It should be supposed that immunological conflict developed according to graft-versus-host reaction because of a high cellularity of human fetal liver cells. At the same time, percentage of lymphomas fell considerably, which could be an expression of graft-versus-tumor reaction.

Biomedical results of xenotransplantation of human umbilical cord cells on PHB/PHV co-polymer carrier were similar (Table 13). Animals' death was prevented by decreasing cell dose. Survivability and life expectancy of tumor mice sharply increased (Table 13).

It was reported that *in vivo* resorption of polyhydroxyalkanoates, including 3- and 4-polyhydroxybutyrate, involved an active participation of phagocytic macrophages and foreign body giant cells (FBGCs) [20].

Table 11. Content (%) of erythrocytes with fetal hemoglobin and reticulocytes in the peripheral blood of AKR/JY mice of different ages (control) following an implantation of artificial human liver in porous TiNi-alloy (experiment), X±m

Mouse age (months)	Fetal hemoglobin		Reticulocytes	
	control	experiment	control	experiment
3	-	-	8.8± 1.75	32.88± 6.58 *
4	1.25± 0.37	3.19±0.37 *	10.5± 0.63	23.80± 3.65 *
5	2.14± 0.49	3.83± 0.97	16.5± 2.74	19.5± 4.4
7	3.67± 1.29	7.00± 2.04	12.43±2.94	6.83± 4.94

Table 12. The reasons of *AKR/JY* mice death in the dynamics of aging (control) following an implantation of human fetal cells in porous TiNi-alloy (experiment)

Group	Lymphomas %	Total death %	Death because of immunological conflict %	Death because of unknown reasons %
Control	70	50	0	10
Experiment	44	89	33	22

Macrophages and FBGCs phagocytize and resorb biodegradable material (Figure 9). Besides phagocytosis, cells initiate the inflammation and nonspecific antitumor immunity. Thus, independently of cells and materials, implanted xenografts allow long-term stimulating mechanisms of natural resistance and of graft-versus-tumor reaction.

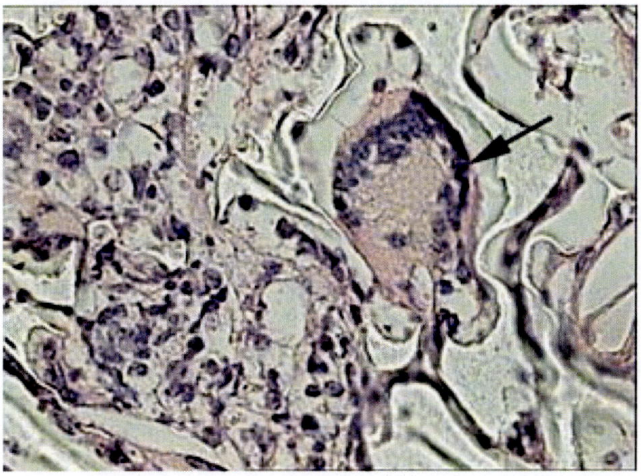

Figure 9. Foreign body polynuclear giant cell in a pore of polymer scaffold seeded by human umbilical cord cells and implanted subcutaneously. Hematoxylin–eosin stain, magnification 600.

Table 13. The causes of AKR/JY mice death in the dynamics of aging (control) following an implantation of human umbilical cord cells in porous polymer matrix (experiment), X ± m

Group	Lymphomas %	Total death %	Average longevity (days)
Control	71	58	181 ± 15
Experiment	60	0	243*

7. Effect of Human Fetal Cell Transplantation on Dynamics of Experimental Toxic Hepatitis

According to the data of the World Public Health Organization, the mortality rate from chronic liver failure is fifth among the other diseases. Different treatment techniques as well as traditional and alternative ones have been developed as of this up-to-date research point. Cell therapy is thought the most prospective method. Nevertheless, concrete mechanisms of therapeutic action and application regimes of cell medication are less studied, especially in cases of chronic pathology.

Some reports about genetic and epigene changes in adult and embryonic stem cells under long-continued cultivation have lately appeared. It's recommended to use those cells, which have passed through the minimal quantity of passages in culture [8]. From the other side, xenotransplantation methods are allowed in clinical trials on the grounds of donor material deficiency [15].

In this connection, clinical attestation of hepatocytes and updating of mechanisms, providing a basis for disease correction with the help of transplantation of xenogeneic embryonic cells in primary culture, are of fundamental and practical interest.

An experiment was conducted in the autumn – winter period on 73 *Wistar* rats of both sexes with mass of 180 – 250 g, kept under standard conditions. Toxic hepatitis was generated in the rats by subcutaneous injection of CCl_4 (carbon tetrachloride) at the rate of

0.4 ml/100 g of animal body mass in oily solution, twice a week for four weeks. In the process of experimental hepatitis modeling, the animal mortality was 6%. Biochemical blood indices and liver histological sections were estimated for hepatitis signs before therapy on the 30th day following hepatitis induction.

Primary cell culture of human embryonic liver of 14-18 week gestation for injection was obtained from the Institute of Cell Culture of FPUE SSC VB "Vektor" (Novosibirsk). The cells were received under therapeutic abortion without any infection marker presence in the mother or fetus according to PCR (polymerase chain reaction) and IEA (immune-enzyme analysis) data. Cell composition of human embryonic liver after unfreezing was: hepatocytes (49-54%), erythroid normoblasts (30-35%), lymphocytes (4-8%), fybroblastoid cells (3-5%), macrophages (3%), and single granulocytes. Used cells were quite well preserved after the unfreezing. Viability was 51±12% (42-85%). However, it's necessary to note that small cells survived, and big hepatocytes (size about 20 μm) died, as a rule. Attempts to separate hepatocytes from other cells sharply decreased viability of the culture.

Transplantation of 10^7 cells/kg was carried out by a single subcutaneous injection in 0.5 ml of phosphate buffer. Single-stage course exposure of direct electric current was begun in the liver region with the help of galvanization apparatus "Potok-1". Current strength in all leads varied from 3A to 5A, and current density was 0.9-1.5 mA/cm^2. Time of exposure was 5 min; course of treatment was 5 days. Animal output from the experiment was done by ether-anesthesia at 7, 14, and 28 days following cell transplantation.

Biochemical indices in blood serum, in liver homogenate were concentration of malonic dialdehyde (MDA) and conjugated bilirubin, catalase and aminotransferase (aspartate aminotransferase, AsAT; alanine aminotransferase, AlAT) activity, thymol assay.

Liver parts were fixed in 10% neutral formalin for 24 hours. Thin paraffin sections were prepared, stained with hematoxylin and eosin by standard method. Number of binuclear and necrotic hepatocyte forms (karyopycnosis, nucleus fragmentation, karyorhexis, vacuolization of nucleus and cytoplasm, karyolysis, and cytolysis) was determined on 100 met liver cells. Data were statistically treated according to Student's t-test.

Table 14. Number (%) of necrotic hepatocytes in dynamics of toxic hepatitis correction, X±m

Research terms, days	Electrophoresis	Cell therapy	Cell therapy+ electrophoresis	Hepatitis control
7	12.6±3.32	8.17±1.19$^{\#}$	7.8±1.16$^{\#}$	15.33±1.76
14	9.0±1.1	7.33±0.88$^{\#}$	7.67±1.63$^{\#}$	14.14±4.04
28	5.4±0.24$^{\#}$	6.4±0.24$^{\#}$	7.17±0.87$^{\#}$	14.6±2.29

Note: here and in the table 15 (*) – significant differences (p<0.05) in comparison with reference level; (#) - in comparison with hepatitis (control).

Single subcutaneous injection of human embryonic hepatocytes culture had a weak influence on dynamics (7th, 14th, and 28th days after cells administration) of redox processes in rat liver homogenate with chronic toxic hepatitis. At the same time, histological sections showed a definite hepatotropic effect of human embryonic cells injection. The number of necrotic hepatocytes was decreased by 2-3 times in dynamics of post-toxic rat liver regeneration (Table

14). Organ repair was stimulated to some extent. As a result, hepatitis serum indices (AlAT, AsAT) were diminished to the level of intact animals by the end of the observation period.

Table 15. Dynamics of biochemical blood serum and liver homogenate indices after 5-days of electrophoresis in rats with chronic toxic hepatitis, X ± m

Research terms, days	Liver homogenate		Blood serum			
	Malonic dialdehyde, μM/l	Catalase, mM/min/g of protein	AsAT, μKAT/l	AlAT, μKAT/l	Un-conjugated bilirubin, μM/l	Thymol test, SH units
Reference level						
0	5.30 ± 0.45	0.14±0.01	0.22±0.03	0.30±0.03	22 ± 6	1.09±0.21
Hepatitis (control)						
7	7.8±0.98	0.21±0.06	0.66±0.05*	0.89±0.07*	46 ± 3*	3.64±0.36*
14	6.5 ± 1.38	0.49±0.07*	0.67±0.02*	0.93±0.02*	46 ± 2*	3.61±0.39*
21	3.74 ± 0.27*	0.46± 0.10*	0.58±0.03*	0.89±0.04*	44 ± 2*	3.45±0.36*
Electrophoresis and hepatitis						
7	13.55±1.8*#	0.15 ± 0.02	0.40±0.01*	0.82±0.05*	22 ± 1#	1.09±0.26#
14	6.71 ± 0.83	0.15± 0.03#	0.54±0.07*	0.76±0.02*#	20 ± 2#	1.67±0.21#
21	2.97 ± 0.71*	0.28± 0.04*	0.55±0,03*	0.63±0.04*#	29 ± 6	2.69±0.30*

Throughout the period of examination (7-28 days after cell injection), the index of integral cell transplant efficiency allowed th expectation that xenogeneic liver cells realized a hepatoprotective effect, which was not connected with braking of lipid peroxidation (LPO) inside of liver. In turn, the course of electrophoresis also improved some homeostasis indices (Table 15) and liver morphology (Table 14) for a long-term period after the short-term LPO intensification due to a seemingly known anti-inflammatory effect of physiotherapy manipulations. Nevertheless, the cell therapy effect was more stable.

The combination of two methods of toxic hepatitis correction hasn't led to a prospective synergic effect. The indices didn't differ from the regime of cell monotherapy. On the other hand, if electrophoresis is considered an additional irritant (Table 15), forwarding the realization of resistant adaptation strategy [6], it's possible to conclude that cells can function under mixed physical-chemical damages.

Apparently, embryonic liver cells work for some time in a substitutive therapy regime, binding endogenous and exogenous toxins, generating albumin and other proteins (transferrin, globulins), which facilitates postnecrotic liver reparation according to a strategy of tolerant adaptation [6] to extreme irritators.

Clinical Application of Hybrid Implants and Cell Manipulations

1. Antibacterial Phototherapy of Adenoid Diseases

At the present time, as a result of the growing resistance of pathogenic microbes to antibiotic therapy, and the great number of complications related to such drugs (allergic reactions, ototoxicity, hepato-, nephro-, cardiotoxicity, etc.), much attention is paid to the search for and development of new methods of antibacterial therapy using physical or combined physical-pharmacological factors. In the late 80s and the early 90s, the high effectiveness of phototherapy on a wide variety of diseases, including those caused by microorganisms – otitis, tonsillitis, pyodermia, and festering wounds – was demonstrated. It has been experimentally proven that some pigments have a pronounced antibacterial effect under exposure to light [3,9,21]. Photopharmacology, a new scientific branch, has been formed.

The aim of this research was to carry out clinical trials of photodynamic therapy for adenoid disease in children.

The course of treatment (one procedure per day for 10 days) was carried out on a group of 42 patients aged below 14 with a chronic form of tonsillitis. Methylene blue and chlorophylliptum were used as photosensitizers. A count of blood indices was done by standard methods. *In vitro* experiments were carried out with pathogenic strain 209 of *Staphylococcus aureus* is characterized by standard morphologic, biochemical, hemolytic and plasmocoagulating properties. A microbe dredge was prepared in sterile conditions in flasks with injection water ($4x10^6$ microbes per ml) in accordance with turbidity standards. The dredge was mixed with water solution of methylene blue in 1:1 proportion in such a way that the final concentration of the pigment was 0.1% and cultivated for 1 minute. The control culture of microbes was diluted with solvent (injection water).

0.1 ml samples of the dredge from each group ($2x10^5$ microbes) were put on agar nutrient medium (ANM) in sterile plastic dishes, 9 cm in diameter, rubbed into agar for two minutes and cultivated for 24 hours at 37°C. Bacterial culture at the initial (exponential) phase of growth was exposed to the action of an autonomous photostimulator (APS) for 3 hours, which emits light in impulse mode in the red (650 nm) parts of the spectrum ("Duny", Tomsk) *ex situ* (by putting the stimulator on the dish lids). To reduce the loss of light the dishes were covered with foil. Simultaneously, ANM was cultivated without APS (control growth) and without microbes as a medium sterility test. The results of the growth were stated by digital photography. Then, with the use of PhotoShop 6.0 software, the number of colony forming units (CFU) of microbes per 1 cm^2 of the surface and per entire dish (survival test), average colony size (reproduction test), and area occupied by bacterial colonies were determined.

To estimate the light influence on myelokaryocytes *in situ* (in normal conditions of cell life support), femur bones of BALB/c mice were cultured with APS ("Duny", Tomsk) for 3 hours in 6 ml of RPMI-1640 medium at 37°C. Morphological signs of apoptosis and necrosis were determined on bone marrow smears stained by azure II – eosin. Apoptosis markers were apoptotic bodies, decreasing cellular size, nuclear chromatin fragmentation and margination and cytoplasmic blebbing. Karyopiknosis and karyorhexis were calculated as necrosis signs.

Spectral research has been conducted at the Chair of Optics in Saratov State University (Head of the Chair, Prof. V.V. Tuchin). To irradiate tonsils a physiotherapy device developed by "Duny" and approved by the Ministry of Health was used. In this device, light is produced by LEDs. Spectra of light of the "Duny" device are shown in Figure 10. It shows that the maximum in the red part of the spectrum lies between 650 and 670 nm and the infrared maximum lies between 940 and 960 nm.

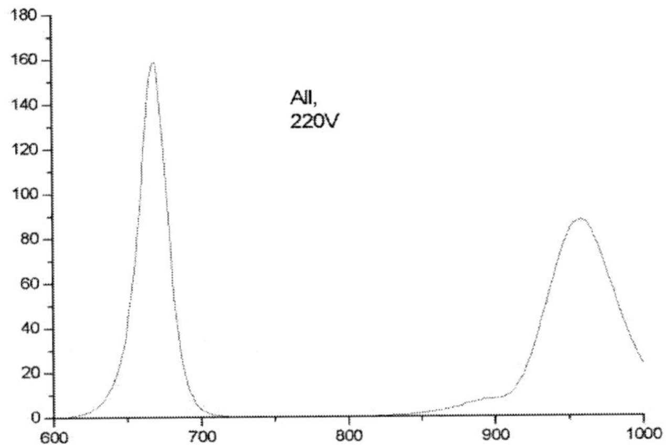

Figure 10. Spectral characteristics of LED physiotherapy device.

Statistical manipulation was done with the use of the Mann-Whitney U-test (Pu) and Student's t-test (P).

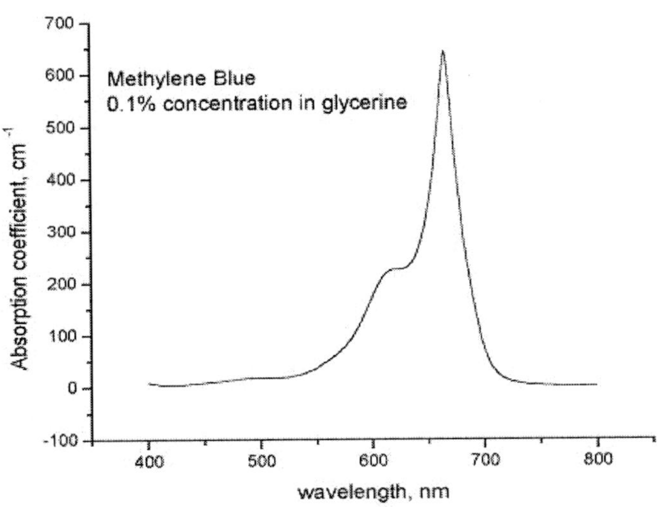

Figure 11. Methylene blue absorption spectrum.

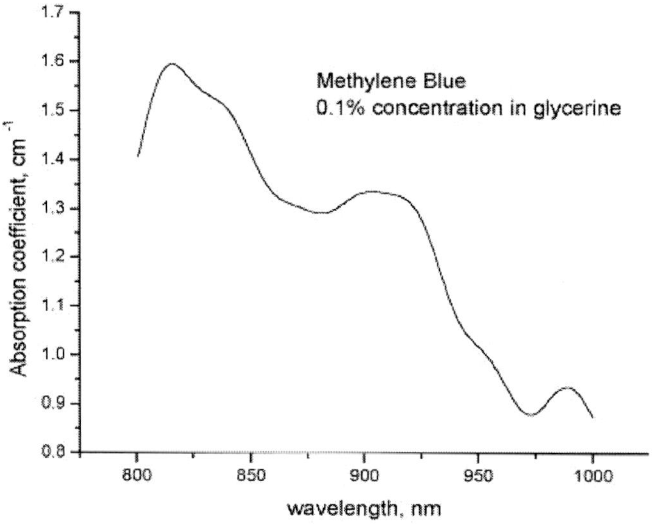

Figure 12. Methylene blue absorption spectrum in IR region.

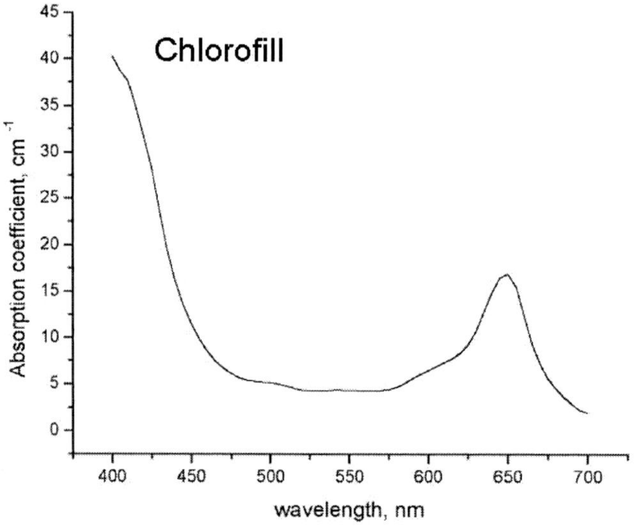

Figure 13. Chlorophylliptum absorption spectrum.

For photodynamic treatment, pigment-containing pharmacological medications approved for clinical use were chosen: methylene blue solution and chlorophylliptum. Methylene blue and chlorophylliptum's absorption spectra are shown in Figures 11-13, where it is clearly seen that methylene blue has an absorption maximum in the 640 to 670 nm region and two peaks in the infrared region from 820 to 830 nm and from 900 to 940 nm. Chlorophylliptum has a maximum in the red part of the spectrum – 640 to 660 nm (Figure 13).

Thus, when comparing spectra of radiation of the LED device and absorption spectra of the medications used, it should be noted that absorption maximum of both methylene blue and chlorophylliptum in the red part of the spectrum coincides with the LED device's radiation maximum, but the methylene blue absorption is considerably more intense than that of chlorophylliptum. Therefore, one may expect the antibacterial photodynamic effect of methylene blue to be noticeably higher than the effect of chlorophylliptum. So, to conduct further investigations, methylene blue only was used. Besides, methylene blue is one of the most efficient singlet oxygen-generating dyes after photoactivation [11].

Irrigation of palatal tonsil lacunas was done directly with a 20 cm^3 syringe with a Hartman needle. After the tonsil irrigation with 0.1% methylene blue solution; the patient's throat was rinsed with 0.9% physiological solution (100 ml). After rinsing, the patient's tonsils were irradiated for 2 minutes with the LED matrix device with a nozzle specially designed for this purpose.

Table 16. Changes of peripheral blood indices in patients with adenoid disease after a course of photodynamic therapy with methylene blue, X + m, P

Index	Before treatment	After treatment
Erythrocytes, T/l	4.27 ± 0.06	4.04 ± 0.06 *
Hemoglobin, g/l	133.7 ± 2.58	121.9 ± 2.15 *
Leukocytes, G/l	5.59 ± 1.64	5.55 ± 1.76
Eosinophils, %	3.65 ± 0.68	2.77 ± 0.39
Band neutrophils, %	2.00 ± 0.30	1.92 ± 0.23
Polymorphonuclear Neutrophils, %	39.46 ± 1.36	42.92 ± 1.82 *
Lymphocytes, %	47.04 ± 1.59	44.27 ± 1.81
Monocytes, %	6.65 ± 0.59	7.85 ± 0.69
Reaction of erythrocytes precipitation mm/h	5.62 ± 0.61	4.42 ± 0.56

Note: *) – significant differences ($p<0.05$) in comparison with level before treatment.

After a course of photodynamic therapy, a minor but reliable reduction of erythrocytes and hemoglobin content ($P<0.05$) is found in patients in 10 days (Table 16). This testifies to the fact that lipid oxidation is intensified because of accelerating photochemical reactions in the presence of a photosensitizer (hemoglobin + methylene blue). A reliable increase in number of polymorphonuclear neutrophils is also noted and this is an indirect evidence of activating the microphage component of the organism's nonspecific resistance. The table demonstrates also that erythrocyte precipitation has a downtrend that is considered a favorable prognostic sign.

As an integral index of inflammatory process intensity in an organism, the dynamics of such parameters as body temperature as well as the changes pulse rate, which depends directly on body temperature and testifies to activation of the sympathetic-adrenal system, one of the two principal stress-realizing systems were investigated (Table 17).

Table 17. Changes of pulse rate and temperature curve in patients with adenoid disease after a course of photodynamic therapy with methylene blue, X + m, P

Index	Before treatment	After treatment
Pulse rate, bpm	77.31 ± 2.15	72.08 ± 1.56 *
Body temperature, °C	37.10 ± 0.07	36.47 ± 0.04 *

Statistical analyses of the results shows that a reliable decrease in body temperature and pulse rate is observed (P<0.05), which correlates with favorable clinical parameters (positive changes in pharyngoscopic picture – Figure 14). The spectrum of pathogenic microorganisms which frequently cause adenoid disease has been studied (Figure 15).

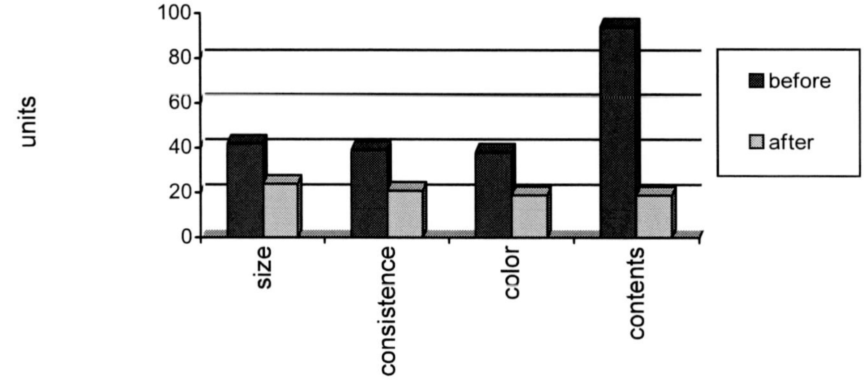

Figure 14. Changes in palatal tonsil state after therapy.

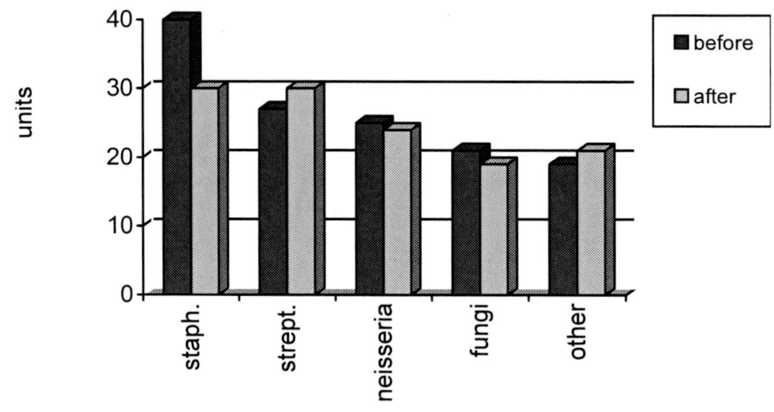

Figure 15. Composition of palatal tonsil microflora before and after treatment.

The diagram clearly shows that the most frequent pathogenic microbe causing adenoid disease is staphylococcus. Indeed, *Staphylococcus aureus* is an active pathogenic agent in the development of bronchitis and pneumonia, and in pustulous eruptions on skin and mucous membranes. The suggested treatment method reduces the level of staphylococcus, which is the absolutely pathogenic microorganism for tonsils. Other saprophytic and conditionally pathogenic microbes practically do not react to photodynamic therapy, which, in its turn, does not cause development of disbacteriosis, in contrast to the classic antibiotic cure. To a bacterial cell, death means permanent loss of reproductive ability (growth and proliferation). A cell is considered dead if it is not able to generate a colony (clone of daughter cells) on a solid nutrient medium. In this connection, we compared clinical results with the effect of red light and methylene blue upon growth and reproduction of *Staphylococcus aureus in vitro* – on agar nutrient medium which, to some extent, can be a model of contamination of skin and mucous membranes of an organism.

The results show that methylene blue has a marked inhibitory effect on growth and proliferation of pathogenic *Staphylococcus aureus* strain 209 in agar. A combination of methylene blue with monochromatic light of the red region increases the effectiveness of chemical therapy (Table 22, II). Figures 16-17 clearly demonstrate the bactericidal effect of red light in combination with chromophore (a variant of photodynamic therapy) on the growth of pathogenic staphylococcus *in vitro*.

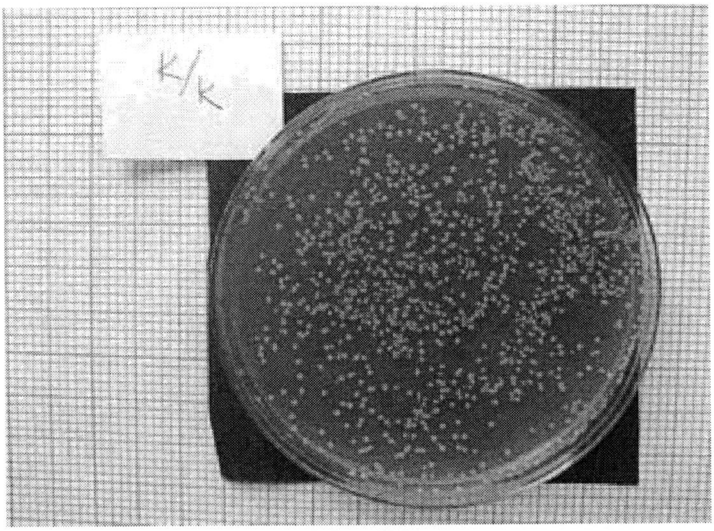

Figure 16. Control growth of staphylococcus in agar culture.

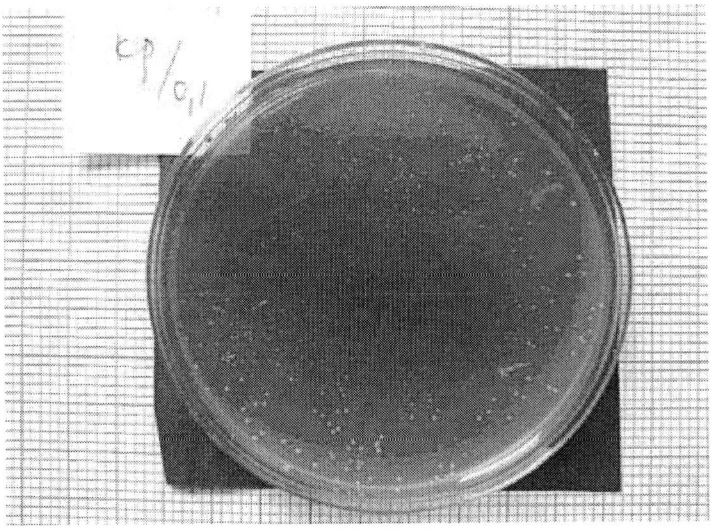

Figure 17. Growth of staphylococcus after combined effect of 0.1 % methylene blue and red light.

It is known that the combination of laser light (energy density 20-40 J/cm^2) and photosensitizers inhibits a growth of pathogenic strains of *Staphylococcus aureus*, including antibiotic resistant ones [3,21] that may be used in a treatment of infectious skin diseases

[19]. In our research, analogous effect was achieved under lesser light energy (no more than 1 J/cm^2) that has significance in overcoming mucous membrane contamination.

Of interest are our findings that impulse red light effect *in situ* on femurs for 3 hours led to a 2-fold accelerating apoptosis of bone marrow cells (Table 18) owing to an increase in apoptotic bodies' formation [14]. The number of erythroid cells, unlike granulocytes, fell in marrow irradiated down to 27 % ($P<0.05$) of the control level (Figure 18). It is necessary to emphasize the full concurrence of experimental results with the blood picture in patients who underwent photodynamic therapy (Table 16).

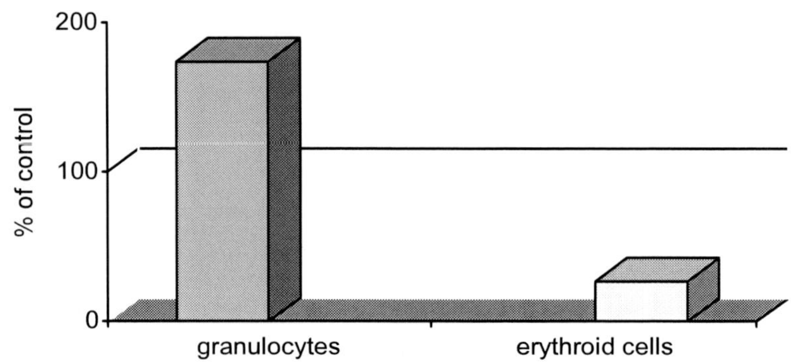

Figure 18. Red light influence on cell number in bone marrow.

Table 18. Morphological signs of cell death in bone marrow after red light influence in situ, X, Pu

Index	Apoptosis signs, %	Necrosis signs, %
Control (n=8)	7.63	6.75
Red light (n=8)	15.5*	8.88

One of the mechanisms of erythroid cell damage may be a direct photochemical process of active light absorption by molecular oxygen in the 638 nm and 762 nm spectra [22] with following acceleration of free radical processes. On the other hand, hemoglobin absorbs red light poorly but its amount in erythroid cells may be enough to initiate cell death by means of photophysical and/or photochemical reactions.

Hence, the results presented testify to the high clinical effectiveness of antibacterial therapy based on photodynamic effects caused by the combined influence of physical and chemical factors on pathogenic microorganisms. Clinical results have a good correlation with results of bacteriological research conducted *in vivo* and *in vitro*.

2. Phototherapy of Dysbacteriosis in Gastrointestinal Tract

Staphylococcus aureus bacteria are Gram-positive spherical cells approximately 1 μm in diameter; they belong to the normal microflora of the gastrointestinal tract. Pathogenic properties of staphylococci are determined by their ability of invasion and toxin production.

Staphylococcus aureus is a typical representative of pathogenic staphylococci; it causes monopathies and systemic changes including food poisoning and colitis [10].

Since staphylococci are very resistant to physical and chemical influences and easily adapt to extremal factors, there is a problem of bacterial decontamination and infection prevention. Antimicrobial chemotherapy based mainly on massive doses of antibiotics has constant difficulties which relate to the microorganisms acquiring resistance to chemical agents and to toxic action of the host's homeostasis. Among other antibacterial measures, UV irradiation has certain significance. However, its application in prevention and therapy of staphylococcus affections is very limited. It is known that low-intensity monochromatic visible light of different wavelengths is able to regulate cell behavior and nucleic acid synthesis [5].

Phototherapy is then a promising method as well as photodynamic therapy (diagnostics), which is based on the combined effect of light and chemical photosensitizers on cells. At the same time, the facts on theeffect of visible light on non-photosynthesizing microorganisms are very contradictory [12] and simply insufficient for pathogenic microbes. In this connection, the study of the effect of visible monochromatic light on the viability and growth of the pathogenic strain of *staphylococcus aureus in vitro* and the state of normal microflora of gastrointestinal tract (GIT) is topical.

The experiments were carried out with a sample of pathogenic strain 209 *Staphylococcus aureus*, which is characterized by typical morphologic, biochemical, hemolytic and plasma-coagulating properties. In sterile conditions in 10 ml flasks a microbe dredge was prepared according to turbidity standards (10^6 microbe cells per ml, 3 ml volume) and irradiated with green light. The light was emitted by autonomous photostimulators (APS) made by "Duny", Tomsk (supply voltage 4.5V, current amplitude 4mA, impulse length 6ms, wavelength 655 ± 20 nm [red range] and 565 ± 20 nm [green range]). The intensity of the green light was 6 times lower than that of the red light.

Before use, the APS was treated with 96 percent ethyl alcohol for 30 minutes. Then it was immersed into the microbe dredge for 3 hours at 37°C (prophylactic irradiation). Then 01 ml samples of dredge of each group (approx. 10^5 microbe cells) were put on agar nutrient medium (ANM) in 9 cm glass dishes, then rubbed into agar for 2 minutes and cultivated for 24 hours at 37°C.

In another series of experiments, microbes (approx. $2x10^5$ microbe cells) were put on ANM in 9 cm plastic dishes as described above. During the first three hours of cultivation, bacteria were irradiated *ex situ* (by putting APS on dish lid) with a green or red light (therapeutic mode). The dishes were covered with aluminum foil to decrease the loss of light.

In the third series of experiments, a microbe dredge was prepared with injcction water in sterile conditions ($4x10^6$ microbe cells per ml) according to turbidity standards. Then the dredge was mixed with methylene blue water solution so that the final concentration of the dye was 0.02% and 0.1% and cultivated for 1 minute. The control group of microbes was diluted with injection water. Then the staphylococci were grown and irradiated with impulse light (photodynamic therapy model) as described above in the second series of experiments. In all series, non-irradiated cultures (control growth) and ANM without microbes (as a medium sterility test) were cultivated. Three dishes were taken for each group.

Results were recorded with a digital camera. Then, using PhotoShop 7.0 software and gray level statistics [7], the number of colony forming units (CFU) of microbes per 1 cm^2 of the surface and per entire dish (survival test), average colony size (reproduction test), and area occupied by bacterial colonies were determined. In the third series of experiments, measurements in the agar culture were carried out within the track of methylene blue (prolonged influence mode plus high concentration of the dye) and outside it (short influence plus low concentration).

Table 19. Results of testing *Staphylococcus aureus* strain 209 growth in agar culture ($x10^5$ microbe cells) after prior influence of green light in impulse mode, X, Pu

\

Object	Colonies per 1 cm^2	Colonies per dish	Colony size
Agar medium purity control	-	1 *	-
Control culture of microbes	10.71 n=31<0,001	681.16 P<0,001	1.04 n=20 <0,001
Microbe culture after green light influence	0	1*	0

Note: * – spore-forming CFU which do not relate to staphylococcus;
n – number of counted areas or colonies.

In vivo trials included 2 groups of patients with intestinal dysbacteriosis of II-III stages: 21 patients were prescribed endogenic photocorrection in the morning, 2-3 hours before meals. The intake of red APS occurred *per os* 3 times for 10 days. The second (control) group included 19 patients with intestinal dysbacteriosis treated by innoculation of living lacto- or bifidobacteria *per os* for 10 days. A control examination of the microorganism's composition was done by the 14th day after therapy. A statistical analysis was carried out with the use of Mann-Whitney U-test (Pu).

To a bacterial cell, death means permanent loss of reproductive ability (growth and proliferation). A cell is considered dead if it is not able to generate a colony (clone of daughter cells) on a solid nutrient medium. *Staphylococcus aureus* strain 209 grows well on agar, reproducing depending on the concentration of inoculated cells and medium properties, on average 10 to 60 colonies per 1 cm^2 of Petri dish surface with the size of colonies originated from one CFU of 0.5 to 1 mm^2 (Figure 16, Table 19).

A three-hour influence of impulse monochromatic light of green range on bacteria in poor nutrient medium (physiological solution) resulted in complete loss of ability to survive and reproduce in favorable conditions by the pathogenic staphylococcus (Table 19). The single spore-forming CFU not belonging to staphylococci, which do not produce spores, could have been a result of imperfect purity of the initial agar medium.

The most important parameters of light when influencing biological objects are power density (intensity) I (W/cm^2), energy density W (J/cm^2), wavelength, mode and duration of influence [5,17]. Since the first two parameters for the light sources used were little, to achieve death of cells it was necessary to increase the duration. So, for example, recalculation of physical parameters of light according to [17] showed that for three-hour green light influence on the liquid culture, total energy density was approx. 2 J/cm^2.

It is well known that starving bacteria are very sensitive to lighting [12], which is confirmed by our data, too. The results obtained can have significance for development of sterilization methods for solutions and biological liquids which do not allow heating or chemical treatment. On the other hand, the resistance of cells to damaging influences rises in full nutrient medium, rich in factors necessary for growth and damage reparation (growth factors, minerals, and physicochemical conditions of cultivation) [10,12]. Thus, further study of green and red spectrums of visible light on staphylococci was carried out in solid nutrient medium, which can within certain limits imitate contamination of skin and mucous membranes of an organism, and the mode of influence can be used for sanitation of pockets of infection in the organism with mild methods, which do not disrupt host homeostasis.

Total area occupied by 3384 bacterial CFU made up approx. 21 cm^2 (33% of total agar surface in Petri dish) (Table 20). Regular form of colonies (S-type) with aureate pigment occurred. Fuzzy edges of colonies in the center of Petri dishes related to formation of condensate because of the great number of microbes in the culture. Despite this, large bacterial population's investigations are more informative for antimicrobial agent study since large populations are inhibited more than the smaller. Moreover, the probability of formation of resistant mutants rises.

Illumination of microbe culture on ANM through plastic with monochromatic red light by APS in impulse mode for 3 hours caused a reliable decrease in area occupied by staphylococcus by 21%. There was a reliable decrease in colony sizes (up to 87% of control size of CFU, $P<0.01$). The number of colonies diminished slightly (Table 20). So, impulse red light limits the number of bacterial cell fissions (bacteriostatic effect) and virtually does not influence the survival rate of staphylococci.

At the same time, irradiation of microbe culture with impulse green light showed pronounced bactericidal effect – the number of staphylococcus colonies decreased to 60% ($P<0.001$) of the control level (Table 20). On the contrary, bacteriostatic effect (decrease of colonies size by 10%, $P>0.05$) held the smaller part the in antimicrobial effect of green light. Nevertheless, the total area occupied by *Staphylococcus aureus* was less than 55% ($P<0.001$) of the control level.

The difference between the effects of green and red light on staphylococci can be explained by the predominance of its absorption and penetration, respectively [17]. Absorption of light in the short-wave region can, in its turn, result in active formation of free radicals and peroxides, which destroy bacterial cells. So, one of the maximums of speed of oxygen absorption by cells lies in the green region (580 nm) [12]. Increase of intensity of oxidation stress and photoreception of visible light by intracellular oxygen molecules [22] underlie the hyper production of active oxygen forms and peroxides, which are mediators of cell death [14].

Of interest is the fact that phototherapy of a human's digestivetract with red APS did not inhibit the growth of normal microflora. Moreover, from Table 21 it is clearly seen that after endogenic phototherapy there was an increased percent of humans with the usual content of bifidobacteria, lactobacteria, and colibacilli. Despite this, phototherapy *in vivo* seemed to be more effective for bifidobacteria and lactobacteria unlike the intestinal inoculation by the same artificial bacterial culture (control). An effect of APS is simultaneous on all types of bacteria whereas bacterial inoculation with microflora cultures is realized consistently and requires a long-term therapy.

Table 20. Results of testing of *Staphylococcus aureus* strain 209 growth in agar culture after 3 hours of green or red light influence in impulse mode, X, Pu

Object	Colony size, mm^2	Colonies per 1 cm^2	Colonies per dish	Colonies area per 1 cm^2, cm^2	Colonies area per dish, cm^2
Agar medium control	0	0	0	0	0
Control microbe culture	0.581 n=23	61.08 n=24	3884	0.33	20.98
Microbe culture after red light influence	0.506* n=29 <0,01	53.26 n=31 >0,2	3387 >0,2	0.26* <0,02	16.53* <0,05
Microbe culture after green light influence	0.526 n=37 >0,05	36.5* n=37 <0,001	2326* <0,001	0.18* <0,001	11.45* <0,001

Note: n – number of counted 1 cm^2 squares.

Intracellular effects of light influence are realized not only through chemical (photooxidation, photoreduction, photodissociation, etc.), but also through physical processes of energy transformation [12]. The energy of an absorbed quantum are transformed into thermal energy (rotational and vibrational) of molecules, mechanical vibrations of macromolecules and acoustic waves in cell membranes [12,17], which resulted in significant changes in its secondary structure.

Total density of light power in a Petri dish is moderate (0.1 J/cm^2), so in the mechanism of death of microbe cells highly sensitive photoreceptors (endogenous photosensitizers) are involved. As a result, photodynamic therapy using the combined effect of light and exogenous lethal photosensitizers (PS) is regarded an alternative to antibiotics because methylene blue (MB) is one of the best PS [9].

Table 21. Composition of microflora (%) of intestinal tract in patients after endogenic phototherapy or bacterial therapy, X

Indices	Endogenic phototherapy		Bacterial therapy (control)	
	Before influence	After influence	Before influence	After influence
Lactobacteria number:				
below normal	75	50	75	90
normal and higher	25	50	25	10
Bifidobacteria number:				
below normal	95	75	70	70
normal and higher	5	25	30	30
Colibacteria number:				
below normal	90	70	80	70
normal and higher	10	30	20	30

In our studies, brief prior incubation of staphylococcus dredge with MB of 0.02% concentration, which is used for antiseptic treatment of organism cavities, decreased the area of microbes spreading in agar culture by 60% (Table 22, II). The decrease of the area of bacterial growth was caused by 30% decrease of CFU number (bactericidal effect) as well as separate microbe colonies size (bacteriostatic effect). Mainly, intensification of bacteriostatic effect was found within the MB track (prolonged, 24-hour influence of the dye on staphylococcus growth), as shown in table 22, I. The size of separate colonies decreased 70% as compared to control microbe culture, which led to an almost 6-fold decrease of bacterial biomass.

Illumination of agar culture through plastic with monochromatic red light in combination with MB for 3 hours (Table 22, Figure 17) led to a reliable decrease of the area of culture (56-60% of only dye level, 10-25% of the size of the control group). Within the track of MB, total inhibitive effect of light was more pronounced due to 3 times higher bacteriostatic effect (Table 22, I, II). Nevertheless, bactericidal effect of red light was more pronounced under brief incubation of staphylococci with the dye (outside the MB track).

Thus, combined action of MB and red light results in almost two times decreasing biomass as compared to only the dye. Under brief exposure to the dye and subsequent irradiation, microbe survival rate decreases (bactericidal effect). In the dye track, red light lowers the number of bacteria due to limited reproduction (bacteriostatic effect).

However, illumination of a microbe culture with green impulse light in conditions of brief exposition with the dye did not affect their number in agar culture. While CFU number slightly decreased, the size of separate colonies slightly increased, which testified to the fact of intensification of microbe reproduction (Table 22, II). Probably, It is probably related to predominant absorption of MB in red region.

At the same time, within the dye track (prolonged dye influence), total area of microbe colonies irradiated with green light was 30% less than after red light irradiation (Table 22, I). Such effect of green light related to maximal decrease of the number of colonies as compared to the control culture of staphylococcus. Seemingly, the absorption spectrum of MB could change after interaction with agar.

High concentration of MB (0.1%) reduced yield of staphylococci in agar culture 4 times as compared to the lower concentration due to both bacteriostatic and bactericidal effects (Table 22, I, II). Red or green light had additional suppressive effect on microbe culture, almost fully correlating with low concentration of the dye (Table 22, I,II). Microbe biomass with combined factors (prolonged dye influence plus green light) was only 2% of the initial value for the control culture, i.e., lethality reached 98%.

However, the efficiency of physical factors decreases with the high concentration of MB. The absence of statistically reliable differences between them for all parameters studied (table 22) testifies to this fact. It is known that a combination of low-energy laser light (energy density of approx. 20-40 J/cm^2) and photosensitizers suppresses growth of pathogenic strains of *Staphylococcus aureus,* including those resistant to antibiotics [3,21], which can be used for treatment of infectious diseases of the skin [19]. In our studies, a similar effect is reached with a microdevice with much lower light intensity, which can have significance for overcoming GIT dysbacteriosis.

Clinical case 1. Patient C., aged 37. Complaints of periodic pains in the epigastric region, frequent liquid stools, abdominal distention, weakness. On fibrogastroscopy the signs of gastroduodenitis were found. On colonoscopy the signs of colitis were revealed. On palpation, pain in the intestine was noted. When examining feces for dysbacteriosis the following was found: pathogenic microbes of intestine family – 0; the total amount of colibacilli – 62 $x10^6/g$; colibacillus with poor enzymic properties – 0%; lactosenegative enterobacteria – 5%; hemolyzing colibacillus – 0%; coccus forms in total sum of microbes - 0%; hemolyzing staphylococcus in respect to all coccus forms – 0%; bifidobacteria – 10^7; lactobacteria – 0; microbes of Proteus genus – 0; microbes of Candida genus – 0. Conclusion: a decrease of a number of colonies of colibacilli, bifidobacteria, and the absence of lactobacteria colonies.

Table 22. Testing of *Staphylococcus aureus* strain 209 growth in agar culture after prior cultivation of microbe dredge with different methylene blue concentrations and 3-hour effect of red or green light in impulse mode, X, Pu

Object of research	Inside methylene blue track (I)					Outside methylene blue track (II)				
	Colony size, mm^2	Colonies per 100 mm^2	Colonies per dish	Colonies area per 100 mm^2	Colonies area per dish, mm^2	Colony size, mm^2	Colonies per 100 mm^2	Colonies per dish	Colonies area per 100 mm^2	Colonies area per dish, mm^2
1 Agar medium control	0	0	0	0	0	0	0	0	0	0
2 Control microbe culture	-	-	-	-	-	1.06 n=57	17.92 n=57	1140	18.25 n=57	1161
3 0.02% methylene blue	0.32 # n=12	9.83 n=12	625	3.15 # n=12	200 #	0.73 * n=29	12.52 * n=29	796 *	7.37 * n=29	469 *
4 0.02% methylene blue + red light	0.31 n=10	5.58 n=10	355	1.76 $ n=10	112 $	0.90 n=34	5.0 $ n=34	318 $	4.47 $ n=33	284 $
5 0.02% methylene blue + green light	0.32 n=12	3.82 $ n=12	243 $	1.22 $ n=12	78 $	0.93 $ n=30	8.48 n=30	539	7.30 n=30	464
6 0.1% methylene blue	0.18 # n=28	2.18 # n=28	139 #	0.78 # n=28	50 #	0.30 * n=25	5.84 * n=25	371 *	1.73 * n=25	110 *
7 0.1% methylene blue + red light	0.16 n=22	1.77 n=22	113	0.48 n=22	31	0.33 n=31	4.81 n=31	306	1.38 n=31	88
8 0.1 % methylene blue + green light	0.20 n=26	1.31 n=26	83	0.42 n=26	27	0.25 n=31	6.63 n=32	422	1.61 n=32	103

Note: n–number of counted squares (1cm^2); (*) – reliable difference with 2; (#) – the same with II; ($) – the same with 3 or 6.

Diagnosis: Chronic colitis, associated with dysbacteriosis of II stage.

The patient was prescribed endogenic photocorrection of the immune system: the intake of a photostimulator per os, on an empty stomach with 200 ml of warm water once every 3 days for 10 days. After therapy the general condition of the patient improved considerably, abdominal pains were reduced, and normalization of stools was noted.

On examining feces for dysbacteriosis the following results were found: pathogenic microbes of the intestine family – 0; the total amount of colibacilli – 100 x10^6/g; colibacillus with poor enzymic properties – 0%; lactosenegative enterobacteria – 5%; hemolyzing colibacillus – 0%; coccus forms in total sum of microbes; 0%, hemolyzing staphylococcus in respect to all coccus forms – 0%; bifidobacteria – 10^9; lactobacteria – 10^7; microbes of Proteus genus – 0; microbes of Candida genus – 0. Conclusion: an increased number of colonies of colibacilli, bifidobacteria, and normal amount of lactobacteria colonies. Control colonoscopy and examination of feces to dysbacteriosis one month after therapy revealed no pathology.

Diagnosis: Chronic colitis, stage of remission.

The given example demonstrated high efficiency when dysbiosis treatment in respect to lacto-, bifido-, and coliflora of the intestine. Moreover, a stable clinical effect after endogenic phototherapy was recorded during 6 months of observation.

Clinical case 2. Patient S., aged 19. Complaints of periodic pains in the epigastric region, frequent liquid stools. On fibrogastroscopy the signs of gastroduodenitis were found. On palpation, pain in the large intestine was noted. When examining feces for dysbacteriosis the followingwas found: pathogenic microbes of the intestine family – 0; the total amount of colibacilli – 10^6/g; colibacillus with poor enzymic properties – 0%; lactosenegative enterobacteria – 5%; hemolyzing colibacillus – 0%; coccus forms in total sum of microbes - 0%; hemolyzing staphylococcus in respect to all coccus forms – 0%; bifidobacteria – 10^7; lactobacteria – 0; microbes of Proteus genus – 0; microbes of Candida genus – 0. Conclusion: a decrease in a number of colonies of colibacilli, and the absence of lactobacteria colonies.

Diagnosis: Chronic colitis, associated with dysbacteriosis of II stage.

The patient was prescribed colibacterin 5 doses per day and lactobacterin 5 doses per day for 15 days.

After therapy abdominal pains were reduced; normalization of stools was not noted.

On examining feces for dysbacteriosis the following results were found: pathogenic microbes of intestine family – 0; the total amount of colibacilli – 2 x10^6/g; colibacillus with poor enzymic properties – 100%; lactosenegative enterobacteria – 5%; hemolyzing colibacillus – 0%; coccus forms in total sum of microbes; 0%, hemolyzing staphylococcus in respect to all coccus forms – 0%, bifidobacteria – 10^7; lactobacteria – 0; microbes of Proteus genus – 0; microbes of Candida genus – 0. Conclusion: a slight decrease in the number of colonies of colibacilli.

Diagnosis: Chronic colitis, associated with dysbacteriosis.

The given example demonstrated low efficiency of dysbiosis treatment with bacterial method.

3. Hybrid Implant Trial in Patients with Diabetic Complications

Currently, cell and tissue transplantation is considered to be a prospective method of treatment of wide range of pathology. Methods of diabetes I and II compensation through transplantation of pancreas insular part have improved. Nevertheless, there are problems with cell therapy (strict patient selection, host immune response, dispersion of transplanted cells in organism, their fast utilization, short-term treatment effect), which could be solved or minimized by means of using carriers – i.e., incubators (scaffold) for cell material. They allow posing a question about the opportunity for creation, growing and implantation of artificial organs, and prolongation of therapeutic effects of transplanted cells.

In this connection, clinical testing of porous constructions as prospective carriers for hybrid implants and artificial organ formation was of doubtless interest.

Subjective and objective examination of the blood system, immune status, and clinical state of 60 patients with pancreatic diabetes (PD) of I and II types after hybrid implantation was done. Hybrid implants were made by means of sowing of porous scaffolds (TiNi-alloy or pure titanium) with autologous blood cells. Then they were briefly (for 24 h) cultured in vitro in *RPMI-1640* medium with 5-10 % of patient's homologous serum. Implantation was carried out into rectus under local anesthesia. Documents for clinical work were: 1) order of Health Ministry of RF (protocol N11 of Committee for new medical equipment from 23.12.1999); 2) decision of Academic council of Russian state medical university (Moscow, 2005) about limited approbation of efficacy and safety of cell therapy by patients with diabetic foot syndrome; 3) RF patent N 2191607 from 27.10.2002, "The carrier of cell culture of artificial internal organs".

Statistical analysis was carried out according to Student's t-test and Wilcoxon T-test.

Transplantation of isolated cells has, in the main, been a function of building material for recipient's cells. Cell introduction into a porous scaffold allows a transplant to form its own microenvironment, to defend itself from host immune cells (and, in that way, to activate them indirectly), to communicate, and to behave as a valid, additional organ for replacement therapy.

Possible mechanisms of replacement therapy with the help of hybrid artificial organs were studied in patients with pancreatic diabetes (PD) of I and II types. In spite of the variety of immune disturbance under pancreatic diabetes, it is known that, as a rule, it is accompanied by depression of organisms, nonspecific resistance, and development of a secondary immunodeficiency state and connected with cell immunity reaction disturbance in people and in animals. There is high tendency of such patients to acquire infectious complications, including tuberculosis, sepsis, gangrene, skin infections and others. In the case of type I PD, an autoimmune component, connected with T-helper activation and/or T-suppressor decrease, is added.

Without going into the complex details of immunity disturbances caused by diabetes, it is necessary to note, that implantation of hybrid artificial organ to patient with type I PD induced a systemic immunomodulating effect, which becomes apparent with optimization and increasing of immune indices (CD 3, 4, 8, 16, 22, 25, 95; Ig A, M, and G).

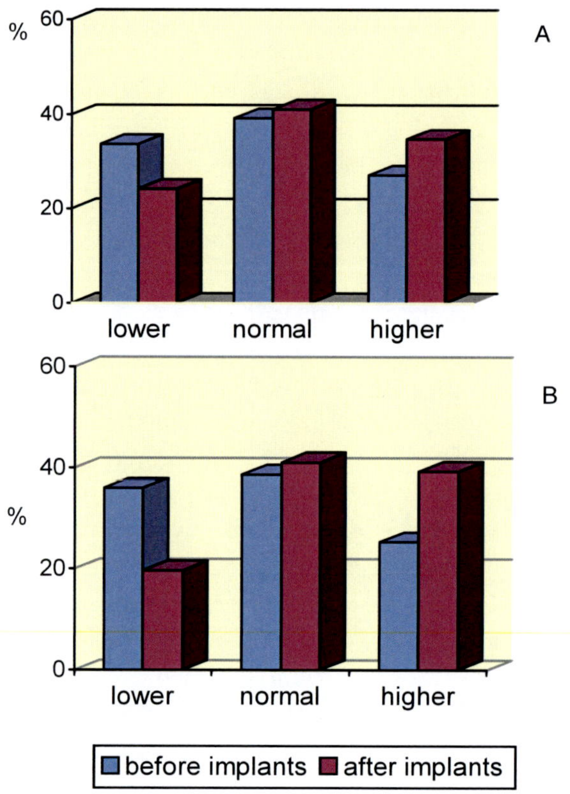

Figure 19. Distribution of immune indices in blood of patients with I (A) and II (B) types of diabetes.

Distribution of immune indices moved into the range of high values (Figure 19). In patients with PD of type II, who appeared in clinics with numerous complications (gangrene of lower extremities, retinopathy, renal disturbance, etc.), implantation had a more evident optimizing effect on immune status. A practically 2-fold decrease in the number of patients with depressed immunity must be especially emphasized (Figure 19).

A

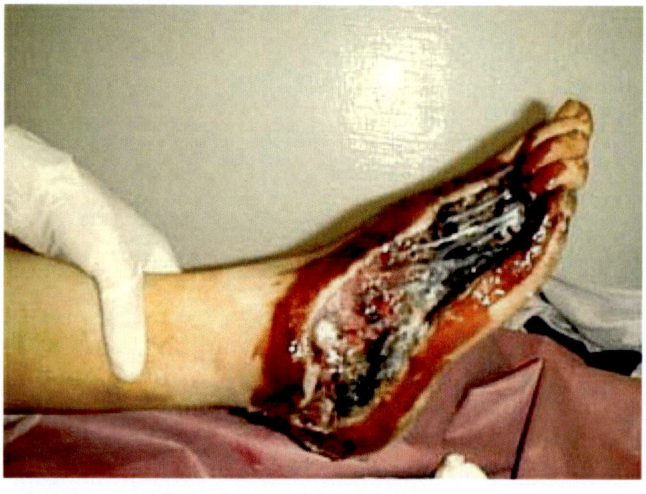

Figure 20. (Continued on next page.)

B

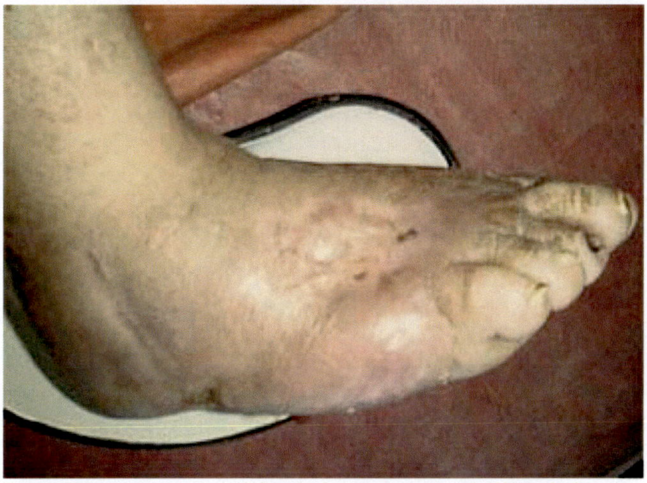

Figure 20. Feet of patients with complicated diabetes before (A) and after (B) correction.

Applied therapy was of a complex, long-term character and included, except implantation, ultrasonic distraction of sphacelated tissues, diet, endogenous Zn and Cr electrophoresis of blood cells as described above (see experimental research). A similar approach allowed for decreased evidence of diabetic complications, improved extremity amputation prevention (Figure 20), a significant decrease in dosage or even gradual insulin withdrawal.

Besides the immunomodulating effect, the basis for this offered method is a decrease in free radical and hypoxic processes. Sulfhydryl (SH) groups of erythrocytes, decreased significantly in patients with PD, are a cell protection mechanism and lipid peroxidation (LPO) marker. SH-groups and glutathione play the main role in gas-transport erythrocyte function, cell energy maintenance, prevention of enzymes from inactivation and other proteins from oxidative denaturation, and protection of membranes from peroxide influence. In other words, this is an integral index of oxidation-reduction, anabolic, catabolic and disintoxication reactions.

SH-group restoration in patients suffering from PD, revealed a marker of tissue hypoxia and organism's intoxication. On the other hand, increasing (up to adult people level) the number of sulfhydryl groups in peripheral blood erythrocytes (Figure 21), conditioned by marrow reticulocyte reaction on implantation, presented as one of the mechanisms of tissue protection from LPO.

Reticulocytes (young erythrocytes) have a high substrate content. At that, cell normalization under the conditions of insulin therapy accrued not earlier than 3-4 weeks after treatment in consequence of the change of defective pools in valid cells. It is interesting that in the case of hybrid implants, the effect becomes apparent exactly in these terms and it is accompanied by active leukocyte's apoptosis (CD95 marker) in peripheral blood.

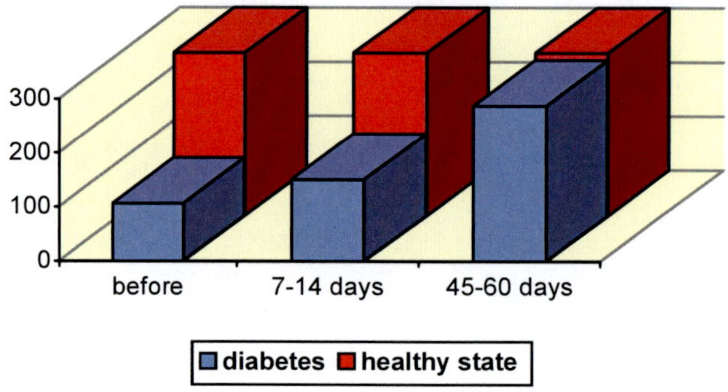

Figure 21. SH-groups level (%) in erythrocytes of healthy volunteers (healthy state) and of patients with diabetes before and after treatment.

The method developed should not be viewed as an alternative to modern methods of PD treatment. All are complex, building on each other. Nevertheless, the results affirm the existence of their proper niche and the promise of hybrid implants in the complex treatment of metabolic disease. Similar structures incorporate into complex hierarchy of interorgan relations and take part in the reconstruction of host organism homeostasis, as a consequence of high compatibility and bioactivity.

Conclusion

The aim of this chapter to present some our efforts in the physical-chemical regulation of cells was achieved. *In vitro* research had shown a principle opportunity to design hybrid implants containing bone marrow and other cells in porous scaffolds. Certain in vitro and in vivo properties of tissues in porous scaffolds are similar in character to those for a whole organism.

Apparently, transplanted cells, for example, fetal liver or umbilical cord cells, can work for some time in a substitutive therapy regime, binding actively endogenous and exogenous toxins, and generating different proteins, which facilitates regeneration or antitumor immune defense. At the same time, hybrid implants, independent of the nature of autologous or xenogeneic cells or of artificial biocompatible materials (metal, ceramics, and polymers), allow long-term stimulation of the mechanisms of natural resistance and graft-versus-tumor reaction.

Stem and committed cells are a heterogeneous population, which is under the control of distant and local regulatory mechanisms. Structure-functional units (microregions), including special cells, stromal elements, microcirculation vessels and terminal nerve endings are in every tissue. Our research showed that not only organic, but also well-defined physical and chemical parameters (surface, electrical impulses, visible light, some microelements, nano-scale particles and directed fluid flow) belonged to the cellular microenvironment, and their local or system changes affected functional activity of stem and stromal cells. Thus, physical-chemical stressors are permitted analogously to regulate an activity of both pathogenic microbes and mammalian cells.

They may have not only a direct effect on the stem and progenitor pool, but also be mediated through microenvironment factors, in which state is unsteady in different individuals and under different vital activities. So, physical-chemical irritants allow positive controlling cell behavior during transplantation that is allowed to correct the course and outcome of different diseases. On the other hand, negative fybrogenic potential of young stromal cells under cell therapy is proved by our flow-through cultivation technique.

The findings of preliminary clinical trials can be considered as promising ones. The high effectiveness of therapy of adenoid disease based on the combined influence of physical (red light) and chemical factors (methylene blue) is evident. Clinical data have a good correlation with bacteriological and cell research conducted *in vivo* and *in vitro*.

Endogenic phototherapy using a photon tablet makes it possible to have a selective in vivo effect on the composition of human intestinal microflora. Under the influence of light with maxima in the range of 560-580 nm and 650-660 nm, a stimulation of lactobacteria and colibacilli growth occurs. In some cases the effect of endogenic phototherapy exceeds the effect after conducting standard bacterial therapy with exogenic intestine inoculation with lacto- and colibacilli. Endogenic photostimulation provides an opportunity not only to influence proper microflora, but also to have an effect on the immune system and erythroid cells.

Our method of correction of diabetic complications with hybrid implants should not be viewed as an alternative to modern ones. All of them are complex, building on each other. Nevertheless, a proper niche and prospects for hybrid artificial organs in porous carriers for complex treatment of metabolic disease exist. Similar structures incorporate into the complex hierarchy of interorgan relations and take part in the reconstruction of host organism homeostasis, as a consequence of high compatibility and bioactivity.

The data presented emphasize the urgent need for detailed and extended experimental and clinical trials of physical-chemical manipulations with microbial and mammalian stem cells.

References

Biologic therapy of cancer: Second Edition. Eds. V.T. DeVita, S.Hellmann, & S.A. Rosenberg. Philadelphia: Lippincott-Raven Publ.; 1995.

Blandova, ZK., Doushkin, VA., Malashenko, AV. et al. The lines of laboratory animals for medical-biological investigations. Moscow: *Science*; 1983 (in Russian).

De Simone, N.A., Christiansen, C. & Dore, D. (1999). Bactericidal effect of 0.95-mW helium-neon and 5-mW indium-gallium-aluminum-phosphate laser irradiation at exposure times of 30, 60, and 120 seconds on photosensitized Staphylococcus aureus and Pseudomonas aeruginosa in vitro. *Phys. Ther.*,79(9), 839-846.

Dexter, T.M., Allen, T.D. & Lajtha, L.G. (1997). Conditions controlling the proliferation of haemopoietic stem cells in vitro. *J.cell.Physiol,* 91, 335-347.

Karu, TI. Photobiology of low-power laser effects. *Health Physics.*, 1989, 56, 691-704.

Kulinskiy, V.I. & Olkhovskiy, I.A. (1992). Two adaptation strategy under unfavorable conditions – resistant and tolerant. Role of hormones and receptors. *Achievements of modern biology,* 112 (5-6), 697-713 (in Russian).

Lee, S., Park, S.H., Pyo, H.B. & Kim S. (1999). Measurement of bone mineral density using trabecular patterns of conventional X-ray films. *Medical Biological Engineering & Computing,*,2 (II), 1066-1067.

Maitra, A; Arking, DE; Shivapurkar, N et al. Genomic alterations in cultured human embryonic stem cells. *Nat. Genet.,* 2005, 37, 1099-1103.

Millson, CE; Wilson, M; Macrobert, AJ; Bedwell, J; Bown, SG. The killing of Helicobacter pylori by low-power laser light in the presence of a photosensitiser. *J. Med. Microbiol.*, 1996, 44(4), 245-252.

Pokrovskiy, V.I. & Pozdneev, O.K. (1999). *Medical microbiology.* Moscow: Nauka (in Russian).

Rochs, R.; Sklenar, H.; Lavery, R.; Roder B. Methylene blue binding to DNA with alternating GC base sequence: a modeling study. *J. Am. Chem. Soc,* 2000, 122, 2860-2866.

Roubin, L.B., Eremeeva, O.V. & Akhobadse V.V. (1971). Light influence on metabolism of unphotosynthesizing microorganisms. *Successes of modern biology,* 71(2), 220-234 (in Russian).

Rummel, S.A. & van Zant G. (1994). Future paradigm for autologous bone marrow transplantation: tumor purging and ex vivo production of normal stem and progenitor cells. *J.Hematotherap,* 3, 213-218.

Samali, A., Gorman A.M. & Cotter T.G. (1996). Apoptosis – the story so far...*Experientia,* 52, 933-941.

Savitz, S.I., Rosenbaum, D.M., Dinsmore, J.H., Wechsler, L.R. & Caplan, L.R. (2002). Cell transplantation for stroke. *Ann. Neurol.,* 52, 266-275.

Sukhikh, GT. Transplantation of fetal cells in medicine: The present and future. *Bull. of experimental biology and medicine,* 1998,126, At.1, 3-13 (in Russian).

Tuchin, VV. Lasers and fiber optic in biomedical investigations. Saratov: Saratov Univ. Publ.House; 1998 (in Russian).

Volova, T. Polyhydroxyalkanoates-Plastic material of the 21st Century (Production, Properties, Application). New York: Nova Science Publishers Inc.; 2004.

Wang, L.,Wang, Y. & Cheng D. (1998). Therapeutic effects of ultrashort wave and He-Ne laser on experimental infection in skin flaps of rabbits. *Chung Kuo Hsiu Fu Chung Chien Wai Ko Tsa Chih.,*12, 55-57.

Williams, SF; Martin, DP. Applications of PHAs in medicine and pharmacy. In: Steinbüchel A, ed. *Series of Biopolymers in 10 vol. Vol.4.* Wiley-VCY Verlag GmbH, 2002, 91-121.

Wilson, M. & Yianni, C. (1995). Killing of methicillin-resistant Staphylococcus aureus by low-power laser light. *J Med Microbiol.,*42, 62-66.

Zakharov, SD., Skopionov, SA., Chudnovskiy, VM. et al. Initial mechanisms of nonspecific action of low intensive laser irradiation on erythrocytes with participation of molecular oxygen. *Proceedings of Academy of Science of USSR. Physical series.,*1990, 54(8),1629-1635 (in Russian).

In: Stem Cell Applications in Disease and Health
Editors: W. Burnsides and R. Ellsley
ISBN: 978-1-60456-255-2

Chapter III

Origin of Adult Neural Stem Cells and Perspectives for Brain Repair

Luca Bonfanti[1a] ***and Ferdinando Rossi***[2b]
[1]Department of Veterinary Morphophysiology
[2]Department of Neuroscience, University of Turin, Italy; [1,2]Istituto Nazionale di Neuroscienze (INN), and Neuroscience Institute of Turin (NIT)

Abstract

The mammalian central nervous system (CNS), as a non-renewable tissue, can be seriously affected by injuries and neurodegenerative diseases. Research carried out during the last decades revealed that neural stem cells persist within the adult brain, providing a source of neuronal and glial cell precursors throughout life. Yet, in mammals, including humans, adult neurogenesis occurs within restricted areas of the brain, giving rise prevalently to neuronal precursors that integrate inside two regions: the hippocampus and the olfactory bulb. Newly-born neurons of the olfactory bulb are generated in the forebrain subventricular zone (SVZ), now considered as the major stem cell reservoir in the CNS. In the SVZ, the neural stem cells have been identified as a subpopulation of astrocytes, originating during the postnatal period, from embryonic radial glia. This counterintuitive finding overlaps with the recent discovery that embryonic radial glia can themselves act as stem cells, capable of producing both neurons and glia during development. In the adult brain, stem cells reside in specialized 'niches' endowed with molecular and cellular signals capable of regulating their symmetric/asymmetric division, as well as the early specification and fate of their

[a] Correspondence: Luca Bonfanti DVM, PhD
Department of Veterinary Morphophysiology
Via Leonardo da Vinci 44 - 10095 Grugliasco (TO), Italy
E-mail: luca.bonfanti@unito.it
[b] Ferdinando Rossi MD, PhD
Department of Neuroscience
Corso Raffaello 30 - 10125 Turin, Italy
Email: ferdinando.rossi@unito.it

progeny. Although radial glia generally disappear early postnatally by transformation into parenchymal astrocytes, it has recently been demonstrated that some radial glial cells persist within the adult forebrain SVZ, hidden inside a dense astrocytic meshwork called 'glial tubes'. Some of these astrocytes occasionally divide to give rise to transit amplifying cells that actively proliferate, thus generating a rapidly expanding neuronal progeny. In parallel with restricted stem cell niches, widespread neural precursor cells in the CNS parenchyma retain the capability/potentiality to divide and possibly stemness properties. Despite true advancements in neural stem cell biology, the very mode, time, and pattern of cell division, as well as the mechanisms involved in its regulation within the stem cell niche and elsewhere are at present under investigation. Recent data indicate that both radial glia during development and astrocytes in the adult neurogenic niches play a dual role, participating both in the genesis of new cells (stem cell function) and in their regulation (paracrine/modulatory function). Also, the reactive behaviour of stem/progenitor cells in different CNS lesion paradigms has started to be elucidated. A further understanding of neural stem cell origin, and of mechanisms allowing the retention of stemness properties, could be fundamental in designing new strategies for brain repair, aimed at modulating in situ the endogenous sources of neuronal and glial progenitor cells. Nevertheless, since brain neurogenic niches show differences in different species and remarkably change through development and postnatally in order to adapt their function to the mature nervous tissue, a thorough knowledge of transformation from radial glia to neurogenic astrocytes across vertebrate and mammalian species is crucial in unraveling neural stem cell biology and regulation within the brain.

Introduction

The central nervous system (CNS) of mammals is a highly complex structure made up of a huge number of neurons (10^{11}), an even higher number of glial cells (10^{12}) and an amount of synapses estimated to be around 10^9/mm^3 (10^{15} in humans; Chklovskii et al., 2004). Nevertheless, what really makes the nervous system complex, beyond the number of its constitutive elements, is the extremely heterogeneous nature of their anatomical and functional relationships, which ultimately enable the performance of integrated neuronal functions. In comparison with other tissues of the body, the CNS has always been known for its incapability to undergo spontaneous repair and for its hostility to integration of exogenously delivered cells. In fact, at the end of the 19th century the Italian histologist Giulio Bizzozero classified the nervous system in the group of 'perennial' tissues, in contrast with 'labile' and 'stable' tissues which are capable of cell renewal and cell replacement throughout life as well as after injury or disease (quoted in Colucci-D'Amato et al., 2006).

For these reasons, we lack efficacious therapies for most neurological disorders that are characterized by loss of neuronal and glial cells. In addition to traumatic and vascular (e.g. stroke) lesions, the CNS is threatened by many neurodegenerative (e.g. Alzheimer, Parkinson, and Huntington disease; amyotrophic lateral sclerosis) and demyelinating diseases (e.g. multiple sclerosis) that ultimately hamper important cognitive and/or motor functions and represent very high costs in terms of human health and social care.

Since many years, biomedical research has been stimulated by the extraordinary flexibility and differentiative plasticity of stem cells, especially in the perspective of exploiting the acquired knowlege for novel therapeutical approaches. Studies carried out in the last few years showed that many adult organs are endowed with stem cell compartments (Robey, 2000), also involving 'perennial' tissues incapable of cell renewal and so unfit to self-repair, such as muscle (Asakura et al., 2002; Chargé et al., 2004; Peng and Huard, 2004; Zammit et al., 2006; Dellavalle et al., 2007) and nervous tissue (Reynolds and Weiss, 1992; Gage, 2000; Gross, 2000; Alvarez-Buylla et al., 2001). The discovery of stem cells within the adult mammalian brain during the last decade opened extraordinary research opportunities in the field of neuroscience, thus leading into new perspectives and/or strategies for the healing of neurodegenerative diseases (Arvidsson et al., 2002; Lie et al., 2004; Emsley et al., 2005; Steiner et al., 2006; Dietrich and Kempermann, 2006).

This new scenario, however, remains very uncertain and complex, due to our scarse knowledge concerning the interaction between stem cells and the host tissues, both in physiological and pathological/experimental conditions (Rossi and Cattaneo, 2002). Therefore, in this chapter the issue of neural stem cells (NSCs) will be primarily analysed under its anatomical, physiological, and biological profiles, then placed in the context of CNS complexity in order to hypothesize realistic perspectives for brain repair, based on stem cell-related therapies, along with related problems.

CNS Development, Neurogenic Sites and Neurogenic Niches

Embryonic and Adult Neurogenesis

The complex structure and cellular heterogeneity of the mammalian CNS is built up from cell divisions occurring in peri-ventricular germinative layers that harbour stem cells (Bayer and Altman, 2004; Figure 1A). These *neural stem cells* (NSCs) provide the source for most neuronal and glial cell precursors. Their progeny populate the CNS parenchyma from the neuraxis toward the pial surface by a combination of radial and tangential migration, following a precise spatial and temporal pattern which allows undifferentiated cell precursors to differentiate in their final destination and to establish appropriate connections (Rakic, 1990; Marin and Rubenstein, 2003).

The primitive germinal layers consist of a *ventricular zone* (VZ) containing direct descendants of the primitive neuroectoderm (neuroepithelium), and a *subventricular zone* (SVZ), which later emerges from the VZ. The SVZ contains rapidly proliferating cells and expands at early postnatal development, eventually giving rise to subsets of neurons and glial cells (for a review see Levison, 2006). Another early generated cell type is *radial glia* (Figure 1A,B). Radial glial cells have their soma within the VZ, close to the ventricular surface, and a long, radial process spanning the thickness of the assemblying CNS, reaching the pial surface with enlarged endfeet. In mammals endowed with a thick cerebral cortex (primates), they maintain the topographical relationships of neuronal precursors ascending from the SVZ in spite of unequal growth of the building cortex (Rakic, 1990,2003), thus playing a critical role in providing positional cues for neuronal differentiation. They are considered as a transient cell population destined to disappear by progressive transformation into astrocytes in the adult (Pixley and De Vellis, 1984; Voigt, 1989). In addition to this 'structural' function, it has been shown that embryonic radial glia can act as stem cells, being capable of asymmetric divisions leading to the genesis of both astrocytes and neurons (Malatesta et al., 2000; Hartfuss et al., 2001; Noctor et al., 2001; see below).

After birth, through a brief postnatal period whose length depends both on species and brain complexity/size (Kornack, 2000), the CNS parenchyma undergoes stabilization, thus making the striking plasticity of the neurogenic processes an exception to the rule. Yet, the dogmatic belief that post-mitotic synaptic changes were the only examples of adult structural plasticity in the CNS, was challenged by research firstly carried out in the sixties in a skeptic milieu, then re-evaluated at the beginning of the nineties[1] with the definitive demonstration that cell proliferation, differentiation and integration can occur even in the postnatal and adult CNS (Lois and Alvarez-Buylla, 1994; Luskin, 1993; reviewed in Gross, 2000). In addition to species-specific variations, the temporal windows of onset and downregulation of neurogenic processes can be heterogeneous throughout the neuraxis. For instance, most neocortical neurons are generated during mid-gestation whereas some neuronal cell populations continue to be added after birth in the cerebellum, hippocampus and olfactory bulb (reviewed in Rakic et al., 2004). As a consequence, besides the exhaustion of germinative layers occurring in most CNS regions at early postnatal stages, in some locations they persist for longer time, giving rise to a *protracted* genesis of neurons (e.g. in the cerebellum) that in some cases shifts into a *persistent* neurogenic activity (e.g. in the hippocampus and olfactory bulb) (Bonfanti, 2008). It is now well accepted that a consistent, endogenous proliferative activity does exist at least within restricted brain sites, referred to as *neurogenic sites* (reviewed in Peretto et al., 1999; Gage, 2000; Alvarez-Buylla and Garcìa-Verdugo, 2002; Bonfanti and Ponti, 2008; Figure 1).

In rodents, neurons are continuously produced within two brain neurogenic sites: the forebrain subventricular zone (SVZ) and the hippocampal subgranular zone (SGZ; reviewed in Alvarez-Buylla and Garcìa-Verdugo, 2002; Kempermann et al., 2004; Bonfanti and Ponti, 2008). In addition to unique opportunities to investigate neural cell birth, migration, and differentiation, these regions contain the NSC '*niches*', namely the place where microenvironmental cues and cell-cell interactions act to regulate stem cell quiescence, proliferation, and early specification (Doetsch, 2003; Alvarez-Buylla and Lim, 2004; Nystul and Spradling, 2006; Conover and Notti, 2008; see below).

Although harbouring all subsequent steps of cell differentiation from stem cell division to cell replacement, persisting neurogenic sites do not faithfully recapitulate development. Indeed, they change pre- and post-natally under their morphological, cellular, and molecular profile in order to adapt to the non-permissive environment of the mature nervous tissue

[1] Pioneering works demonstrated that neuronal cell renewal does occur in the periphery (Graziadei and Monti-Graziadei, 1979) and that proliferative activity also persists in certain brain areas (Altman and Das, 1965; Altman, 1969; Kaplan and Hinds, 1977). Nevertheless, a substantial change in the concept of brain structural plasticity has been prompted by the discovery of massive neurogenesis and cell migration within the forebrain subventricular zone (SVZ) of newborn (Luskin, 1993) and adult rodents (Lois and Alvarez-Buylla, 1994), as well as consistent genesis of new neurons in the adult hippocampus (Cameron and McKay, 2001; Kempermann, 2002). The first isolation of adult neural stem cells from the mammalian brain (Reynolds and Weiss, 1992) provided a cell biological basis for the continuous genesis of neurons, astrocytes and oligodendrocytes within the mature brain.

(Alves et al., 2002; Tramontin et al., 2003; Peretto et al., 2005). The complex postnatal modifications occurring in brain neurogenic sites have not yet been completely understood and likely hide the cues to unravel the mechanisms allowing persistent neurogenesis to occur in the "perennial" nervous tissue. Thus, investigating the origin and the postnatal modifications of NSCs might be the better way to figure out strategies that could be exploited in the perspective of brain healing.

Forebrain SVZ

In the postnatal and adult forebrain the VZ is replaced by an actively proliferating SVZ, that is considered as the major CNS stem cell reservoir (Peretto et al., 1999; Gage, 2000; Alvarez-Buylla and Garcìa-Verdugo, 2002). The anatomical arrangement of adult SVZ, as well as its extension, depend on the species and their cerebral ventricle conformation (reviewed in Garcia-Verdugo et al., 2002; Bonfanti and Ponti, 2008; see below). In laboratory rodents, the middle/posterior part of the SVZ remains in contact with the ventricles whereas the anterior part - due to the postnatal closure of the olfactory ventricle - forms the rostral migratory stream (RMS), namely a tangential migratory pathway for the newly-born neuroblasts through the olfactory peduncle and bulb axes. At the ventricular level, where the stem cell niche is more active, the SVZ cytoarchitecture involves four main cell types (Figure 1C): i) newly generated neuroblasts (*type A* cells), which migrate in the form of tangentially-oriented chains through the RMS (Lois and Alvarez-Buylla, 1994; Lois et al., 1996), ii) astrocytes (*type B* cells) organized into a dense meshwork and forming longitudinally-oriented channels called glial tubes (Lois et al., 1996; Jankovski and Sotelo, 1996; Peretto et al., 1997), iii) a third element with high proliferative capacity (*type C* cells; Doetsch et al., 1999a,b,2002), and iv) *ependymal cells*, cuboidal/columnar cells which express motile cilia on their apical surface and form an epithelial layer along the walls of the lateral ventricle. Ultrastructurally, type C cells show a cytoplasm with features intermediate between neuroblasts (highly electrondense) and astrocytes (light, watery), and are considered as 'transit amplifying cells' derived from NSC asymmetric division and expanding the neuronal population (Doetsch et al., 2002; see below).

Hippocampal SGZ

Unlike the SVZ, the hippocampal neurogenic site during development loses a direct contact with the ventricular cavities. Immature neurons are generated in the subgranular zone (SGZ) of the hippocampal dentate gyrus, and develop into mature dentate gyrus granule neurons (Gage, 2000; Figure 1C). At least two morphologically distinct types of astrocytes exist in the adult SGZ (type B cells; Seri et al., 2004), one with radial glia-like morphology (*radial astrocytes*) and the other with morphology of typical mature astrocytes (*horizontal astrocytes*). Radial astrocytes extend a major radial projection into the granule cell layer and have extensive basal processes that form basket-like structures that nestle clusters of neuroblasts (*type D* cells) and insulate them from contacting other cells and processes in the hilus. The type D cells progress through three maturational stages as they translocate to the granule cell layer of the dentate gyrus to become granule neurons (Seri et al., 2004; Figure 1C).

Sub-Callosal Zone (SCZ)

Beside the SVZ and SGZ, a large number of proliferating cells are found between the hippocampus and the corpus callosum in the adult rodent brain, in a region called the subcallosal zone (SCZ; Seri et al., 2006). This region corresponds to an extension of the posterior horn of the lateral ventricle – an area where the ventricular walls collapse during development – and could be considered a caudal and dorsomedial extension of the SVZ. Unlike the SVZ, it is not associated to an overt open ventricular cavity but to isolated cavities filled with cerebrospinal fluid that might be connected to the main compartment of the lateral ventricle. The SCZ retains cells that have properties of stem cells *in vitro,* but *in vivo* function as a local source of oligodendrocytes (Seri et al., 2006).

Common Elements in the NSC Niche

Both the forebrain SVZ (with its posterior extension, SCZ) and the hippocampal SGZ are considered to contain a stem cell niche. Common elements can be found in the glial/neuronal anatomical relationships of adult germinal niches. Astrocytes form tubes in the SVZ and baskets in the SGZ, insulating the neuronal progeny from coming into contact with the surrounding parenchyma. According to a recent theory, gaps in the SVZ glial tubes or the SGZ astrocytic baskets and/or contact with myelinated axons in the SCZ are conducive to the formation of oligodendrocytes and permissive to the migration of oligodendrocyte precursors (Seri et al., 2006), thus reinforcing the hypothesis that neuronal-glial interaction can affect the fate of NSC progeny. Finally, as it will be described in the next section, adult NSCs have been identified in vivo with elements of the glial (astrocytic) lineage, belonging to a subpopulation of glial tube's astrocytes in the SVZ and to radial astrocytes in the SGZ. Thus, glial cells of the neurogenic zones play a dual role being both stem cells and partner cells capable of regulating neurogenesis at the niche level.

Discovery and Identification of Neural Stem Cells: A Brief Overview

Retrospective Isolation of NSCs

The fact that stem cells are functional states rather than physical entities (Blau et al., 2001), along with the lack of reliable markers for resident NSCs does hamper their visualization and identification in vivo, forcing the biologists to isolate them *in vitro*, in a retrospective manner. In the nervous system, multipotency consists of the capacity to form the three main cell types: neurons, astrocytes, and oligodendrocytes. About fifteen years ago,

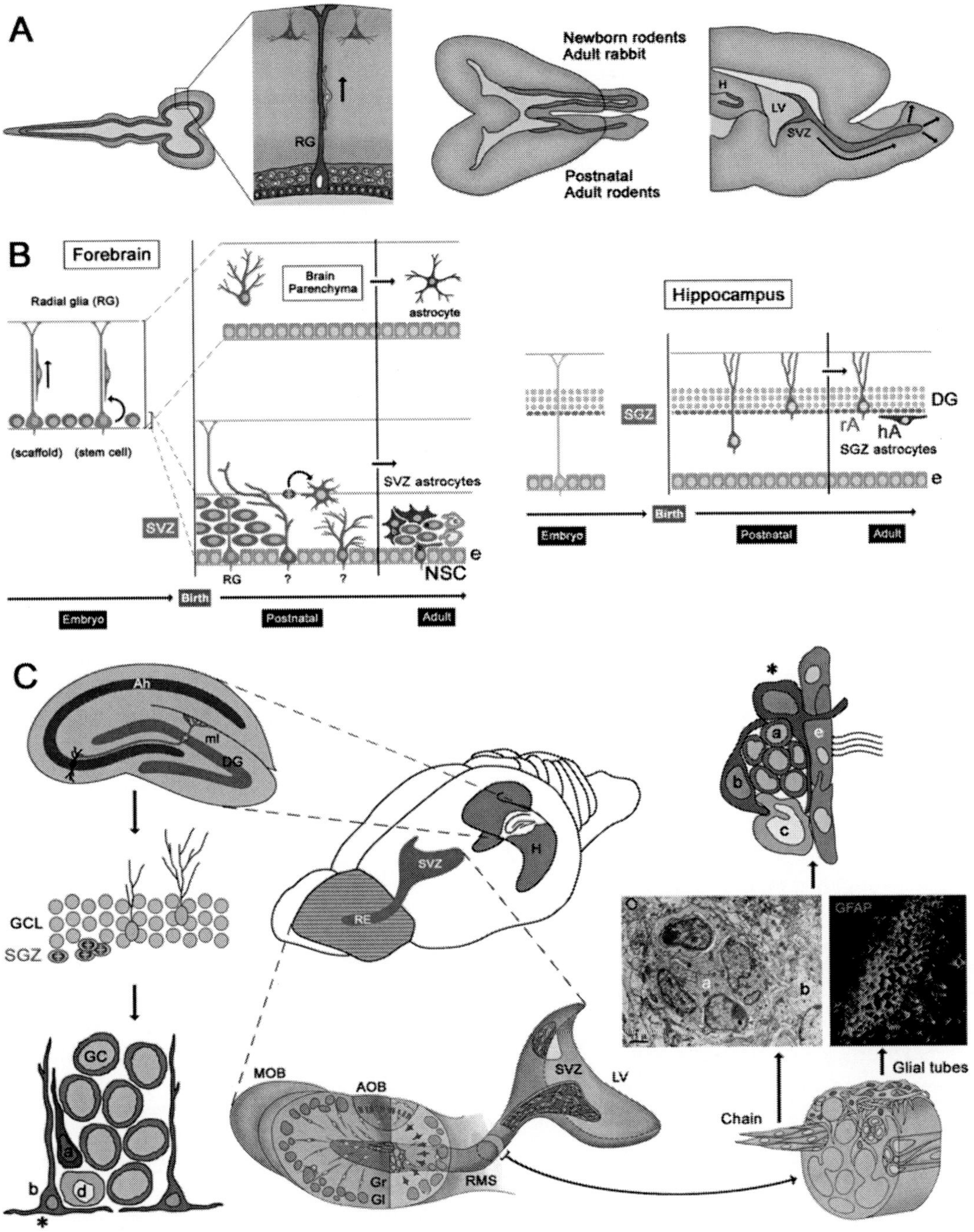

Figure 1. *Origin of adult neural stem cells and neuroanatomical correlates of adult neurogenesis in mammals.* (**A**) Origin and persistence of CNS germinative layers (dark grey). During embryonic development (left), proliferating ventricular/subventricular zones line the primitive ventricular cavities (inner light grey) and generate neuronal precursors which migrate radially (arrow) along radial glia (RG) through the brain parenchyma. At perinatal stages (middle, top) the SVZ persists only in the hippocampus (H) and in the anterior forebrain, associated to the lateral (LV) and olfactory ventricle. In rodents (middle bottom and right), the olfactory ventricle closes soon after birth, leaving a chord of tissue called *rostral extension*. Neuronal precursors generated within the SVZ migrate tangentially in

the *rostral migratory stream* (long arrow) and radially in the olfactory bulb (arrows; see also C). (**B**) Origin of 'astrocytic' neural stem cells in neurogenic zones. During embryonic development radial glia play a dual role, including that of stem cell capable of generating neurons and glia. After birth, radial glia transform into mature astrocytes in the CNS parenchyma, and into SVZ and SGZ astrocytes. NSC, neral stem cell; rA, radial astrocyte; hA, horizontal astrocyte, of the dentate gyrus (DG). Modified from Gritti & Bonfanti, *Neuron Glia Biol*., 2008. (**C**) Anatomy and cytoarchitecture of adult neurogenic sites. Genesis of new neurons occurs in the olfactory bulb (OB) and hippocampus (H). Their germinative layers are respectively the SVZ (bottom and right) and the subgranular zone (SGZ, left). The SVZ is composed of two main compartments: astrocytic *glial tubes*, and tangential *chains* of neuronal precursors sliding within the glial ensheathment. In electron microscopy they correspond to type A cells (a) and type B cells (b). A third cell type (C cell) is present in the SVZ stem cell niche, but not in the SGZ; e, ependyma. The SVZ chains dissolve in the olfactory bulb, whereby single neuroblasts move radially through the main (MOB) and accessory olfactory bulb (AOB). The SGZ is located in the inner layer of the dentate gyrus (DG, left top) and gives rise to granule cells (GC), through type D cells (d). Even in the absence of glial tubes and chains, cell types similar to those described in the SVZ form a stem cell niche in the SGZ (left, bottom). Asterisks indicate putative stem cells. (A,C, Modified from Bonfanti and Ponti, Vet. J., 2008).

stem cells were firstly isolated from the adult CNS (Reynolds and Weiss, 1992), and since.that seminal paper cultured NSCs have been deeply studied in parallel with adult neurogenesis in vivo

Both the SVZ and SGZ of adult rodents harbour a population of multipotent neural progenitor cells that can be isolated and expanded in culture under the effect of trophic factors such as the epidermal growth factor (EGF), the fibroblast growth factor 2 (FGF2) or a combination of them, ultimately giving rise to neurons, astrocytes and oligodendrocytes (Gage, 2000; Galli et al., 2003). Protocols based on the use of growth factors for isolation and expansion of NSCs derived from the embryonic and adult CNS have been developed, allowing extensive growth and propagation of NCSs in floating cell clusters termed "*neurospheres*" (Reynolds and Weiss, 1992; Vescovi et al., 1993; Gritti et al., 2000,2002; Figure 2). This mixed cellular milieu offers a microenvironment allowing the renewal of *bona fide* stem cells and immature progenitors, which represent a small proportion (5-10%) of the whole cell population (Singec et al., 2006). Thus, a central goal in the field has been the improvement of the NeuroSphere Assay (NSA; Reynolds and Rietze, 2005) and the development of alternative approaches to select and clonally expand the bona fide stem cell population in vitro. Similarly to other renewing tissues, the neurogenic regions of the CNS comprise a hierarchy of different types of cells, i.e. bona fide stem cells, transit amplifying progenitors and mature differentiated cells.

The NSA represents a selective serum-free in vitro assay, which has been successfully applied not only to neural tissues of different origin (rodent or human, fetal or adult, normal or pathological), but also to non-neural tissues as the mammary epithelium and the cardiac muscle (Messina et al., 2004, Wooward et al., 2005). By taking advantage of the functional features of the different cell populations retrieved within the stem cell compartment, the NSA is selectively enriched for the stem cell component. In fact, stem cells in culture are characterized by the capacity to proliferate and self-renew extensively, thus giving rise to long-term expanding stem cell lines. Conversely, transit amplifying progenitors, although displaying limited proliferative capacity, do not self-renew and are lost throughout extensive sub-culturing (Potten and Loeffler, 1990).

Lately, these cultured NSCs have been suggested to derive from the type C cells of the SVZ (Doetsch et al., 2002). Since in vivo type C cells represent transit-amplifying progenitors rather than bona fide stem cells, the concept has been proposed that tissue culturing and/or mitogens can re-awake a latent stem cell program in what can be considered "potential stem cells" (Potten and Loeffler, 1990). Upon removal of the mitogens, the progeny of NSCs promptly differentiate into the three main cell types of the CNS (astrocytes, oligodendrocytes and neurons; Doetsch et al., 1999a; Weiss et al., 1996; Kilpatrick and Bartlett, 1993; Vescovi et al., 1993; Morshead et al., 1994), a result that could also be obtained with proliferating cells isolated from the adult hippocampal formation (Gage et al., 1995).

The most important concept regarding this method is that it represents a selective system by which, in a heterogeneous primary culture, committed progenitors and/or differentiated mature cells rapidly die and thus are eliminated, whereas the undifferentiated NSCs are positively selected and forced to access a state of active proliferation (Gritti et al., 2000; Reynolds and Rietze, 2005). Of note, not all the cells within a neurosphere are true stem cells. Only a fraction of this progeny retain stem cell features, while the remainder of the cells undergoes spontaneous differentiation. Indeed, a neurosphere represents a highly heterogeneous entity made up by stem cells, differentiating progenitors, and even terminally differentiated neurons and glia, and, accordingly, can be envisioned as the in vitro counterpart of the in vivo neurogenic compartment.

Importantly, neurospheres can be sub-cultured by mechanical dissociation and by re-plating under the same in vitro conditions. As in the primary culture, at every sub-culturing passage, differentiating/differentiated cells rapidly die while the NSCs proliferate, giving rise to secondary spheres that can then be further sub-cultured. This procedure can be sequentially repeated several times and, since each stem cell gives rise to many stem cells by the time a sphere is formed, it results in the progressive enrichment and stable expansion of NSCs in culture. Thus, in order to avoid any ambiguity in applying the definition of NSC to a candidate cell, particular caution is necessary in performing and interpreting this assay. The definition of a neural cell as a stem cell (or of a population of cells as containing a stem cell) should be applied only to a founder cell (or population of cells) that self-renew *extensively* and can be propagated in *long-term cultures*. In this view, the generation of clonal neurospheres by the NSA needs to be monitored, by both clonogenic and population analysis, to document stem cell activity and to identify bona fide stem cells (Gritti and Conti, 2008).

Historically, hemopoietic stem cells have been the first to provide a useful prototypic functional model in order to devise conceptual and practical strategies to tackle investigations on other stem cell types (Morrison et al., 1997). When comparing hemopoietic stem cells and NSCs with regard to their functions, the two systems appear to sit at opposite sides of the functional spectrum. Although enormous amounts of new blood cells are generated every day from stem cells of the hemopoietic tissue, one of the most significant hurdles in the field of hemopoietic stem cells is their resilience to undergo extensive proliferation in culture. On the other hand, in spite of the relative overall quiescence and structural/cytoarchitectural "rigidity" of the brain parenchyma, the study of NSCs received much of its initial impulse from *in vitro* findings that first demonstrated and exploited the amazing proliferation capacity that NSCs display in culture. Furthermore, a mirror-like situation regards the reliable antigenic fingerprint based on the expression of multiple, lineage-specific markers available to distinguish various hemopoietic precursors, in contrast with the paucity of specific markers for NSCs (Galli et al., 2003).

Multipotent NSCs have been isolated from fetal (Uchida et al., 2000), adult (Sanai et al., 2004) and post-mortem (Palmer et al., 2001) human nervous tissues, and this enhances the interest in the study of mammalian neurogenesis. In addition, it has been possible to establish continuous mouse transgenic/genetically-modified (Galli et al., 2003) or human NSC lines (Carpenter et al., 1999; Vescovi et al., 1999) that expand merely by growth factor stimulation and under chemically defined conditions. In particular, human NSC lines represent a renewable source of normal nervous cells that might facilitate basic studies on human neurogenesis, drug discovery and, remarkably, may virtually eliminate the need of foetal human tissue for therapeutic neural transplantation.

Based on the assumption that specialized niches are essential to prevent stem cells from differentiating (Alvarez-Buylla and Lim, 2004; Doetsch, 2003; Spradling, 2001; Scadden, 2006; Nystul and Spradling, 2006), ex vivo homogenous propagation of tissue-restricted stem cells has indeed represented a major challenge for biologists. With regard to NSCs, their expansion as neurospheres finds support from the hypothesis that their heterogeneous composition would recapitulate a niche-like microenvironment (Alvarez-Buylla and Lim, 2004; Garcion et al., 2004; Campos, 2004). This idea has been recently challenged by a novel and powerful strategy for lineage restricted tissue-specific stem cell lines derivation from mouse embryonic stem (ES) cells and their long-term stable propagation (Conti et al., 2005; Figure 2). Transiently generated ES-derived neural precursors, usually fated to rapidly differentiate to neurons and glia, can be competently expanded as adherent clonal stem cell lines via exposure to EGF and FGF-2. Immunocytochemical analyses of these cultures reveal that almost all of the cells express markers of neural stem and progenitor cells (i.e. nestin and sox2), but lack the expression of markers of glial (GFAP and S-100) and of neuronal (β3-tubulin and Map2) lineages. Notably, in these growth conditions, cells divide symmetrically with no accompanying differentiation. Shifting the cultures to neuronal differentiation conditions reveals that these cells can efficiently generate mature neurons, astrocytes and oligodendrocytes, thus indicating their neural stem flavour (Conti et al., 2005; Glaser et al., 2007). These properties, together with single cell clonal analysis, are suggestive of the generation of self-renewing NSC cultures, named *Neural Stem (NS) cells* (Conti et al., 2005), whose expansion can occur in the absence of a complex cellular niche. Accordingly, NS cell expansion in monolayer conditions seems to restrain spontaneous differentiation and permit proliferation of homogeneous NSCs.

Antigenic characterization of NS cells revealed a stringent similarity to forebrain radial glia, a population recently demonstrated to behave as a fetal source of both neurons and glia during CNS development (Gotz et al., 2002; Hartfuss et al., 2001; Fishell and Kriegstein, 2003; Malatesta et al., 2000; Pinto and Gotz, 2007). Indeed, analysis of NS cells revealed that these cells are homogenously immunopositive for nestin, SSEA1/Lex1, Pax6, prominin, RC2, vimentin, 3CB2, GLAST (glutamate-aspartate transporter), and BLBP (brain lipid binding protein), a set of markers diagnostic for neurogenic radial glia (Pollard and Conti, 2007). These results further indicate that the acquisition of radial glial properties endows the cells with a "niche" that traps them in a state of symmetric cell division. It is therefore possible that NS cells represent the resident stem cell population within neurospheres.

In conclusion, while in the natural setting of the nervous tissue stem cell division is securely regulated by the cellular microenvironment of the niche, stem cell propagation in vitro may not be attributable solely to intrinsically programmed mechanisms within progenitor cells, but also to changes of the transcriptional state induced by the artificial environment (Smith, 2001). Thus, we ought to remember that: i) challenging a cell in vitro only unveils its developmental potential and not its actual in vivo fate, and ii) the extent by which a cell can self-renew in vitro may depend on the culture conditions adopted and/or on distinct regions from which it is isolated (Gritti and Conti, 2008). Thus, keeping in mind the different possibilities described in this paragraph, the artificial nature of cell culture should be clear, thus underlining the requirement for carefulness in extrapolating in vitro results to normal development or physiology without corresponding in vivo data. In other words, researchers should be extremely cautious about the possible differences between in vivo and in vitro systems, particularly on the possibility that unusual biological and molecular properties induced by the culture environment may lead to the acquisition of stem cell features.

In Vivo Identification of NSCs (Neurons From Astrocytes)

A consistent number of reports support the hypothesis that adult SVZ NSCs belong to the astrocytic lineage (Doetsch et al., 1999b; Laywell et al., 2000; Skogh et al., 2001; Imura et al., 2003; Garcia et al., 2004; Namba et al., 2005; reviewed in Alvarez-Buylla et al., 2001). Experiments carried out in vitro and in vivo demonstrated that i) astrocytes and transit amplifying cells (type B and C cells) can produce neurospheres that contain multipotent progenitor cells (Doetsch et al., 2002); ii) labeled neurons migrating to and integrated in the olfactory bulb were detected after targeting SVZ astrocytes in transgenic GFAP-TVA mice which express the receptor of the avian leukosis virus under the control of the GFAP promoter (Doetsch et al., 1999b); iii) quiescent astrocytes within the SVZ were spared by treatment with high doses of anti-mitotic drugs that kill virtually all transit amplifying cells and neuroblasts, and are able to regenerate a functional neurogenic system.

These in vivo studies not only have identified in the SVZ a subset of astrocytes as bona fide NSCs but have also traced a lineage relationship between the different SVZ cell types. In fact, slow-dividing type B cells give rise to type C cells that, in turn, proliferate with remarkable frequency giving rise to type A proliferating neuroblasts (Doetsch et al., 1999b), the precursors of olfactory bulb interneurons.

The fact that astrocytes of the adult forebrain SVZ retain stem cell potential has also been shown in cultures from transgenic mice expressing herpes simplex virus thymidine kinase from the GFAP promoter, in which dividing GFAP-expressing cells can be selectively eliminated by anti-viral agents (Imura et al., 2003). In a similar model, the same conclusion has been reached in vivo after observation of depletion of newly generated neurons in the olfactory bulb (Garcia et al., 2004).

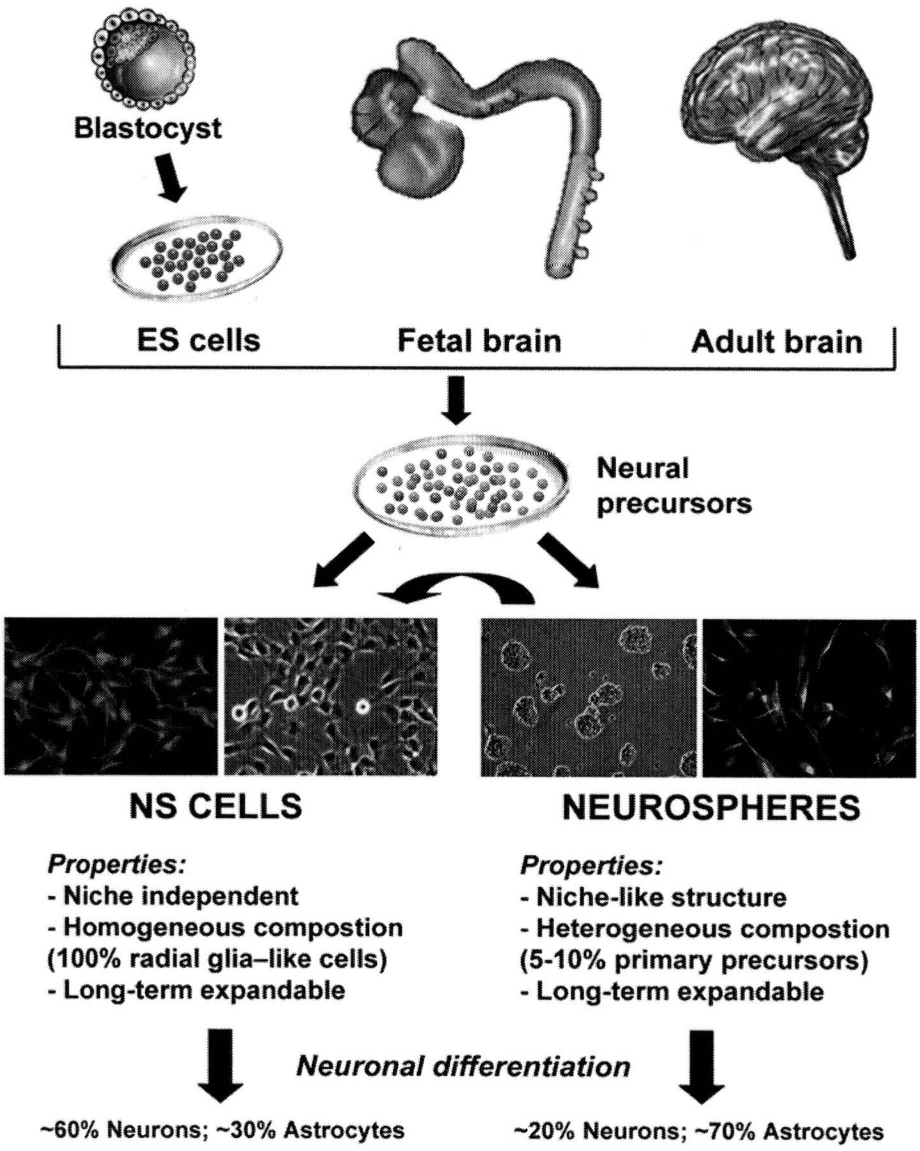

Figure 2. *Schematic illustration of the sources of ES cells, neurospheres and NS cells.* Neurospheres and NS cell lines can be produced from ES cell lines (derived from the inner cell mass of the blastocysts), from the germinative areas of the fetal and from the SVZ of the adult brain. Whilst ES cells can be considered pluripotent being capable to originate to endoderm, mesoderm and ectoderm derivatives, neurospheres and NS cells can be considered "tripotent" since can give rise to neuroectoderm derivatives (neurons, astrocytes and oligodendrocytes). Modified from Gritti & Conti, in: *Postnatal and Adult Neurogenesis, L. Bonfanti (Ed.), Research Signpost, 2008.*

The transition from quiescent to 'activated' SVZ astrocyte coincides with an increased expression of the EGF receptor (EGFR), thought to be implicated in the shift from a relatively quiescent cell to a 'transit amplifying' cell capable of expanding the neuronal progenitor population (Doetsch et al., 2002). Hence, substantial evidence indicates GFAP-expressing progenitor cells as a predominant source of constitutive adult neurogenesis, suggesting that NSCs could retain cytological aspect and functions of highly specialized cells in the CNS in contrast with the common idea that stem cells are highly undifferentiated elements (Alvarez-Buylla et al., 2001; Gritti and Bonfanti, 2008).

Similarly, SGZ radial astrocytes are thought to function as giving rise to neurons, astrocytes and oligodendrocytes in the hippocampus (Seri et al., 2004). In the SGZ, type D cells are thought to function as 'transit amplifying' cells of the hippocampal NSC niche. Steiner et al. (2006) characterized the highly proliferative intermediate precursor (type-2 cell) of the hippocampus as the level where the transition between the glial-like state and neuronal differentiation does occur. The common features shared by NSCs of the two main brain neurogenic sites can find an explanation in their developmental origin.

The Dual Role of Radial Glia and the Origin of Adult NSCs

Although classically considered as a transient scaffold for neuronal migration (Rakic, 1990,2003; see above), radial glial cells have been recently re-considered by studies demonstrating that they could act as stem cells during embryogenesis (reviewed in Pinto and Gotz, 2007; Rakic, 2003; Figure 1B). Indeed, some radial glia are endowed with self-renewal capacity and generate astrocytes, neurons (Noctor et al., 2001) and oligodendrocytes (Merkle et al., 2004). This might occur in vivo in two different ways: i) newly born neuroblasts are generated by asymmetric division and then migrate along the radial fiber of their mother cell (Noctor et al., 2001), or ii) radial glia cells transform into migrating neurons by translocating the nucleus along the radial process, leaving a stem cell within the VZ (Miyata et al., 2001). Although prevalently assessed in cerebral cortex development, similar properties of radial glia as neural progenitors have been demonstrated to exist within wide areas of the CNS, with remarkable regional- and lineage-dependent heterogeneity (Merkle et al., 2007; reviewed in Pinto and Gotz, 2007).

Most of radial glial cells actually disappear at the end of neurogenesis by progressive transformation into parenchymal astrocytes of the mature CNS (Rakic, 2003). Postnatal modifications involve both morphological changes (displacement of the cell soma from the ventricle, retraction of the radial process from the pial surface, increased ramification of cell processes) and molecular modifications, resulting in the expression of specific sets of intermediate filament proteins (reviewed in Bonfanti and Peretto, 2007; Pinto and Gotz, 2007). The radial glia/astrocyte transition can also be viewed as a sort of transdifferentiation process that can display different traits depending on regional specializations (Chanas-Sacré et al., 2000). While in most of CNS parenchyma radial glial cells transform into stellate astrocytes, in restricted CNS regions (e.g. Bergmann fibres in the cerebellum and Muller cells in the retina) they retain some morphological and molecular features resembling those displayed at the precursor cell state.

An additional scenario can be envisaged within the neurogenic sites, where the large majority of radial glial cells do transform into astrocytes (glial tubes in the SVZ, radial astrocytes in the SGZ), but a subset of them retains some morphological, molecular, and functional features of primary precursors. In the SVZ, radial glia change their orientation and degree of ramification thus forming the unique astrocytic meshwork of the glial tubes (Peretto et al., 2005). During the morphogenesis of the hippocampus, radial glial cells translocate from the SVZ to the dentate gyrus (Eckenhoff and Rakic, 1984), then persisting as radial astrocytes in the SGZ (Seri et al., 2004). Among the adult SVZ and SGZ cell types, glial cells have some cytological and molecular features of radial glia, since they contain glycogen granules, the intermediate filaments vimentin and nestin, and glutamate transporters (Bonfanti and Peretto, 2007; Pinto and Gotz, 2007). In the SVZ, some of them retain a thin cellular process protruding in the ventricle through the ependymal monolayer (Doetsch et al., 2002). The mechanisms underlying the morphological, molecular and functional transition that glial cell compartments undergo within the neurogenic niches are largely unknown, yet recent studies confirmed that radial glia-derived SVZ astrocytes retain stem cell properties.

The observations relative to the persistence of radial glia within the SVZ were followed by more direct proof of a link between embryonic radial glia and adult NSCs. This was obtained by employing a lox-Cre-based approach to specifically and permanently label a restricted population of radial glia in newborn mice (Merkle et al., 2004). Radial glia in the lateral wall of the lateral ventricle genetically tagged with a replication incompetent, EGFP-expressing adenovirus at postnatal day 0, was shown to give rise to GFAP$^+$ astrocytes of the SVZ as well as to multiple classes of brain cells including neurons, ependymal cells, and oligodendrocytes. The fact that radial glia become the forebrain NSCs is also supported by other observations in the work of Merkle and colleagues, namely, i) the RMS contained marked migratory neuroblasts produced after the disappearance of radial glia; ii) marked glial and neuronal cells also labelled for exogenously-administered bromodeoxyuridine were found in the olfactory bulb; iii) the progeny of marked radial glial cells isolated from adult animals generated self-renewing, multipotent neurospheres.

Radial glia-like cells whose cell bodies translocate from the SVZ to the dentate gyrus (Eckenhoff and Rakic, 1984) persist as the *radial astrocytes* of the SGZ (Seri et al., 2004; Figure 1B), taking part in the hippocampal stem cell niche. These cells show functional properties similar to those of the SVZ, although some intermediate transit cells can be slightly different (Seri et al., 2001, 2004; Namba et al., 2005). By comparing glial cells of the two main neurogenic sites, SVZ and SGZ, another common pattern can be found in the lineage radial glia-astrocyte/stem cell, although revealing that radial glia transform differently within different tissue environments.

Unlike the hippocampal SGZ, the high density of astrocytic cells packed within the glial tube structure, further complicated by the intermix of their processes, could explain how the existence of a remnant of radial glial cells in this adult neurogenic site was not detected soon after its discovery (Luskin, 1993; Lois et al., 1994; Jankovski and Sotelo, 1996; Lois et al., 1996; Peretto et al., 1997). Due to such a complexity, the exact morphology and spatial organization of individual glial cells therein is at present unknown, and immunocytochemistry indicates that the glial compartment of the adult SVZ is highly heterogeneous. For instance, subsets of SVZ astrocytes express vimentin, nestin (Peretto et al., 1999), platelet-derived growth factor receptor-α (PDGFR α; Jackson et al., 2006), EGF receptor (Doetsch et al., 1999b), carbohydrate antigen known as Lewis X (LeX; Bartsch and Mai, 1991; Capela and Temple, 2002). Combined morphological and functional studies estimated that about 12% of SVZ cells are GFAP$^+$, but only 1% of SVZ cells generate neurospheres (Doetsch et al., 1999b), thus indicating that not all SVZ astrocytes are stem cells. Nevertheless, none of the markers investigated so far can be considered restricted to the stem cell elements, and only slight differences can be observed among these glial cells at the ultrastructural level (Doetsch et al., 1997), thus leaving unresolved the issue of NSCs in vivo identification.

All this evidence proves that radial glial cells not only serve as progenitors for many neurons and glial cells soon after birth, but also give rise to adult SVZ stem cells that continue to produce neurons throughout adult life. Yet, NSC features seem to be retained solely within persistent neurogenic sites since astrocytes from other (non-neurogenic) CNS regions can generate stem cells in vitro only when isolated at early developmental stages, whereas they lose this property after the second postnatal week in mice (Laywell et al., 2000). This confirms that adult neurogenic sites other than harbouring stem/progenitor cells do possess intrinsic cellular/molecular factors capable of favouring the retention of stem properties as well as of allowing them to be extrinsecated. This fact does not exclude that quiescent stem properties could possibly be retained elsewhere (see below).

Neurogenesis declines with aging in the SVZ and, more strikingly, in the SGZ in terms of newly generated neurons, accompanied by changes in the niche cytoarchitecture. A significant decrease in cell proliferation occurs in the mouse SVZ between 2 and 22 months of age, due to a decrease in neuroblast and transit amplifying cell numbers, in contrast with a relatively constant amount of astrocytes (Luo et al., 2006). Although only residual pockets of neurogenesis remain in the dorsolateral horn of the SVZ in old mice, the key components of the younger neurogenic niche are maintained.

The production of new neuronal precursors from radial glia and the persistence of radial glia during adulthood are features even more evident in other vertebrates, whereby they sustain regeneration and brain repair (see below).

Local Progenitors in Non-Neurogenic Parenchyma

Even in the context of an expanding knowledge on brain plasticity and repair, apart from two restricted neurogenic sites, most of the CNS remains a perennial, non renewable tissue. Yet, in addition to SVZ and SGZ, some other loci with potential and actual capabilities to generate glial and/or neuronal cells have been identified in mammals, with remarkable species-related differences (reviewed in Gould, 2007; Bonfanti and Ponti, 2008).

Adult Gliogenesis Is More Widespread than Neurogenesis

The presence of mitotically active glial cells has been observed since long time in radioautographic studies (Walker and Leblond, 1958; Adrian and Walker, 1962). In recent years it was confirmed that glial elements are highly plastic cells under many profiles, both as part of homeostatic mechanisms and in response to injury and disease (reviewed in Horner and Jacobson, 2008).

A continuous production of glial progenitors proliferating *in situ* and differentiating into astrocytes and oligodendrocytes has been shown to occur in the corpus callosum (Gensert and Goldman, 1996, 2001), spinal cord (Horner et al., 2000) and substantia nigra (Lie et al., 2002) of adult rodents. In the adult rat spinal cord, which is not endowed with spontaneous neurogenic capacities, a process of slow but widespread and persistent gliogenesis does occur (Horner et al., 2000). Multipotent adult NSCs that if transplanted in the hippocampus can give rise to both neurons and glia, when grafted in the spinal cord assume a restricted glial fate (Shihabuddin et al., 2000). Thus, the spinal cord environment seems to be particularly gliogenic.

Similar conclusions can be drawn for the cerebellum. Cells endowed with stem properties have been recently isolated from the postnatal (Lee et al., 2005) and even adult parenchyma of the rodent cerebellum (Klein et al., 2005). These cells can generate local phenotypes following transplantation to early postnatal cerebella, when neurogenesis is still active. Beyond this period, however, the cerebellar milieu lacks clear inductive capacities and there is no evidence for either spontaneous, compensatory or inducible neuronal neogeneration (Grimaldi and Rossi, 2006).

The cell intrinsic/extrinsic properties that create a gliogenic environment in the adult CNS are not well understood. The largest class of mitotically active cells in the adult brain and spinal cord were found to express the nerve/glial antigen 2 (NG2) proteoglycan (Horner et al., 2002; Dawson et al., 2003). NG2-expressing cells are a widespread, heterogeneous population of glial progenitor cells most of which retain the capability to divide during adulthood (Belachew et al., 2003). They are also called *synantocytes* (Berry et al., 2002; Butt et al., 2002,2005) or *polydendrocytes* (Nishiyama, 2007), and are morphologically, antigenically, and functionally distinct from mature astrocytes, oligodendrocytes, and microglia, being at present the best example of parenchymal progenitors (Horner et al., 2002; Nishiyama, 2007).

NG2+ cells are also present within both main neurogenic zones (SVZ and SGZ). During the postnatal period, these resident cells are intrinsically multipotent in vitro and can generate neurons in vivo (Aguirre and Gallo, 2004; Aguirre et al., 2004; Belachew et al., 2003), whereas the multipotency of parenchymal NG2-expressing cells remains to be evaluated (Nishiyama, 2007).

Thus, it has not been resolved if the stem-like pools of progenitors isolated from various parenchymal regions at post-developmental periods are of physiologically relevance or artifacts of experimental manipulation (Ourednik et al., 2001). The existence of spatial differences in the developmental competence and normal fates of diverse population of CNS progenitors in vivo (that are not preserved by the in vitro assays used to isolate these cells) suggests that it is important to understand the difference between fate and potential (Jessell, 2000).

In the context of gliogenesis an exception is represented by *microglia*, the resident immune cells of the CNS. Unlike astrocytes and oligodendrocytes - derived from a common neuroepithelial precursor - the origin of microglial cells is still hotly debated, and their functions are quite distinct from macroglia. Although dogma holds that microglia are derived from blood monocytes late in development and continue to populate the CNS via vascular infiltration postnatally, it is now argued that during embryonic and fetal development, microglial precursors are of a peripheral mesodermal origin which is separate from the monocyte lineage (reviewed in Chan et al., 2007). In adults, the resident microglial population undergoes slow turnover in addition to the traditionally established repopulation via vascular infiltration of blood monocytes (Lawson et al., 1992). However, some recent findings indicate that expansion of the microglia population in response to different pathological stimuli is primarily sustained by CNS resident elements rather than blood-borne recruited cells (Midner et al., 2007; Ajami et a., 2007). In healthy tissue, microglia have classically been referred to as "resting", idly awaiting a pathogenic signal, but mounting evidence suggests they play a more active role in nervous tissue monitoring, and a rethinking of their the dynamic nature without the pathological context actually is among future perspectives (reviewed in Hanisch and Kettenmann, 2007; see below).

A different scenario can be found when exploring the neurogenic potential of some mammalian species, as described in the next paragraph.

Spontaneous Parenchymal Neurogenesis

Although proliferative activity in non-neurogenic regions is thought to give rise primarily to glial cells, several in vitro data indicate that neuron competent precursors naturally exist in widely divergent tissues of the adult brain (Palmer et al., 1995,1999; Weiss et al., 1996; Kondo and Raff, 2000; Dawson et al., 2003; Tamura et al., 2004,2007). In addition, spontaneous in vivo parenchymal-derived neurogenesis has been described to occur in some mammals (Bedard et al., 2002; Dayer et al., 2005; Luzzati et al., 2006; Bonfanti, 2006; Ponti and Bonfanti, 2008).

A small rate of local neurogenesis has been shown to occur in the rat neocortex (Dayer et al., 2005; Tamura et al., 2007), whereas the migration of new neuroblasts from the SVZ to the brain parenchyma remains quite controversial (Gould et al., 1999a;2001; Kornack and Rakic 2001; Koketsu et al., 2003; Dayer et al., 2005) or limited to young animals (Luzzati et al., 2003; Ponti et al., 2006a).

Besides the common pattern reported in the CNS of all mammals studied so far, intriguing interspecies differences have been found concerning the occurrence of alternative sources of newly born cells (Bernier et al., 2002; Bédard et al., 2002,2005; Luzzati et al., 2003,2006; Ponti et al., 2006b), thus making the picture of adult neurogenesis more complex when observed under a comparative profile. The occurrence of newly-formed neurons within the mature brain parenchyma independently from the SVZ neurogenic site has been described in the rabbit caudate nucleus (Luzzati et al., 2006). This region represents the most medial-dorsal part of corpus striatum, and it is generally considered non-neurogenic during the postnatal life under physiological condition. By using BrdU injections coupled to confocal analysis with the neuronal marker NeuN performed at two months survival time from the treatment, the genesis of a low amount of new mature neurons (0.7% of surviving BrdU+ cells) was demonstrated. Additional phenotypic analyses showed that about 1/6 of these cells differentiate into the calretinin striatal interneurons. Although the number of fully differentiated new neurons found is extremely low, the percentage of newly generated calretinin cells reaches non negligible values if referred to the whole population of striatal calretinine interneurons (Luzzati et al., 2006, 2007). In addition, the immunocytochemical analysis of early neuronal markers (i.e. PSA-NCAM and doublecortin) showed that in the caudate nucleus of rabbits neuroblasts associate to form chains that are similar to those of the RMS (*striatal chains*, Luzzati et al., 2006) but independent from the SVZ neurogenic site. Phenotypic analyses of the newlyborn striatal cell precursors indicated that most of them express the astroglial marker BLBP, a protein is abundant in radial glia, and the early neuronal marker doublecortin. This is the first example of a neurogenic sequence leading to produce new neurons in the mature brain parenchyma, indicating a striking ability of the rabbit striatum to maintain adult neurogenesis under physiological conditions.

The hypothesis that in lagomorphs the adult CNS parenchyma could be particularly supportive for neurogenesis is strengthened by recent results obtained in the rabbit cerebellum. In rodents, no spontaneous genesis of cerebellar neurons occurs after the exhaustion of the external germinal layer (EGL), a transitory actively-proliferating zone which is thought to disappear long before puberty. In rabbits, a protracted genesis of newly generated neuronal precursors does persist in the cerebellar cortex at peripuberal and adult ages (Bonfanti, 2006; Bonfanti and Ponti, 2008; Ponti et al., unpublished). Neuroblasts generated within a proliferative subpial layer replacing the EGL are arranged to form thousands of tangential chains reminiscent of those responsible for cell migration in the forebrain SVZ. Subpial chains cover the cerebellar surface up to the 5th month of life, then disappearing after puberty. By contrast, newly-born cells continue to exist within the adult cerebellar cortex. These cells are GABA-ergic, Pax2+ neurons of the molecular layer and Sox2+/Olig2+ glial cell precursors reminiscent of synantocytes/polydendrocytes, thus suggesting that neuronal and glial progenitors originating from the neuroepithelium persist within the adult cortical parenchyma, even in the absence of a germinal layer.

Thus, recent research on neurogenesis carried out in different CNS regions and/or in different mammalian species has challenged the concept of "non-neurogenic" parenchyma, that should be replaced by a more general definition, such as "CNS parenchyma outside classic neurogenic sites."

The Regulation of Neurogenesis

In the perspective of future uses of NSCs for therapeutic approaches, it is essential to know how their cell biology and functions are regulated in the context of neurogenesis, at different levels. Neurogenesis is modulated by both physiological stimuli (Kuhn et al., 1996; Kempermann et al., 1997; van Praag et al., 1999; Gould et al., 1999b,c; Mirescu and Gould 2006; addressed in this section of the chapter) and pathophysiological conditions (Bengzon et al., 1997; Parent et al., 1997; addressed in the next section). Basically, the regulation of neurogenesis can occur at two extremes: i) at the cellular/molecular level within the stem cell niche and ii) at the behavioural level, through interaction with the environment in which the animals live.

During development, neurogenesis precedes gliogenesis, being characterized by the dual role of radial glia in the genesis and migration of neurons. Yet, the role of glial cells in neurogenesis is not exhausted after birth, since astrocytes (and ependymal cells in the SVZ) are key components of the adult neurogenic niche after they complete their postnatal assembly (Tramontin *et al.*, 2003; Peretto *et al.*, 2005). In both neurogenic zones, some astrocytes are more kin to traditional astroglia, whereas a subpopulation retains stem-like/radial glia-like properties (see also Gritti and Bonfanti, 2008). Thus, similarly to their radial glia precursors, also astrocytes play a dual role, by serving both as primary precursors and/or supporting cells, adding an additional intriguing feature to the process of adult neurogenesis (reviewed in Chiang et al., 2008). After birth, also ependymal cells can contribute to the regulation of neurogenesis, and newly generated neurons might exert influences on both glial and neuronal cells, contributing to maintain the correct proportions between the stem cell pool and their progeny (reviewed in Genzen and Bordey, 2008).

Hence, the regulation of neurogenesis and NSC activity is a highly articulated process whose multiple steps have partially begun to be dissected and elucidated.

Glial Regulation

Astrocytes

Astrocytes in the niche contact all essential cell types within the niche itself. In the adult mammalian neurogenic niches quiescent and active NSCs, neuronal progenitors, and immature neurons are closely associated with astrocytes, blood vessels, and basal lamina (Ma et al., 2005). Although the astrocytic identity of NSCs has been proven (Doetsch et al., 1999b; Alvarez-Buylla et al., 2001; Merkle et al., 2004) the dynamics of cell division and subsequent cell expansion in vivo remain obscure. Anatomically, the close proximity of astrocytes with every cellular and structural component in the neurogenic niches may enable them to function as local sensors, and their unique cellular continuity through gap junctions may enable them to integrate and propagate local signals. Functionally, astrocytes express various signalling molecules that actively modulate the dynamic process of adult neurogenesis. By expressing secreted and membrane-bound molecules, including cell adhesion molecules, growth factors, neurotrophic factors and cytokines, astrocytes may regulate neurogenic niches in a spatial- and temporal-specific manner (reviewed in Ma et al., 2005).

Factors that are important for cell proliferation, cell patterning and neurogenesis in the embryonic stage are maintained in the adult SVZ and SGZ neurogenic niches. For instance, *Wnt*, *Shh*, and *Notch* signalling are important in maintaining proliferation of adult neural progenitors in vivo (Chambers et al., 2001; Doetsch et al., 2002; Lai et al., 2003). Epidermal growth factor (EGF) and basic fibroblast growth factor (FGF-2), which are used as mitogens to propagate adult NSCs in vitro, are able to induce proliferation of adult neuronal progenitors and quiescent adult NSCs in vivo (Doetsch et al., 2002).

In addition to secreted factors, signalling via cell-cell contact also promotes neurogenesis, as astrocyte-conditioned medium does not fully account for neurogenic activities from astrocytes. Besides, the majority of identified neurogenic factors are not associated with all, but only certain subsets of astrocytes in the SVZ and SGZ, thus further accentuating the complex regulation of adult neurogenesis by specialized local astrocytes (see Ma et al., 2005).

The neuronal progeny of adult SVZ NSCs exhibit astonishing ability to migrate in a highly directed manner, up to several millimetres in rodents and several centimetres in humans (reviewed in Bonfanti and Ponti, 2008; Curtis et al., 2007a). Along the RMS, neuroblasts migrate to the olfactory bulb through the meshwork of astrocytes called glial tubes (Lois et al., 1996; Jankovski and Sotelo, 1996; Peretto et al., 1997) which not only create a physical route for the long-distance migration, but also communicate with migrating cells and regulate their migration speed. In turn, migrating neuroblasts in the RMS release neurotransmitter GABA, thus contributing to regulate migration (see below).

Transplantation experiments showing that adult progenitors isolated from non-neurogenic regions can give rise to neurons when transplanted into neurogenic niches and that they change phenotypes of their neuronal progeny when transplanted into another neurogenic niche support the hypothesis the complex signalling present in the niches regulate neuronal fate specification of adult neural progenitors. However, recent works suggest that indeed NSCs residing in the different neurogenic niches represent specified and non-equivalent populations of primitive cells endowed with an intrinsic and restricted neurogenic program which is maintained during brain development until adulthood in a regionally dependent manner (Merkle et al., 2007; Young et al., 2007).

During development, reciprocal interactions between differentiating glial cells and neurons are important for establishing neuronal connections. The molecular mechanisms regulating synaptic integration of adult-born neurons remain to be fully characterized. However, several astrocyte-secreted synaptogenic factors that promote dendritic development and synapse formation in embryonic systems, including Wnts, thrombospondins, and tumour necrosis factor-α (TNFα), might play an important role also in the adult environment (reviewed in Ma et al., 2005). Dendritic spine morphology and organization are regulated through local astrocyte contact and ephrin ligands (for further details see Ma et al., 2005).

Ependymal Cells

Ependymal cells are ciliated cuboidal/columnar cells forming an epithelial monolayer (ependyma) along the walls of the brain and spinal cord ventricular cavities (Bruni, 1998). Besides being a protective barrier between the brain parenchyma and cerebrospinal fluid they function as important 'partner cells' in the neurogenic niche (reviewed in Genzen and Bordey, 2008). Similarly to astrocytes, ependymal cells are derived from radial glia, but they are post-mitotic under physiological conditions (Spassky et al., 2005). Although it is believed that adult ependymal cells might display limited regenerative capacity after damage, the hypothesis that they might represent an alternative source of NSCs has lost ground. Even though ependymal cells can proliferate in vitro they do not maintain the capacity to self-renew or produce neurons (Chiasson et al., 1999). Indeed, these cells are derived from radial glia but are post-mitotic (Spassky et al., 2005).

Although ependymal cells do not act as NSCs, they are capable to release molecules that modulate cell proliferation and/or NSC self-renewal in the SVZ. The peptides Noggin and Pigment Epithelium-Derived factor (PEDF), which are synthesized and released by ependymal cells, have been identified as key molecules for the regulation of SVZ neurogenesis. Noggin binds to and inhibits bone morphogenetic protein (BMP). As BMP inhibits neurogenesis (and promote gliogenesis), Noggin antagonizes this BMP signalling, therefore promoting neurogenesis in this stem cell niche (Lim et al., 2000), although a more widespread expression of Noggin, BMP4, and BMP7 was observed in the entire SVZ, including regions that do not contain ependymal cells (Peretto et al., 2004). Also PEDF, a protein synthesized also by endothelial cells, has been shown to promote the self-renewal of NSCs (Ramírez-Castillejo et al., 2006).

In addition to secreting molecules, ependymal cells may influence neurogenesis through other mechanisms. The process of coordinated ciliary motion is necessary for normal CSF flow through the cerebral ventricles and may affect neuroblast migration in the RMS by establishing a gradient of chemo-repulsive guidance molecules, such as Slit (Sawamoto et al., 2006). Since ependymal cells build a protective barrier between the cerebrospinal fluid and the brain parenchyma, they may be also involved in alterations of neurogenesis and/or gliogenesis that are present in the context of CNS injury and inflammation (Hagberg and Mallard, 2005; see below).

Neuronal Regulation

The proliferation of neural progenitors is dynamically influenced by many physiological and pathological changes, including hormones, growth factors, physical activity, drugs, ischemia, and seizures (Lim and Alvarez-Buylla, 1999; Lie et al., 2004; Ma et al., 2005).

Neurotransmitters are a class of secreted molecules that are generally associated with neuronal communication, yet they might be important signals during nervous system development. For example, transmitter-gated receptors are operative before synapse formation, suggesting that their action is not restricted to synaptic transmission. Such a 'trophic' effect appears to be mediated through a paracrine, diffuse, non-synaptic mode of action that precedes the more focused, rapid mode of operation characteristic of synaptic connections (Represa and Ben-Ari, 2005).

The amino acid γ-aminobutyric acid (GABA) has been implicated in the regulation of nearly all the key developmental steps, from cell proliferation to circuit refinement. Although in the mature brain GABA functions primarily as an inhibitory neurotransmitter, during CNS development it can also act as a trophic factor to influence events such as proliferation, migration, differentiation, synapse maturation and cell death (Owens and Kriegstein, 2002; Represa and Ben-Ari, 2005). Recent studies indicate that GABA can exert similar effects also in neurogenic sites, possibly acting as a regulator in adult neurogenesis. For example, a negative regulation of newly born olfactory bulb interneurons production via regulated, paracrine GABAergic signalling has suggested (Liu et al., 2005). As GFAP-expressing cells generate neuroblasts in the SVZ, GABA released from neuroblasts provides a feedback mechanism to control the proliferation of GFAP-expressing progenitors, and, in turn, neuroblast production, by activating $GABA_A$ receptors. On the whole, GABA might be considered a key regulator of the neuroblast pool size in the SVZ and on the state of the synaptic network activity in the hippocampus (Bordey, 2007).

Migration of neurons from their birthplace to their final destination is another extremely important step in brain maturation, and cortical migration disorders are the most common brain developmental alteration observed in human patients. During embryonic cortical development GABA works in cooperation with the neurotransmitter glutamate and activation of specific GABA and glutamate receptors differently regulate cell migration, depending on the type of migration (radial, tangential, or chain migration), the type of neurons, and the brain area (reviewed in Manent and Represa, 2007). Paracrine GABA and glutamate acting, respectively, on $GABA_A$ and NMDA receptors have been demonstrated to modulate the migration of hippocampal pyramidal cells in a cross-modulated system that may also be involved the construction of the hippocampal network (Manent et al., 2006). The role of glutamate in adult forebrain neurogenesis remains obscure. On the contrary, it is known that $GABA_A$ receptor activation in adult SVZ neuroblasts reduces their speed of migration in the proximal RMS, probably through a stabilization of their processes that results in reduced motility (Bordey, 2007).

Environmental Regulation

Although the neurobiological bases of neurogenesis in adulthood remain enigmatic, a role in learning and memory has long been recognized across many vertebrates (Lindsey and Tropepe, 2006), in songbirds (Nottebohm, 2002) and in mammals (see Gould et al., 1999b). This implies that, in addition to cellular/molecular regulation, many external, environmental factors can be involved in the modulation of this process.

Hippocampal neurogenesis is modulated by many factors, thus it has been used as a model for quantitative studies aimed at unravelling how cell production and survival are regulated. For example, an enriched environment increases the survival of newly born cells (Kempermann et al., 1997; Gould et al., 1999b,c) whereas stress conditions exert the opposite effect (Mirescu and Gould, 2006). Hormonal levels (Tanapat et al., 1999) and physical activity (Van Praag et al., 1999) also affect hippocampal neurogenesis, although the exact mechanisms are still obscure (Gould et al., 1999b; Mirescu and Gould, 2006; Meshi et al., 2006). Similar to the olfactory system (see below), adult neurogenesis in the dentate gyrus is modulated during learning of hippocampal-dependent tasks, thus it may play a role in learning and memory (Gould et al., 1999b,c; Leuner et al., 2006; Lledo et al., 2006).

Newborn cells in the adult olfactory bulb that had survived to the early wave of cell death can live up to one year or more (reviewed in Lledo et al., 2006). There is increasing evidence for the role of sensory input in control of adult neurogenesis in the olfactory system. In anosmic mice, the lack of metabolic and electrical activity dramatically reduces the survival of newborn neurons acting selectively on those cells which have achieved morphological maturation, whereas this manipulation does not influence proliferation and migration of immature cells (Petreanu and Alvarez Buylla, 2002). Deprivation of the sensory input by naris occlusion reduces survival of newly generated cells (Corotto et al., 1994; Mandairon et al., 2006) and a critical time window for sensory experience-dependent survival of new granule cells has been established at days 14-28 after birth (Yamaguchi and Mori, 2005).

Following protocols of reversible olfactory deprivation in which newborn animals are unilaterally occluded for 20 days, which reduces the number of cells in the bulb, and then allow to access to normal olfactory stimulation for the following 40 days, olfactory bulb cell numbers can recover (Cummings et al., 1997). Interestingly, this effect was found to depend, at least partially, by increase of newly generated cells that survived in the previously deprived olfactory bulb, suggesting that the probability of survival of newly generated neurons may be enhanced by renewed access to environmental stimulation.

Computation modeling suggests activity-dependent incorporation of new granule cells in the olfactory bulb optimizes olfactory discrimination, a theory supported by recent studies showing that survival of adult born granule neurons is transiently increased by olfactory enrichment and correlates to improvement in olfactory memory (Rochefort et al., 2002). A positive modulation of olfactory bulb neurogenesis has also been described following a protocol of olfactory discrimination learning (Alonso et al., 2006). Although in some cases contradictory, all studies indicate a functional relationship between newborn cell survival and learning (see also Giachino et al., 2005; Miwa and Storm, 2005).

In parallel with studies on the cellular/molecular/environmental regulation of neurogenesis, in the last few years researchers have started to tackle the issues concerning how NSC activity can be modulated after CNS injuries and disesease. The initial, obvious aim of such an approach was that of replaying/modulating the neurogenic process in the perspective of brain repair and cell replacement. Nevertheless, the subtle and complex molecular/cellular machinery regulating the biology of NSCs in brain physiology does not work in most lesion paradigms and neurodegenerative diseases. Some progresses have been recently made in the understanding if and how stem cells and their progeny do behave in different pathological contexts. The next paragraph contains a summary of the present knowledge in this emerging field.

Neurogenesis and Pathological Conditions

Neurological injuries as well as chronic neurodegenerative diseases, characterized by neuronal and/or glial loss accompanied by neurological disfunction or cognitive impairment, are largely untreatable, thus representing a huge social and economical burden. In this context, the discovery of NSCs and of neurogenic processes representing a direct outcome of their activity in vivo, prompted new perspectives for brain repair. In the last few years, in parallel with research carried out on adult neurogenesis and that aimed at unraveling the pathogenesis of these diseases, many studies addressed the mutual influences between neurogenesis and pathological condition.

Although this field of research is still at its beginning, it is clear that different responses in stem cell activity and glio-/neuro-genesis are dependent on the specific situations involved, i.e. the site and type of lesion (e.g. mechanical lesions, hypoxia/ischemia, neurodegenerative diseases, demyelination), the age, and the animal species involved (Goings et al., 2004; Picard-Riera et al., 2004; Chen et al., 2007; Curtis et al., 2007b; Das and Basu, 2007; Kim and Szele, 2007).

SVZ Activation in Different Lesion/Disease Paradigms

Endogenous SVZ NSCs become "activated" after neuronal injury (reviewed in Romanko et al., 2004; Kim and Szele, 2007). Such an activation implies changes in cell proliferation, migration from the SVZ to the surrounding parenchyma, and differentiation into glial/neuronal elements depending on different lesion/disease paradigms. Most situations have ultimately the prevalent effect of expanding the neuronal/glial precursor cell population existing within neurogenic sites, a process occurring parallel to an upregulation of developmentally-related molecules that can partially mimic but not recreate the environment typical of embryonic neurogenesis. Unfortunately, the identification of the real nature, origin, and fate of these cells is not easy to assess, and it is known that several methodological artefacts can affect the detection of newly born cells in vivo, particularly within injured tissue (Gould and Gross, 2002; Rakic, 2002).

Mechanical lesions are highly variable and difficult to be reproduced experimentally. Generally, they result in hemorrhage, edema, followed by gliosis and ultimately cell death. How these changes in brain tissue homeostasis do affect cells in the SVZ is poorly understood, due to heterogeneity of experimental models and results obtained, as well as species differences, even between rat and mouse (reviewed in Dizon and Szele, 2006). In general, cell proliferation seems to be increased in rats and decreased in mice, although this could not be related to effective neurogenesis, the total number of cells in the SVZ being also regulated by cell migration and cell death.

Despite the presence of the glial tube astrocytic meshwork, we know that cells generated in the SVZ can emigrate outside their physiological route of the RMS since the first experiments of growth factor (EGF) intraventricular injection (Craig et al., 1996). Under prolonged exposure to EGF these cells invade the surrounding mature parenchyma, including striatum, septum, corpus callosum, and cortex (Doetsch et al., 2002).

Many studies suggest that newly generated cells also migrate away from the proliferative regions in postnatal and adult lesioned brains, both after trauma and ischemia (Zhang et al., 2001,2007; Arvidsson et al., 2002; Parent et al., 2002a,b,2006; Gotts and Chesselet, 2005; Felling et al., 2006; Lee et al., 2006; Yamashita et al., 2006). Cells generated within the SVZ have been proposed to reach the striatum or the corpus callosum within two-three weeks (Parent et al., 2002a,b; Arvidsson et al., 2002; Yamashita et al., 2006). In lysolecithin-induced demyelinating lesions of the corpus callosum, glial progenitor cells recruited from the subjacent SVZ were shown to migrate to the lesion site and differentiate into oligodendrocytes and astrocytes (Nait-Oumesmar et al., 1999; Picard-Riera et al., 2002). A similar result has been recently observed in postmortem brains of humans affected by multiple sclerosis (Nait-Oumesmar et al., 2007).

The genesis of new neurons has been proposed to occur in the neocortex of adult mice after targeted cell death of corticothalamic neurons that does not affect any of the surrounding cell types (Magavi et al., 2000). In this paradigm[2], the dying neurons are phagocitically removed without inflammation or gliosis, what could provide an instructive environment for restorative neurogenesis. These authors suggest that after targeted apoptosis in the murine neocortex new neurons can integrate in the pre-existing neuronal circuits as a result of both migration from the subjacent SVZ and local proliferation/differentiation (reviewed in Magavi and Macklis, 2001; see also below). By the same paradigm neogeneration of cortico-spinal neurons has been also reported (Chen et al., 2004). However, in spite of the interest of this model, compensatory neogenesis of cerebral cortical neurons still avaits a stringent demonstration in conditions matching a real pathological situation.

These results clearly show that different types of injury/disease can trigger specific differentiative pathways in the newly generated progenitor cells, yet leaving open the question if they come from the same multipotent stem cells or from co-existing unipotent

[2] Degeneration is obtained by photoactivation of retrogradely transported nanospheres carrying the chromophore chlorin e6 previously injected into the axonal field of the targeted projection neurons and retrogradely transported.

progenitors (see Kim and Szele, 2007). Both these situations could be active at the same time within different portions of the SVZ, its heterogeneity remaining largely unesplored (see for example Hack et al., 2005; Merkle et al., 2007).

SGZ Activation in Different Lesion/Disease Paradigms

Dentate gyrus neurogenesis is stimulated by several types of injuries leading to granule neurons cell death: adrenalectomy, excitotoxic or mechanical lesions (Cameron and Gould, 1994; Gould and Tanapat, 1997). The same outcome is observed after ischemia (Liu et al., 1998; Bernabeu and Sharp, 2000; Arvidsson et al., 2002; reviewed in Parent et al., 2002a; Kokaia and Lindvall, 2003) and epileptic seizures (Bengzon et al., 1997; Parent et al., 1997; Gray and Sundstrom, 1998).

Two brain injury models in particular, experimental epilepsy and stroke, have provided significant insight into the consequences of injury-induced neurogenesis in the hippocampus. Dentate gyrus neurogenesis in adult rodent epilepsy models suggest that seizure-induced neurogenesis involves aberrant neuroblast migration and integration that may contribute to persistent hippocampal hyperexcitability. On the other hand, both global ischemia (a model that replicates coronary artery occlusion) and focal ischemia (stroke) increases the production of new dentate gyrus granule neurons (Liu et al., 1998; Arvidsson et al., 2002; Kee et al., 2001; Yagita et al., 2001; Komitova et al., 2002), although we lack evidence that they replace those that have died. Injection of growth factors (EGF and FGF-2) after ischemia leads to regeneration of pyramidal neurons in rats starting from progenitor cells located in the posterior SVZ (Nakatomi et al., 2002). This report, however, still awaits definitive confirmation. The recruitment of cell progenitors from the caudal SVZ into the injured hippocampus has also been described to occur after prolonged seizures (Parent et al., 2006).

These findings underscore the need for a better understanding of injury-induced neurogenesis in the adult and suggest that the manipulation of endogenous neural precursors is a potential strategy for brain reparative therapies.

Perspectives for Brain Repair

Studies in the stem cell field are of great interest for unraveling basic cellular mechanisms (proliferation, differentiation, cancer genesis) and exciting therapeutic perspectives (cell therapy, regenerative medicine) (Gage and Verma, 2003). Notwithstanding the headway achieved, many cellular and tissue mechanisms controlling the activity of these cells are still unclear, particularly in the nervous system. Cell replacement therapies for neurologic diseases require that newly-added cells adopt specific phenotypes and become functionally integrated into the adult CNS network. Although stem cells, which also reside in the mature CNS, are able to differentiate into different cell types, their actual phenotypic choice will be ultimately determined by the extrinsic cues present in their microenvironment. As a consequence, crucial to develop efficient therapeutic strategies is to elucidate whether the adult CNS milieu provides instructive information to direct the specification and adaptive integration of multipotent cells. Although the adult injured CNS may be endowed with previously overlooked self-repair potentialities, the real capabilities of the adult brain to incorporate new elements have still to be determined. Thus, new answers to unresolved questions could reside in the dualism between intrinsic (cellular) factors and extrinsic cues such as interactions in the "stem cell niche" and in the host tissue.

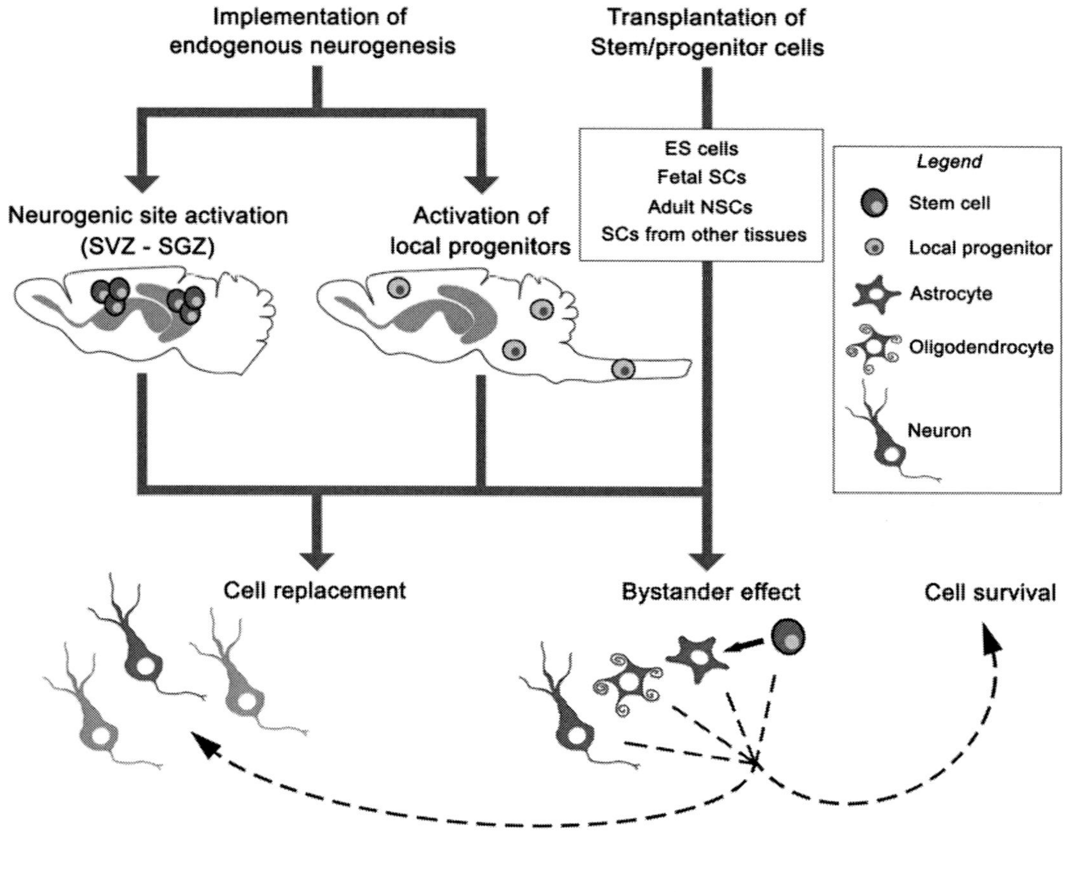

Figure 3. *Prospect of possible neural stem cell (NSC) exploitations for brain repair*. Different sources (endogenous and exogenous) of stem cell/progenitor cells can be used to obtain either cell replacement of lost elements and/or a supporting/trophic (bystander) effect for cell survival. The successful therapeutic effect of all these pathways depends on the interaction with the nervous tissue environment.

Cell Replacement in the Adult CNS

Specification of Stem/Progenitor Cells

Neuronal replacement in the adult CNS may be obtained through cell transplantation or mobilization of resident stem/progenitor cells (summarized in Figure 3). In both instances, the process is accomplished through two major steps. The newly added stem cell has first to be specified towards a specific phenotype and, then, it has to become anatomically and functionally integrated in the recipient network. In the neurogenic regions of the adult CNS, the whole process occurs spontaneously. When these regions are hit by injury, reactive mechanisms boost the neurogenic machinery so to counteract neuronal loss (Goldman, 2005; see also references in the sections above). There is evidence that attempts to set up similar compensatory mechanisms may also take place in normally non-neurogenic sites (e.g. Magavi et al., 2000; Arvidsson et al., 2002; Kokaya and Lindvall, 2003; Curtis et al., 2003; Jin et al., 2004; Lindvall and Kokaia, 2006; Okano et al., 2007). However, the actual extent of these phenomena, particularly in terms of effective replacement of lost neurons remains uncertain. Although it is clear that adult neurogenesis is a physiologically relevant phenomenon only in restricted sites and concerns a limited number of phenotypes, it is plausible that it may be reactivated following injury to favour repair. However, the available evidence indicates that this may happen only in very particular conditions (e.g. Magavi et al., 2000) or with an almost abortive outcome (e.g. Arvidsson et al., 2002). By contrast, numerous reports indicate that the injured CNS milieu retains poor neurogenic abilities (Martens et al., 2002; Zhao et al., 2003; Frielingsdorf et al., 2004; Grimaldi and Rossi, 2006) and, rather, develops strong gliogenic aptitudes (see Rossi and Cattaneo, 2002). If spontaneous mechanisms are not sufficient to sustain repair, latent neurogenic potentialities present in the adult nervous tissue might be enhanced by the application of specific stimulating substances. Over the last few years, intraventricular application of growth factors or neurotrophic molecules has been employed to promote neuronal neogeneration in normally non-neurogenic locations (reviewed in Goldman, 2004, 2007). Following application of brain-derived neurotrophic factor (BDNF; Zigova et al., 1998; Benraiss et al., 2001; Pencea et al., 2001) numerous proliferating cells appear in the regions surrounding the cerebral ventricles. Some of such newlyborn cells settle in the striatum and acquire features suggestive of medium-sized spiny neurons. In addition, new elements expressing markers for immature neurons occur in several other regions. Long-lasting infusion of epidermal growth factor (EGF) and basic fibroblast growth factor (bFGF) also enhance the proliferation of resident progenitors in the SVZ of the lateral ventricles (Craig et al., 1996; Kuhn et al., 1997). These factors influence the phenotypic choices of newlyborn elements: EGF promotes neuronal production in the olfactory bulb, whereas bFGF increases gliogenesis in the olfactory bulb and the dentate gyrus. Infusion of the same substances in the fourth ventricle stimulates the proliferation of resident progenitor cells of the brainstem and spinal cord (Martens et al., 2002) and of the cerebellum (Grimaldi and Rossi, 2006). Here, however, these cells exclusively differentiate into glia. Therefore, application of bioactive molecules can effectively enhance the proliferation of endogenous progenitor cells. Nevertheless, evidence that these cells can acquire mature neuron phenotypes and become stably incorporated in adult circuits is still scant.

Similar approaches applied to experimental models of injury or degeneration have yielded some encouraging results in the striatum (Fallon et al., 2000), hippocampus (Nakatomi et al., 2002) and septum (Calzà et al., 2003), but they resulted uneffective in the cerebellum (Grimaldi and Rossi, 2006). Thus, if latent neurogenic potentialities exist in the adult CNS, they may be not homogeneously distributed throughout the neuraxis.

While the existence of resident progenitors in the adult CNS parenchyma is firmly established, little is known about their actual developmental potentialities in vivo. Several reports indicate that such proliferating cells express region-specific properties and, hence, might be actually restricted in their phenotypic choices (Hitoshi et al., 2002; Kim et al., 2006;

Armando et al., 2007). As a consequence, therapeutic employment of resident progenitors of the mature CNS still requires a more careful characterization and deeper understanding of their biological properties and of the signaling mechanisms that may determine their fate choices. Alternative strategies in this direction could involve the setting of ex vivo procedures that replicate instructive signaling needed to generate specific neuronal phenotypes (e.g. Wichterle et al., 2002; Smidt and Burbach, 2007). Indeed, there is evidence that the best outcomes in terms of specific differentiation and integration are obtained with committed progenitors rather than naïve stem elements (MacLaren et al., 2006). Finally, it should be emphasized that even when in the case of weak neurogenic properties of the recipient milieu, other mechanisms may help to replace lost neurons by selecting specific phenotypes among pools of committed precursors (e.g. Carletti and Rossi, 2005).

Integration of New Neurons in Mature Circuits

The lack of neurogenic capabilities in a given CNS region may be circumvented by the ex vivo production of committed progenitors. On the contrary, their integration into the recipient circuitry is necessarily conditioned by the specific properties of the recipient tissue. During this process the newly-born neuron must migrate towards its final placement, develop its mature phenotypic traits and establish functional connections with the appropriate partners. Hence, integration is an ontogenetic phenomenon that is accomplished within a mature environment. The latter must provide the newly-added elements with specific signaling, guidance cues and conductive substrates.

Full anatomo-functional incorporation of new elements in preexisting networks spontaneously occurs in active neurogenic sites (Ming and Song, 2005; Emsley et al., 2005). By contrast, there is scant indication that the same may occur in other regions of the adult CNS. Neuronal replacement and integration in the mature cerebral cortex by transplanted (Macklis, 1993) or resident progenitors (Magavi et al., 2000; Chen et al., 2004) seems to be strictly dependent on the maintenance of the tissue architecture. For instance, a good degree of anatomical incorporation can be obtained after targeted photolytic ablation of cortical neurons, a procedure that causes minimal disruption of the tissue texture (Macklis, 1993). On the contrary, settling of donor neurons in the same sites is defective after excitotoxic injuries, that provoke reactive gliosis and disruption of the cortical structure. Therefore, a first prerequisite for the homing of new neurons in non-neurogenic sites is the preservation of tissue structure, which is not an obvious condition following injury or degeneration.

Even when the recipient architecture is maintained, full integration is not automatically guaranteed. In fact, the adult tissue may lack appropriate supportive cues and/or express inhibitory molecules that hinder the navigation of new cells or processes. Indeed, there is evidence that juvenile neurons can actively modify the adult cellular/molecular microenvironment to favour their penetration and settling. This phenomenon, called "adaptive rejuvenation" (Sotelo et al., 1994), involves the re-expression of developmentally-regulated molecules by glial scaffolds as well as the induction of structural plasticity and synaptogenesis of adult axons, in order to integrate new elements in the host circuitry (Sotelo and Alvarado-Mallart, 1991; Grimaldi et al., 2005). Newlyborn neurons can be also endowed with strong adaptive properties that allow them to exploit unusual migratory pathways, when the physiological routes are impassable or no longer available. For instance, embryonic

Purkinje cells normally migrate to the cerebellar cortex along a radial path from the ventricular neuroepithelium. In the adult this route is obstructed by inhibitory cues present in the internal granule cell layer, which is not yet present at the ontogenetic stage when Purkinje cells migrate to the cortex. Thus, in adult individuals transplanted Purkinje cells enter the cerebellar cortex via the pial surface and then migrate inwardly along Bergmann fibres (Sotelo and Alvarado-Mallart, 1991; Grimaldi et al., 2005). This path, which is also conductive for granule cells but not for neocortical neurons (Rossi et al., 1992, 1994; Grimaldi et al., 2005), allows donor Purkinje cells to establish specific connectivity with local partners. However, because of the unusual migratory route, such Purkinje cells have inverted orientation and fail to send their axons to their natural targets in the deep cerebellar nuclei (Sotelo and Alvarado-Mallart, 1991; Grimaldi et al., 2005). This example is paradigmatic in showing that specific cell replacement and circuit rewiring in non-neurogenic sites likely requires targeted manipulations of the recipient microenvironment to allow full integration of newly-added elements.

On the whole, the observations reported above indicate that efficient neuronal replacement may be feasible. However, this goal requires some deep understanding of the biological features of the newly-added neurons (either endogenous or exogenous) and also of the recipient tissue. Indeed, it has to be emphasized that while major efforts have been produced to investigate cellular physiology and molecular biology of neural stem cells, little attention has been paid to the properties of the recipient tissue, particularly when related to specific pathological conditions. Because of the complexity of the adult nervous tissue and the multifarious nature of its diseases, this step is likely the most challenging. Nevertheless, it is clear that even the best cells may turn out to be unuseful if we do not succeed in mastering the recipient environment.

Since the discovery of NSCs, it has becoming more and more evident that large-scale sources of NSCs can represent a fundamental resource for both basic research and advances toward the treatment of neurodegenerative disorders. In fact, the extensive growth of neural stem and progenitor cells in vitro, would offer a large population of progenitors that, beside offering an appropriate tool to study the essential biology of tissue-specific stem/progenitors and their molecular and biochemical properties, would enclose considerable potentialities in the therapy for a broad variety of clinical conditions.

Nevertheless, in light of the CNS complexity as well as of the complex cellular/molecular interplay governing division, maintenance and differentiation of stem cells and their progeny, it is clear that significant progress will require the close coordination of in vivo and in vitro approaches. In this scenario, in vitro systems of NSCs can allow a deep analysis at cellular level providing useful information to be further validate in the embryo and adult in order to identify relevance to normal physiology. Nevertheless, these cellular systems can present unique opportunities for genetic or chemical screens, and for biochemical strategies. For biomedicine these issues become less important than whether the cells are useful in stimulating repair, are a delivery vehicle or cell replacement.

Activation of CNS Dormant Cell Progenitors: A Future Perspective

The identification of stem/progenitor cells residing within the mature non-neurogenic nerve tissue introduces new perspectives in the study of adult neurogenesis and CNS structural plasticity. Firstly, it extends the concept of *potential* stem cells (quiescent multipotent progenitors) to the whole nervous system, in addition to the existence of *actual* stem cells within the neurogenic niches (Loeffler and Potten, 1997). Consequently, there is a great interest in the long-term perspective of the potential mobilization of these progenitors as an endogenous source for brain repair, possibly by-passing the problems linked to cell transplantation and to the topographically highly restricted location of neurogenic sites (Lie et al., 2004; Luzzati et al., 2007; Bonfanti and Ponti, 2008).

Research summarized in this chapter actually indicates that besides adult neurogenesis and gliogenesis in brain neurogenic sites, undifferentiated progenitors and/or quiescent stem cells could also persist elsewhere in the mature CNS, cycling at slow rates and maintaining the capacity of 'wake' in particular physiological or lesion-induced conditions (Palmer et al., 1995, 1999; Weiss et al., 1996; Gensert and Goldman, 2001; Lie et al., 2002). Nevertheless, the remarkable interest in the study of the potentialities of parenchymal cell progenitors is somehow slowed down by the complexity of the factors/environmental conditions required for their modulation. Unraveling in depth the regulation of endogenous neurogenesis/gliogenesis could lead to its implementation through re-instruction of neural precursor cells to generate new neurons and glial cells, a trend that is rapidly growing as an emerging field in neuroscience.

Among the issues remaining to be tackled is the heterogeneity of parenchymal cell progenitors and to what extent they can be activated in order to contribute to brain repair. One appealing hypothesis about the origin, dispersion and putative function of multipotent progenitors in the mature brain parenchyma has been proposed by Ourednick and collaborators (Ourednik et al., 2001). By grafting a clone of NSCs of human derivation into the developing brain of fetal bonnet monkeys they found that the progeny of this single traceable clone get segregated into two separate subpopulations. One migrating radially from the germinal zone to the appropriate cortical layer and differentiating into neurons, the other represented by undifferentiated cells dispersing through the SVZ and the mature striatal and cortical parenchyma. The authors suggest that this latter population may provide a local resident pool for self repair and plasticity, thus representing the stem-like cells extracted by several investigators. Such a hypothesis, obtained with an experimental paradigm, is reinforced by the recent finding of neuronal and glial progenitors in the marginal zone of the developing cerebral cortex (Costa et al., 2007), showing distinct properties from those of the VZ and SVZ germinal layers and appearing to be regulated by local environmental cues. The integration of new neurons in the pre-existing neuronal circuits of the murine neocortex after targeted cell death, as a result of proliferation/differentiation of local cell progenitors (Magavi et al., 2000; described above), could be linked to such type of cells "trapped" within the mature parenchyma.

In disease states or after injury, also microglia become hyperplastic and hypertrophic, altering their phenotypes to an "activated" form, and performing important protective inflammatory roles such as the phagocytosis of debris and dead cells. Microglia also play a role in chronic neuroinflammation which may perturb or even initiate pathology in diseases such as Parkinson's or Alzheimer's (Rogers et al., 2007). On the other hand, bone marrow-derived microglia-like cells can infiltrate the CNS of Alzheimer disease trangenic animal models and help to clear the β-amyloid (Malm et al., 2005), and inhibition of bone marrow-derived microglial recruitment in the CNS increases the amount of β-amyloid deposits (Simard et al., 2006). Thus, the nature of the microglial response to pathological states, whether neuroprotective or neurotoxic, is still under debate (Bessis et al., 2007; Glezer et a., 2007).

Recently, the effect of exogenously delivered stem cells, e.g. that could reach the CNS parenchyma through the blood flow, has also been taken into account. It has been discovered that NSCs intravenously transplantated in animal models of CNS inflammation can integrate into *atypical ectopic* (perivascular) niches, thus exerting a bystander effect of neuroprotection and trophic support (Pluchino et al., 2003,2005; Martino and Pluchino, 2006). Such atypical ectopic niches are formed around perivascular areas of the inflamed CNS and contain transplanted neural precursor cells, blood-borne (encephalitogenic) inflammatory cells as well as CNS-resident cells (such as inflammation-reactive astrocytes and microglia; reviewed in Martino and Pluchino, 2006).

This approach could be exploited in the future for certain neurological disorders characterized by recurrent and/or chronic inflammation, which invariably leads to destruction of both CNS-resident as well as transplanted cells. Indeed, a new perspective is opening up, in which transplanted neural stem/precursor cells by exerting bystander immunomodulatory and/or neurotrophic properties may have particular therapeutic relevance (see also Calabrese et al., 2007).

Conclusion

The concept of replacing lost neurons to repair brain damage has now been extensively investigated and tested for several decades. During this time, remarkable advancements have been obtained, particularly in the knowledge of the biological properties of stem/progenitor cells that can be used as donors. In this context, genetic engineering and ex vivo manipulations have greatly enhanced our ability to isolate cells from developing or mature tissues, to maintain and expand them in vitro, and to modify their properties to the advantage of specific therapeutic targets. However, the challenge is now shifting towards some different aspects. The repair of nervous tissue, and particularly the rewiring of neural circuits, relies on the ability of newly-born donor cells to interact with the novel environment in order to appropriately integrate into the recipient architecture and functional networks. In spite of the previously unexpected plastic/adaptive properties of the adult CNS tissue that have been unraveled by many recent studies, functional incorporation of new neurons in the recipient circuits remains a difficult task. Nevertheless, the persistence of neurogenic niches in the adult brain provides a most important opportunity to elucidate the cellular/molecular mechanisms underlying the process of neuronal differentiation and integration in a mature environment. Thus, a deep understanding of the basic biological mechanisms that regulate physiological neurogenesis will be a crucial step to produce more efficient donor cells, but also to design targeted manipulations of the host environment that may facilitate the homing of new elements. Therefore, we are convinced that in the next few years studies in this direction are going to provide some major breakthroughs in the field of cell therapies for CNS damage.

Interestingly, in a future perspective some hints for brain repair strategies could come from a comparative vision of neurogenesis. Firstly, the increasing evidence for diverse neurogenic sites and potentialities among different mammals suggests that neurogenesis in humans has not been thoroughly revealed, thus leaving open the possibility that it could be more widespread and articulated with respect to our present knowledge (Sanai et al., 2004; Quinones-Hinojosa et al., 2006; Curtis et al., 2007a; reviewed in Bonfanti and Ponti, 2008; Gould, 2007). For instance, the studies carried out on rabbits support the existence of spontaneous mammalian neurogenesis and gliogenesis outside the classic SVZ and SGZ neurogenic sites, as a result of local cell progenitor activity (Bonfanti and Ponti, 2008; Luzzati et al., 2007). Secondly, also the radial glial origin of neural stem cells raises particular interest in a comparative perspective. In fact, it is now clear that radial glia is a common primary progenitor cell in many vertebrates but its capability to retain stem cell properties highly varies among classes and species, these variations being related to a different morphological adaptation of radial glial cells to the mature CNS environment. For instance, morphologically recognizable radial glia do persist in birds endowed with high neurogenic potential (Alvarez-Buylla et al., 1990), as well as in amphibians that show astonishing regenerative potential (Parish et al., 2007; for review see Garcia-Verdugo et al., 2002).

Thus, in an evolutionary perspective, the postnatal modifications of embryonic radial glia and their related retention of stemness properties could hold the key for unraveling new strategies for brain repair.

Acknowledgments

This work was supported by grants from Compagnia di San Paolo (Progetto NEUROTRANSPLANT 2004.2019), Regione Piemonte (CIPE 2004 - A14), M.U.R.S.T. (PRIN), and University of Turin. We wish to thank G. Zanutto for expertise in graphics.

References

Adrian, E.K. and Walker, B.E. (1962). Incorporation of thymidine-H3 by cells in normal and injured mouse spinal cord. *J. Neuropathol. Exp. Neurol.*, 21, 597-609.

Aguirre, A.A. and Gallo, V. (2004). Postnatal neurogenesis and gliogenesis in the olfactory bulb from NG2-expressing progenitors of the subventricular zone. *J. Neurosci.*, 24, 10530-10541.

Aguirre, A.A.; Chittajallu, R.; Belachew, S.; and Gallo, V. (2004). NG2-expressing cells in the subventricular zone are type C-like cells and contribute to interneuron generation in the postnatal hippocampus. *J. Cell Biol.*, 165, 575-589.

Ajami, B.; Bennett, J.L.; Krieger, C.; Tetzlaff, W.; and Rossi, F.M. (2007). Local self-renewal can sustain CNS microglia maintenance and function throughout adult life. Nat. Neurosci., 10, 1538-1543.

Alonso, M.; Viollet, C.; Gabellec, M.M.; Meas-Yedid, V.; Olivo-Marin, J.C.; and Lledo, P.M. (2006). Olfactory discrimination learning increases the survival of adult-born neurons in the olfactory bulb. *J. Neurosci.*, 26, 10508-10513.

Altman, J. and Das, G.D. (1965). Post-natal origin of microneurones in the rat brain. *Nature*, 207, 953-956.

Altman, J. (1969). Autoradiographic and histological studies of postnatal neurogenesis. III. Dating the time of production and onset of differentiation of cerebellar microneurons in rats. J. Comp. Neurol., 137, 433-458.

Alvarez-Buylla, A.; Theelen, M. and Nottebohm, F. (1990). Proliferation 'hot spots' in adult avian ventricular zone reveal radial cell division. *Neuron*, 5, 101-109.

Alvarez-Buylla, A.; Garcìa-Verdugo, J.M. and Tramontin, A.D. (2001). A unified hypothesis on the lineage of neural stem cells. *Nat. Rev. Neurosci.*, 2, 287-293.

Alvarez-Buylla, A. and Lim, D.A. (2004). For the long run: maintaining germinal niches in the adult brain. *Neuron*, 41, 683-686.

Alvarez-Buylla, A. and Garcìa-Verdugo, J.M. (2002). Neurogenesis in adult subventricular zone. *J. Neurosci.*, 22, 629-634.

Alves, J.A.J.; Barone, P.; Engelender, S.; Froes, M.M.; and Menezes, J.R.L. (2002). Initial stages of radial glia astrocytic transformation in the early postnatal anterior subventricular zone. *J. Neurobiol.*, 52, 251-265.

Armando, S.; Lebrun, A.; Hugnot, J.P.; Ripoll, C.; Saunier, M.; and Simonneau, L. (2007). Neurosphere-derived neural cells show region-specific behaviour in vitro. *Neuroreport,* 18, 1539-1542.

Arvidsson, A; Collin, T.; Kirik, D.; Kokaia, Z.; and Lindvall, O. (2002). Neuronal replacement from endogenous precursors in the adult brain after stroke. *Nat. Med.*, 8, 963-970.

Asakura, A.; Seale, P.; Girgis-Gabardo, A.; and Rudnicki, M.A. (2002). Myogenic specification of side population cells in skeletal muscle. *J. Cell. Biol.*, 2002, 159, 123-134.

Bartsch, D. and Mai, J.K. (1991). Distribution of the 3-fucosyl-N-acetyl-lactosamine (FAL) epitope in the adult mouse brain. *Cell Tissue Res.*, 263, 353-366.

Bayer, S.A. and Altman, J. (2004). Development of the telencephalon: neural stem cells, neurogenesis, and neuronal migration. In: *The Rat Nervous System*. Paxinos, G. Ed., 3rd Edition, Academic Press, San Diego, 27-73.

Bédard, A.; Cossette, M.; Lévesque, M.; and Parent, A. (2002). Proliferating cells can differentiate into neurons in the striatum of normal adult monkey. *Neurosci. Lett.*, 328, 213-216.

Belachew, S.; Chittajallu, R.; Aguirre, A.A.; Yuan, X.; Kirby, M.; Anderson, S.; and Gallo, V. (2003). Postnatal NG2 proteoglycan-expressing progenitor cells are intrinsically multipotent and generate functional neurons. *J. Cell Biol.*, 161, 169-186.

Bengzon, J.; Kokaia, Z.; Elmer, E.; Nanobashvili, A.; Kokaia, M.; and Lindvall, O. (1997). Apoptosis and proliferation of dentate gyrus neurons after single and intermittent limbic seizures. *Proc. Natl. Acad. Sci. USA*, 94, 10432-10437.

Benraiss, A.; Chmielnicki, E.; Lerner, K.; Roh, D.; and Goldman, S.A. (2001). Adenoviral brain-derived neurotrophic factor induces both neostriatal and olfactory neuronal recruitment from endogenous progenitor cells in the adult forebrain. *J. Neurosci.*, 21, 6718–6731.

Bernabeu, R. and Sharp, F.R. (2000). NMDA and AMPA/kainate glutamate receptors modulate dentate neurogenesis and CA3 synapsin-I in normal and ischemic hippocampus. *J. Cereb. Blood Flow Metab.*, 20, 1669-1680.

Bernier, P.J.; Bédard, A.; Vinet, J.; Lévesque, M.; and Parent, A. (2002). Newly generated neurons in the amygdala and adjoing cortex of adult primates. *Proc. Natl. Acad. Sci. USA*, 99, 11464-11469.

Berry, M.; Hubbard, P. and Butt, A.M. (2002). Cytology and lineage of NG2-positive glia. *J. Neurocytol.*, 31, 457-467.

Bessis, A.; Béchade, C.; Bernard, D.; and Roumier, A. (2007). Microglial control of neuronal death and synaptic properties. *Glia*, 55, 233-238.

Blau, H.M.; Brazelton, T.R. and Weimann, J.M. (2001). The evolving concept of a stem cell: entity or function? *Cell*, 2001, 105, 829-841.

Bonfanti L. (2006). PSA-NCAM in mammalian structural plasticity and neurogenesis. *Prog. Neurobiol.*, 80, 129-164.

Bonfanti, L. (2008). Neurogenesis during mammalian postnatal brain development: anatomical and functional considerations. In: Bonfanti L, editor. *Postnatal and Adult Neurogenesis*. Research Signpost, Trivandrum, India.

Bonfanti, L. and Peretto, P. (2007). Radial glial origin of the adult neural stem cells in the subventricular zone. *Prog. Neurobiol.*, 83, 24-36.

Bonfanti, L. and Ponti, G. (2008). Adult neurogenesis and the New Zealand White Rabbit. *Vet. J.*, 175, 310-331.

Bordey, A. (2007). Enigmatic GABAergic networks in adult neurogenic zones. *Brain Res. Rev.*, 53, 124-134.

Bruni, J.E. (1998). Ependymal development, proliferation, and functions: a review. *Microsc. Res. Tech.*, 41, 2-13.

Butt, A.M.; Kiff, J.; Hubbard, P.; and Berry, M. (2002). Synantocytes: new functions for novel NG2 expressing glia. *J. Neurocytol.*, 31, 551-565.

Butt, A.M., Hamilton, N., Hubbard, P., Pugh, M.; and Ibrahim, M. (2005). Synantocytes: the fifth element. *J. Anat.*, 207, 695-706.

Calabrese, C.; Poppleton, H.; Kocak, M.; Hogg, T.L.; Fuller, C.; Hamner, B.; Oh, E.Y.; Gaber, M.W., Finklestein, D.; Allen, M.; Frank, A.; Bayazitov, I.T.; Zakharenko, S.S.; Gajjar, A.; Davidoff, A.; and Gilbertson, R.J. (2007). A perivascular niche for brain tumor stem cells. *Cancer Cell*, 11, 69-82.

Calzà, L.; Giuliani, A.; Fernandez, M.; Pirondi, S.; D'Intino, G.; Aloe, L.; and Giardino, L. (2003). Neural stem cells and cholinergic neurons: Regulation by immunolesion and treatment with mitogens, retinoic acid, and nerve growth factor. *Proc. Natl. Acad. Sci. USA,* 100, 7325-7330.

Cameron, H.A. and Gould, E. (1994). Adult neurogenesis is regulated by adrenal steroids in the dentate gyrus. *Neuroscience*, 61, 203-209.

Cameron, H.A. and McKay, R.D. (2001). Adult neurogenesis produces a large pool of new granule cells in the dentate gyrus. *J. Comp. Neurol.*, 435, 406-417.

Campos, L.S. (2005). Beta1 integrins and neural stem cells: making sense of the extracellular environment. *Bioessays*, 27, 698-707.

Capela, A. and Temple, S. (2002). LeX/ssea-1 is expressed by adult mouse CNS stem cells, identifying them as nonependymal. *Neuron*, 35, 865-875.

Carletti, B. and Rossi, F. (2005). Selective rather than inductive mechanisms favour specific replacement of Purkinje cells by embryonic cerebellar cells transplanted to the cerebellum of adult Purkinje cell degeneration (*pcd*) mutant mice. *Eur. J. Neurosci.*, 22, 1001-1012.

Carpenter, M.K.; Cui, X.; Hu, Z.Y.; Jackson, J.; Sherman, S.; Seiger, A.; and Wahlberg, L.U. (1999). In vitro expansion of a multipotent population of human neural progenitor cells. *Exp. Neurol.*, 158, 265-278.

Chambers, C.B.; Peng, Y.; Nguyen, H.; Gaiano, N.; Fishell, G.; and Nye, J.S. (2001). Spatiotemporal selectivity of response to Notch1 signals in mammalian forebrain precursors. *Development*, 128, 689-702.

Chan, W.Y.; Kohsaka, S. and Rezaie, P. (2007). The origin and cell lineage of microglia: new concepts. *Brain Res Rev.*, 53, 344-354.

Chanas-Sacré, G.; Thiry, M.; Pirard, S.; Rogister, B.; Moonen, G.; Mbebi, C.; Verdiere-Sahuque, M.; and Leprince, P. (2000). A 295-kDA intermediate filament-associated protein in radial glia and developing muscle cells in vivo and in vitro. *Dev. Dyn.*, 219, 514-525.

Chargé, S.B. and Rudnicki, M.A. (2004). Cellular and molecular regulation of muscle regeneration. *Physiol Rev.*, 84, 209-238.

Chen, J.; Magavi, S.S. and Macklis, J.D. (2004). Neurogenesis of corticospinal motor neurons extending spinal projections in adult mice. *Proc. Natl. Acad. Sci. USA*, 101, 16357-16362.

Chen, Q.; Nakajima, A.; Choi, S.H.; Xiong, X.; Sisodia, S.S.; and Tang, Y.P. (2007). Adult neurogenesis is functionally associated with AD-like neurodegeneration. *Neurobiol Dis.*, E-pub.

Chiang, C.H.; Ma, D.K.; Ming, G.; and Song, H. (2008). Glial influences on neural stem cell niches. In: Bonfanti L, editor. *Postnatal and Adult Neurogenesis*. Research Signpost, Trivandrum, India.

Chiasson, B.J.; Tropepe, V.; Morshead, C.M.; and van der Kooy, D. (1999). Adult mammalian forebrain ependymal and subependymal cells demonstrate proliferative potential, but only subependymal cells have neural stem cell characteristics. *J. Neurosci.*, 19, 4462-4471.

Chklovskii, D.B.; Mel, B.W. and Svoboda, K. (2004). Cortical rewiring and information storage. *Nature*, 431, 782-788.

Colucci-D'Amato, L.; Bonavita, V. and di Porzio, U. (2006). The end of the central dogma of neurobiology: stem cells and neurogenesis in adult CNS. *Neurol. Sci.*, 27, 266-270.

Conover, J.C. and Notti, R.Q. (2008). The neural stem cell niche. *Cell Tissue Res.*, 331, 211-224.

Conti, L.; Pollard, S.M.; Gorba, T.; Reitano, E.; Toselli, M.; Biella, G.; Sun, Y.; Sanzone, S.; Ying, Q.L.; Cattaneo, E.; and Smith, A. (2005). Niche-independent symmetrical self-renewal of a mammalian tissue stem cell. *PLoS Biol.*, 3, e283.

Corotto, F.S.; Henegar, J.R. and Maruniak, J.A. (1994). Odor deprivation leads to reduced neurogenesis and reduced neuronal survival in the olfactory bulb of the adult mouse. *Neuroscience*, 61, 739-744.

Costa, M.R.; Kessaris, N.; Richardson, W.D.; Gotz, M.; and Hedin-Pereira, C. (2007). The marginal zone/layer I as a novel niche for neurogenesis and gliogenesis in developing cerebral cortex. *J. Neurosci.*, 27, 11376-11388.

Craig, C.G.; Tropepe, V.; Morshead, C.M.; Reynolds, B.A.; Weiss, S.; and van der Kooy, D. (1996). In vivo growth factor expansion of endogenous subependymal neural precursor cell populations in the adult mouse brain. *J. Neurosci.* 16, 2649-2658.

Cummings, D.M. and Henning, H.E; (1997). Brunjes, PC. Olfactory bulb recovery after early sensory deprivation. *J. Neurosci.*, 17, 7433-7440.

Curtis, M.A.; Penney, E.B.; Pearson, A.G.; van Roon-Mom, W.M.C.; Butterworth, N.J.; Dragunow, M.; Connor, B.; and Faull, R.L.F. (2003). Increased cell proliferation and neurogenesis in the adult human Huntington's disease brain. *Proc. Natl. Acad. Sci. USA*, 100, 9023-9027.

Curtis, M.A.; Kam, M.; Nannmark, U.; Anderson, M.F.; Axell, M.Z.; Wikkelso, C.; Holtås, S.; van Roon-Mom, W.M.C.; Björk-Eriksson, T.; Nordborg, C.; Frisén, J.; Dragunow, M.; Faull, R.L.F.; and Eriksson, P.S. (2007a). Human neuroblasts migrate to the olfactory bulb via a lateral ventricular extension. Science, 315, 1243-1249.

Curtis, M.A.; Faull, R.L. and Eriksson, P.S. (2007b). The effect of neurodegenerative diseases on the subventricular zone. *Nat. Rev. Neurosci.*, 8, 712-723.

Das, S. and Basu, A. (2007). Inflammation: A new candidate in modulating adult neurogenesis. *J. Neurosci.* Res., Epub.

Dayer, A.G.; Cleaver, K.M.; Abouantoun, T.; and Cameron, H.A. (2005). New GABAergic interneurons in the adult neocortex and striatum are generated from different precursors. *J. Cell Biol.*, 168, 415-427.

Dawson, M.R.L.; Polito, A., Levine, J.M., and Reynolds, R. (2003). NG2-expressing glial progenitor cells: an abundant and widespread population of cycling cells in the adult rat CNS. *Mol. Cell. Neurosci.*, 24, 476-488.

Dellavalle, A.; Sampaolesi, M.; Tonlorenzi, R.; Tagliafico, E.; Sacchetti, B.; Perani, L.; Innocenzi, A.; Galvez, B.G.; Messina, G.; Morosetti, R.; Li, S.; Belicchi, M.; Peretti, G.; Chamberlain, J.S.; Wright, W.E.; Torrente, Y.; Ferrari, S.; Bianco, P.; and Cossu, G. (2007). Pericytes of human skeletal muscle are myogenic precursors distinct from satellite cells. *Nat. Cell Biol.*, 9, 255-267.

Dietrich, J. and Kempermann, G. (2006). Role of endogenous neural stem cells in neurological disease and brain repair. *Adv. Exp. Med. Biol.*, 557, 191-220.

Dizon, M.L.V. and Szele, F.G. (2006). The subventricular zone responds dynamically to mechanical brain injuries. In: Levison, SW, (Ed.) *Mammalian subventricular zones*, Springer.

Doetsch, F.; Garcìa-Verdugo, J.M. and Alvarez-Buylla, A. (1997). Cellular composition and three-dimensional organization of the subventricular germinal zone in the adult mammalian brain. *J. Neurosci.*, 17, 5046-5041.

Doetsch, F.; Garcìa-Verdugo, J.M. and Alvarez-Buylla, A. (1999a). Regeneration of a germinal layer in the adult mammalian brain. *Proc. Natl. Acad. Sci. USA*, 96, 11619-11624.

Doetsch, F.; Caille, I.; Lim, D.A.; Garcìa-Verdugo, J.M.; and Alvarez-Buylla, A. (1999b). Subventricular zone astrocytes are neural stem cells in the adult mammalian brain. *Cell*, 97, 703-716.

Doetsch, F.; Petreanu, L.; Caille, I.; Garcìa-Verdugo, J.M.; and Alvarez-Buylla, A. (2002). EGF converts transit-amplifying neurogenic precursors in the adult brain into multipotent stem cells. *Neuron*, 36, 1021-1034.

Doetsch, F. (2003). A niche for adult neural stem cells. *Curr. Opin. Genet. Dev.*, 13, 543-550.

Eckenhoff, M.F. and Rakic, P. (1984). Radial organization of the hippocampal dentate gyrus: a Golgi, ultrastructural, and immunocytochemical analysis in the developing rhesus monkey. *J. Comp. Neurol.*, 223, 1-21.

Emsley, J.G.; Mitchell, B.D.; Kempermann, G.; and Macklis, J.D. (2005). Adult neurogenesis and repair of the adult CNS with neural progenitors, precursors, and stem cells. *Prog. Neurobiol.*, 75, 321-341.

Fallon, J.; Reid, S.; Kinyamu, R.; Opole, I.; Opole, R.; Baratta, J.; Kork, M.; Endo, T.L.; Duong, A.; Nguyen, G.; Kerkhehabadi, M.; Twardzik, D.; and Loughlin, S. (2000). *In vivo* induction of massive proliferation, directed migration, and differentiation of neural cells in the adult mammalian brain. *Proc. Natl. Acad. Sci. USA*, 97, 14686-14691.

Felling, R.J.; Snyder, M.J.; Romanko, M.J.; Rothstein, R.P.; Ziegler, A.N.; Yang, Z.; Givogri, M.I.; Borganzone, E.R.; and Levison, S.W. (2006). Neural stem/progenitor cells participate in the regenerative response to perinatal hypoxia/ischemia. *J. Neurosci.*, 26, 4359-4369.

Frielingsdorf, H.; Schwarz, K. and Mohapel, P. (2004). No evidence for new dopaminergic neurons in the adult mammalian substantia nigra. *Proc. Natl. Acad. Sci. USA*, 101, 10177-10182.

Fishell, G. and Kriegstein, A.R. (2003). Neurons from radial glia: the consequences of asymmetric inheritance. *Curr. Opin. Neurobiol.*, 13, 34-41.

Gage, F.H.; Coates, P.W.; Palmer, T.D.; Kuhn, H.G.; Fisher, L.J.; Suhonen, J.O.; Peterson, D.A.; Suhr, S.T.; and Ray, J. (1995). Survival and differentiation of adult neuronal progenitor cells transplanted to the adult brain. *Proc. Natl. Acad. Sci. USA*, 92, 11879-11883.

Gage, F.H. (2000). Mammalian neural stem cells. *Science*, 287, 1433-1438.

Gage, F.H. and Verma, I.M. (2003). Stem cells at the dawn of the 21st century. *Proc. Natl. Acad. Sci. USA*, 100, Suppl. 1, 11817-11818.

Galli, R.; Gritti, A.; Bonfanti, L.; and Vescovi, A.L. (2003). Neural stem cells: an overview. *Circ. Res.*, 92, 598-608.

Garcia, A.D.; Doan, N.B.; Imura, T.; Bush, T.G.; and Sofroniew, M.V. (2004). GFAP-expressing progenitors are the principal source of constitutive neurogenesis in adult mouse forebrain. *Nat. Neurosci.*, 2004, 7, 1233-1241.

Garcia-Verdugo, J.M.; Ferron, S.; Flames, N.; Collado, L.; Desfilis, E.; and Font, E. (2002). The proliferative ventricular zone in adult vertebrates: a comparative study using reptiles, birds, and mammals. *Brain Res. Bull.*, 57, 765-775.

Garcion, E.; Halilagic, A.; Faissner, A.; and ffrench-Constant, C. (2004). Generation of an environmental niche for neural stem cell development by the extracellular matrix molecule tenascin C. *Development*, 131, 3423-3432.

Gensert, J.M. and Goldman, J.E. (1996). In vivo characterization of endogenous proliferating cells in adult rat subcortical white matter. *Glia*, 17, 39-51.

Gensert, J.M. and Goldman, J.E. (2001). Heterogeneity of cycling glial progenitors in the adult mammalian cortex and white matter. *J. Neurobiol.*, 48, 75-86.

Genzen, J.R. and Bordey, A. (2008). Ependymal cells and the neurogenic niche. In: Bonfanti L, editor. *Postnatal and Adult Neurogenesis*. Research Signpost, Trivandrum, India.

Giachino, C.; De Marchis, S.; Giampietro, C.; Parlato, R.; Perroteau, I.; Schutz, G.; Fasolo, A.; and Peretto, P. (2005). cAMP response element-binding protein regulates differentiation and survival of newborn neurons in the olfactory bulb. *J. Neurosci.*, 25, 10105-10118.

Glaser, T.; Pollard, S.M.; Smith, A.; and Brustle, O. (2007). Tripotential differentiation of adherently expandable neural stem (NS) cells. *PLoS ONE*, 2, e298.

Glezer, I.; Simard, A.R. and Rivest, S. (2007). Neuroprotective role of the innate immune system by microglia. *Neuroscience*, 147, 867-883.

Goings, G.E.; Sahni, V. and Szele, F.G. (2004). Migration patterns of subventricular zone cells in adult mice change after cerebral cortex injury. *Brain Res.*, 2004, 996, 213-226.

Goldman, S.A. (2004). Directed mobilization of endogenous neural progenitor cells: the intersection of stem cell biology and gene therapy. *Curr. Opin. Mol. Ther.*, 6, 466-372.

Goldman, S.A. (2005). Stem and progenitor cell-based therapy of the human central nervous system. *Nat. Biotech.*, 23, 862-871.

Goldman, S.A. (2007). Disease targets and strategies for the therapeutic modulation of endogenous neural stem and progenitor cells. *Clin. Pharmacol. Ther.*, 82, 453-460.

Gotts, J.E. and Chesselet, M.F. (2005). Migration and fate of newly born cells after focal cortical ischemia in adult rats. *J. Neurosci. Res.*, 80, 160-171.

Gotz, M.; Hartfuss, E. and Malatesta, P. (2002). Radial glial cells as neuronal precursors: a new perspective on the correlation of morphology and lineage restriction in the developing cerebral cortex of mice. *Brain Res. Bull.*, 57, 777-788.

Gould, E. (2007). How widespread is adult neurogenesis in mammals? *Nat. Rev. Neurosci.*, 2007, 8, 481-488.

Gould, E. and Tanapat, P. (1997). Lesion-induced proliferation of neuronal progenitors in the dentate gyrus of the adult rat. *Neuroscience*, 80, 427-436.

Gould, E.; Reeves, A.J., Graziano, M.S.; and Gross, C.G. (1999a). Neurogenesis in the neocortex of adult primates. *Science*, 286, 548-552.

Gould, E.; Tanapat, P.; Hastings, N.B.; and Shors, T.J. (1999b). Neurogenesis in adulthood: a possible role in learning. *Trends Cogn. Sci.*, 3, 186-192.

Gould, E.; Beylin, A.; Tanapat, P.; Reeves, A.; and Shors, T.J. (1999c). Learning enhances adult neurogenesis in the hippocampal formation. *Nat. Neurosci.*, 2, 260-265.

Gould, E.; Vail, N.; Wagers, M.; and Gross, C.G. (2001). Adult-generated hippocampal and neocortical neurons in macaques have a transient existence. *Proc. Natl. Acad. Sci. USA*, 98, 10910-10917.

Gould, E. and Gross, C.G. (2002). Neurogenesis in adult mammals: some progress and problems. *J. Neurosci.*, 22, 619-623.

Gray, W.P. and Sundstrom, L.E. (1998). Kainic acid increases the proliferation of granule cell progenitors in the dentate gyrus of the adult rat. *Brain Res.*, 790, 52-59.

Graziadei, P.P.C. and Monti-Graziadei, G.A. (1979). Neurogenesis and neuron regeneration in the olfactory system of mammals. I. Morphological aspects of differentiation and structural organization of the olfactory neurons. *J. Neurocytol.*, 8, 1-18.

Grimaldi, P. and Rossi, F. (2006). Lack of neurogenesis in the adult rat cerebellum after Purkinje cell degeneration and growth factor infusion. *Eur. J. Neurosci.,* 23, 2657-2668.

Grimaldi, P; Carletti, B; Rossi, F. Neuronal replacement and integration in the rewiring of cerebellar circuits. *Brain Res. Rev.*, 2005, 49, 330-342.

Gritti, A.; Galli, R. and Vescovi, A.L. (2000). Cultures of stem cells of the CNS. In: Fedoroff S, editor. *Protocols for neural cell culture*. Humana Press, Totowa, NJ.

Gritti, A.; Bonfanti, L.; Doetsch, F.; Caille, I.; Alvarez-Buylla, A.; Lim, D.A.; Galli, R.; Verdugo, J.M.; Herrera, D.G.; and Vescovi, A.L. (2002). Multipotent neural stem cells reside into the rostral extension and olfactory bulb of adult rodents. *J. Neurosci.*, 22, 437-445.

Gritti, A. and Conti, L. (2008). Ex-vivo isolation and culturing of somatic and ES-derived neural stem cells. In: Bonfanti L, editor. *Postnatal and Adult Neurogenesis*. Research Signpost, Trivandrum, India.

Gritti, A. and Bonfanti, L. (2008). Neuronal–glial interactions in central nervous system neurogenesis: the neural stem cell perspective. *Neuron Glia Biol.* (in press).

Gross, C.G. (2000). Neurogenesis in the adult brain: death of a dogma. *Nat. Rev. Neurosci.*, 1, 67-73.

Hack, M.A.; Saghatelyan, A.; de Chevigny, A.; Pfeifer, A.; Ashery-Padan, R.; Lledo, P.M.; and Gotz, M. (2005). Neuronal fate determinants of adult olfactory bulb neurogenesis. *Nat. Neurosci.*, 8, 865-872.

Hagberg, H. and Mallard, C. (2005). Effect of inflammation on central nervous system development and vulnerability. *Curr. Opin. Neurol.*, 18, 117-123.

Hanisch, U.K; (2007). Kettenmann, H. Microglia: active sensor and versatile effector cells in the normal and pathologic brain. *Nat. Neurosci.*, 10, 1387-1394.

Hartfuss, E.; Galli, R.; Heins, N.; and Gotz, M. (2001). Characterization of CNS precursors subtypes and radial glia. *Dev. Biol.*, 2001, 229, 15-30.

Hitoshi, S.; Tropepe, V.; Ekker, M.; and van der Kooy, D. (2002). Neural stem cell lineages are regionally specified, but not committed, within distinct compartments of the developing brain. *Development*, 129, 233-244.

Horner, P.J.; Power, A.E.; Kempermann, G.; Kuhn, H.G.; Palmer, T.D.; Winkler, J.; Thal, L.J.; and Gage, F.H. (2000). Proliferation and differentiation of progenitor cells throughout the intact adult rat spinal cord. *J. Neurosci.*, 20, 2218-2228.

Horner, P.J.; Thallmair, M. and Gage, F.H. (2002). Defining the NG2-expressing cell of the adult CNS. *J. Neurocytol.*, 31, 469-480.

Horner, P.J. and Jacobson, B.E. (2008). Gliogenesis in the adult spinal cord. In: Bonfanti L, editor. *Postnatal and Adult Neurogenesis*. Research Signpost, Trivandrum, India.

Imura, T.; Kornblum, H.I. and Sofroniew, M.V. (2003). The predominant neural stem cell isolated from postnatal and adult forebrain but not early embryonic forebrain expresses GFAP. *J. Neurosci.*, 23, 2824-2832.

Jackson, E.L.; Garcia-Verdugo, J.M.; Gil-Perotin, S.; Roy, M.; Quinones-Hinojosa, A.; Vandenberg, S.; and Alvarez-Buylla, A. (2006). PDGFR alpha-positive B cells are neural stem cells in the adult SVZ that form glioma-like growths in response to increased PDGF signaling. *Neuron*, 51, 187-199.

Jankovski, A. and Sotelo, C. (1996). Subventricular zone-olfactory bulb migratory pathway in the adult mouse: cellular composition and specificity as determined by heterochronic and heterotopic transplantation. *J. Comp. Neurol.*, 371, 376-396.

Jessell, T.M. (2000). Neuronal specification in the spinal cord: inductive signals and transcriptional codes. *Nat. Rev. Genet.*, 1, 20-29.

Jin, K.; Peel, A.L.; Mao, X.O.; Xie, L.; Cottrell, B.A.; Henshall, D.C.; and Greeberg, D.A. (2004). Increased hippocampal neurogenesis in Alzheimer's disease. *Proc. Natl. Acad. Sci. USA*, 101, 343-347.

Kaplan, M.S. and Hinds, J.W. (1977). Neurogenesis in the adult rat: electron microscopic analysis of light radioautographs. *Science*, 197, 1092-1094.

Kee, N.J.; Preston, E. and Wojtowicz, J.M. (2001). Enhanced neurogenesis after transient global ischemia in the dentate gyrus of the rat. *Exp. Brain Res.*, 136, 313-320.

Kempermann, G.; Kuhn, H.G. and Gage, F.H. (1997). More hippocampal neurons in adult mice living in an enriched environment. *Nature*, 386, 493-495.

Kempermann, G. (2002). Why new neurons? Possible functions for adult hippocampal neurogenesis. *J. Neurosci.*, 22, 635-638.

Kempermann, G.; Jessberger, S.; Steiner, B.; and Kronenberg, G. (2004). Milestones of neuronal development in the adult hippocampus. *Trends Neurosci.*, 27, 447-452.

Kilpatrick, T.J. and Bartlett, P.F. (1993). Cloning and growth of multipotential neural precursors: requirements for proliferation and differentiation. *Neuron*, 10, 255-265.

Kim, H.T.; Kim, I.S.; Lee, I.S.; Snyder, E.Y.; and Park, K.I. (2006). Human neurospheres derived from the fetal central nervous system are regionally and temporally specified but are not committed. *Exp. Neurol.*, 199, 222-235.

Kim, Y. and Szele, F.G. (2007). Activation of subventricular zone stem cells after neuronal injury. *Cell Tissue Res.*, Epub ahead of print.

Klein, C.; Butt, S.J.; Machold, R.P.; Johnson, J.E.; and Fishell, G. (2005). Cerebellum- and forebrain-derived stem cells possess intrinsic regional character. *Development*, 132, 4497-508.

Kokaia, Z. and Lindvall, O. (2003). Neurogenesis after ischaemic brain insults. *Curr. Opin. Neurobiol.*, 13, 127-132.

Koketsu, D.; Mikami, A.; Miyamoto, Y.; and Hisatsune, T. (2003). Nonrenewal of neurons in the cerebral neocortex of adult macaque monkeys. *J. Neurosci.* 23, 937-942.

Komitova, M.; Perfilieva, E.; Mattsson, B.; Eriksson, P.S.; and Johansson, B.B. (2002). Effects of cortical ischemia and postischemic environmental enrichment on hippocampal cell genesis and differentiation in the adult rat. *J. Cereb. Blood Flow Metab.*, 22, 852-860.

Kondo, T. and Raff, M. (2000). Oligodendrocyte precursor cells reprogrammed to become multipotential CNS stem cells. *Science*, 289, 1754-1757.

Kokaia, Z. and Lindvall, O. (2003). Neurogenesis after ischaemic brain insults. *Curr. Opin. Neurobiol.*, 13, 127-132.

Kornack, D.R. (2000). Neurogenesis and the evolution of cortical diversity: mode, tempo and partitioning during development and persistence in adulthood. *Brain Behav. Evol.*, 55, 336-344.

Kornack, D.R. and Rakic, P. (2001). Cell proliferation without neurogenesis in adult primate neocortex. *Science*, 294, 2127-2130.

Kuhn, H.G.; Dickinson-Anson, H. and Gage, F.H. (1996). Neurogenesis in the dentate gyrus of the adult rat: age-related decrease of neuronal progenitors proliferation. *J. Neurosci.*, 16, 2027-2033.

Kuhn, H.G.; Winkler, J.; Kempermann, G.; Thal, L.J.; and Gage, F.H. (1997). Epidermal growth factor and fibroblast growth factor-2 have different effects on neural progenitors in the adult rat brain. *J. Neurosci.*, 17, 5820-5829.

Lai, K.; Kaspar, B.K.; Gage, F.H.; and Schaffer, D.V. (2003). Sonic hedgehog regulates adult neural progenitor proliferation in vitro and in vivo. *Nat. Neurosci.*, 6, 21-27.

Lawson, L.J.; Perry, V.H. and Gordon, S. (1992). Turnover of resident microglia in the normal adult mouse brain. *Neuroscience*, 1992, 48, 405-415.

Laywell, E.D.; Rakic, P.; Kukekov, V.G.; Holland, E.C.; and Steindler, D.A. (2000). Identification of a multipotent astrocytic stem cell in the immature and adult mouse brain. *Proc. Natl. Acad. Sci. USA*, 97, 13883-13888.

Lee, A.; Kessler, J.D.; Read, T.A.; Kaiser, C.; Corbeil, D.; Huttner, W.B.; Johnson, J.E.; and Wechsler-Reya, R.J. (2005). Isolation of neural stem cells from the postnatal cerebellum. *Nat. Neurosci.*, 8, 723-729.

Lee, SR; Kim, HY; Rogowska, J; Zhao, BQ; Bhide, P; Parent, JM; Lo, EH. Involvement of matrix metalloproteinase in neuroblast cell migration from the subventricular zone after stroke. *J. Neurosci.*, 2006, 26, 3491-3495.

Leuner, B., Gould, E. and Shors, T.J. (2006). Is there a link between adult neurogenesis and learning? *Hippocampus*, 16, 216-224.

Levison, S.W. (2006). *Mammalian subventricular zones*. (Ed.), Springer.

Lie, D.C.; Dziewczapolski, G.; Willhoite, A.R.; Kaspar, B.K.; Shults, C.W.; and Gage, F.H. (2002). The adult substantia nigra contains progenitor cells with neurogenic potential. *J. Neurosci.*, 22, 6639-6649.

Lie, D.C.; Song, H.; Colamarino, S.A.; Ming, G.L.; and Gage, F.H. (2004). Neurogenesis in the adult brain: new strategies for central nervous system diseases. *Annu. Rev. Pharmacol. Toxicol.* 44, 399-421.

Lim, D.A. and Alvarez-Buylla, A. (1999). Interaction between astrocytes and adult subventricular zone precursors stimulates neurogenesis. *Proc. Natl. Acad. Sci. USA*, 96, 7526-7531.

Lim, D.A.; Tramontin, A.D.; Trevejo, J.M.; Herrera, D.G.; Garcìa-Verdugo, J.M.; and Alvarez-Buylla, A. (2000). Noggin antagonizes BMP signaling to create a niche for adult neurogenesis. *Neuron*, 28, 713-726.

Lindsey, B.W. and Tropepe, V. (2006). A comparative framework for understanding the biological principles of adult neurogenesis. *Prog. Neurobiol.* 80, 281-307.

Lindvall, O. and Kokaia, Z. (2006). Stem cells for the treatment of neurological disorders. *Nature*, 441, 1094-1096.

Liu, J.; Solway, K.; Messing, R.O.; and Sharp, F.R. (1998). Increased neurogenesis in the dentate gyrus after transient global ischemia in gerbils. *J. Neurosci.*, 18, 7768-7778.

Liu, X.; Wang, Q.; Haydar, T.F.; and Bordey, A. (2005). Nonsynaptic GABA signaling in postnatal subventricular zone controls proliferation of GFAP-expressing progenitors. *Nat. Neurosci.*, 8, 1179-1187.

Lledo, P.M.; Alonso, M. and Grubb, M.S. (2006). Adult neurogenesis and functional plasticity in neuronal circuits. *Nat. Rev. Neurosci.* 7, 179-193.

Loeffler, M. and Potten, C.S. (1997). Stem cells and cellular pedigrees – a conceptual introduction. In: *Stem cells*. Potten, C.S., 1997, Academic Press, USA, 1-27.

Lois, C. and Alvarez-Buylla, A. (1994). Long-distance neuronal migration in the adult mammalian brain. *Science*, 264, 1145-1148.

Lois, C.; Garcìa-Verdugo, J. and Alvarez-Buylla, A. (1996). Chain migration of neuronal precursors. *Science*, 271, 978-981.

Luo, J., Daniels, S.B., Lennington, J.B., Notti, R.Q. and Conover, J.C. (2006) The aging neurogenic subventricular zone. Aging Cell 5, 139-152.

Luskin, M.B. (1993). Restricted proliferation and migration of postnatally generated neurons derived from the forebrain subventricular zone. *Neuron*, 11, 173-189.

Luzzati, F.; Peretto, P.; Aimar, P.; Ponti, G.; Fasolo, A.; and Bonfanti, L. (2003). Glia-independent chains of neuroblasts through the subcortical parenchyma of the adult rabbit brain. *Proc. Natl. Acad. Sci. USA*, 100, 13036-13041.

Luzzati, F., De Marchis, S., Fasolo, A., Peretto, P. (2006). Neurogenesis in the caudate nucleus of the adult rabbit. *J. Neurosci.* 26, 609-621.

Luzzati, F., De Marchis, S., Fasolo, A. and Peretto, P. (2007). Adult neurogenesis and local neuronal progenitors in the striatum. *Neurodegener. Dis.*, 4, 322-327.

Ma, D.K., Ming, G.L. and Song, H. (2005) Glial influences on neural stem cell development: cellular niches for adult neurogenesis. *Curr. Opin. Neurobiol.* 15, 514-520.

Macklis, J.D. (1993). Transplanted neocortical neurons migrate selectively in regions of neuronal degeneration produced by chromophore-targeted laser photolysis. *J. Neurosci.*, 13, 3848-3863.

MacLaren, R.E.; Pearson, R.A.; MacNeil, A.; Douglas, R.H.; Salt, T.E.; Akimoto, M.; Swaroop, A.; Sowden, J.C.; and Ali, R.R. (2006). Retinal repair by transplantation of photoreceptor precursors. *Nature*, 444, 203-207.

Magavi, S.S., Leavitt, B.R. and Macklis, J.D. (2000). Induction of neurogenesis in the neocortex of adult mice. *Nature* 405, 951-955.

Magavi, S.S. and Macklis, J.D., (2001). Manipulation of neural precursors in situ: induction of neurogenesis in the neocortex of adult mice. *Neuropsychopharmacology* 25, 816-835.

Malatesta, P., Hartfuss, E. and Gotz, M., (2000). Isolation of radial glial cells by fluorescent-activated cell sorting reveals a neuronal lineage. Development 127, 5253-5263.

Malm, T.M.; Koistinaho, M.; Parepalo, M.; Vatanen, T.; Ooka, A.; Karlsson, S.; and Koistinaho, J. (2005). Bone-marrow-derived cells contribute to the recruitment of microglial cells in response to beta-amyloid deposition in APP/PS1 double transgenic Alzheimer mice. *Neurobiol. Dis.*, 18, 134-142.

Mandairon, N.; Sacquet, J.; Jourdan, F.; and Didier, A. (2006). Long-term fate and distribution of newborn cells in the adult mouse olfactory bulb: Influences of olfactory deprivation. *Neuroscience*, 141, 443-451.

Manent, J.B.; Jorquera, I.; Ben-Ari, Y.; Aniksztejn, L.; and Represa, A. (2006). Glutamate acting on AMPA but not NMDA receptors modulates the migration of hippocampal interneurons. *J. Neurosci.*, 26, 5901-5909.

Manent, J.B. and Represa, A. (2007). Neurotransmitters and brain maturation: early paracrine actions of GABA and glutamate modulate neuronal migration. *Neuroscientist*, 13, 268-279.

Marin, O. and Rubenstein, J.L. (2003). Cell migration in the forebrain. *Annu. Rev. Neurosci.*, 26, 441-483.

Martens, D.J.; Seaberg, R.M. and van der Kooy, D. (2002). In vivo infusions of exogenous growth factors into the fourth ventricle of the adult mouse brain increase the proliferation of neural progenitors around the fourth ventricle and the central canal of the spinal cord. *Eur. J. Neurosci.*, 16, 1045-1057.

Martino, G. and Pluchino, S. (2006). The therapeutic potential of neural stem cells. *Nat. Rev. Neurosci.* 7, 395-406.

Merkle, F.T.; Tramontin, A.D.; Garcìa-Verdugo, J.M.; Alvarez-Buylla, A. (2004). Radial glia give rise to adult neural stem cells in the subventricular zone. *Proc. Natl. Acad. Sci. USA*, 101, 17528-17532.

Merkle, F.T.; Mirzadeh, Z. and Alvarez-Buylla, A. (2007). Mosaic organization of neural stem cells in the adult brain. *Science*, 317, 381-384.

Meshi, D.; Drew, M.R.; Saxe, M.; Ansorge, M.S.; David, D.; Santarelli, L.; Malapani, C.; Moore, H.; and Hen, R. (2006). Hippocampal neurogenesis is not required for behavioral effects of environmental enrichment. *Nat. Neurosci.*, 9, 729-731.

Messina, E.; De Angelis, L.; Frati, G.; Morrone, S.; Chimenti, S.; Fiordaliso, F.; Salio, M.; Battaglia, M.; Latronico, M.V.; Coletta, M.; Vivarelli, E.; Frati, L.; Cossu, G.; and Giacomello, A. (2004). Isolation and expansion of adult cardiac stem cells from human and murine heart. *Circ. Res.*, 95, 911-921.

Mildner, A.; Schmidt, H.; Nitsche, M.; Merkler, D.; Hanish, U.K.; Mack, M.; Heikenwalder, M.; Bruck, W.; Priller, J.; and Prinz, M. (2007). Microglia in the adult brain arise from Ly-6C(hi)CCR2(+) monocytes only under defined host conditions. *Nat. Neurosci.*, 12, 1544-1553.

Ming, G.L. and Song, H. (2005). Adult neurogenesis in the mammalian central nervous system. *Annu. Rev. Neurosci.*, 28, 223-250.

Mirescu, C. and Gould, E. (2006). Stress and adult neurogenesis. *Hippocampus*, 16, 233-238.

Miwa, N. and Storm, D.R. (2005). Odorant-induced activation of extracellular signal-regulated kinase/mitogen-activated protein kinase in the olfactory bulb promotes survival of newly formed granule cells. *J. Neurosci.*, 25, 5404-5412.

Miyata, T.; Kawaguchi, A.; Okano, H.; and Ogawa, M. (2001). Asymmetric inheritance of radial glial fibers by cortical neurons. *Neuron*, 31, 727-741.

Morrison, S.J.; Wright, D.E.; Cheshier, S.H., and Weissman, I.L. (1997). Hematopoietic stem cells: challenges to expectations. *Curr. Opin. Immunol.*, 9, 216-221.

Morshead, C.M.; Reynolds, B.A.; Craig, C.G.; McBurney, M.W.; Staines, W.A.; Morassutti, D.; Weiss, S.; and van der Kooy, D. (1994). Neural stem cells in the adult mammalian forebrain: a relatively quiescent subpopulation of subependymal cells. *Neuron*, 13, 1071-1082.

Nait-Oumesmar, B.; Decker, L.; Lachapelle, F.; Avellana-Adalid, V.; Bachelin, C.; and Van Evercooren, A.B. (1999). Progenitor cells of the adult mouse subventricular zone proliferate, migrate and differentiate into oligodendrocytes after demyelination. *Eur. J. Neurosci.* 11, 4357-4366.

Nait-Oumesmar, B.; Picard-Riera, N.; Kerninon, C.; Decker, L.; Seilhean, D.; Hoglinger, G.U.; Hirsch, E.C.; and Van Evercooren, A.B. (2007). Activation of the subventricular zone in multiple sclerosis: evidence for early glial progenitors. *Proc. Natl. Acad. Sci. USA*, 104, 4694-4699.

Nait-Oumesmar, B.; Picard-Riera, N.; Kerninon, C. and Van Evercooren, A.B. (2007). The role of SVZ-derived neural precursors in demyelinating diseases: From animal models to multiple sclerosis. J. Neurol. Sci., Epub ahead of print.

Nakatomi, H.; Kuriu, T.; Okabe, S.; Yamamoto, S.; Hatano, O.; Kawahara, N.; Tamura, A.; Kirino, T.; and Nakafuku, M. (2002). Regeneration of hippocampal pyramidal neurons after ischemic brain injury by recruitment of endogenous neural progenitors. *Cell*, 110, 429-441.

Namba, T.; Mochizuki, H.; Onodera, M.; Mizuno, Y.; Namiki, H.; and Seki, T. (2005). The fate of neural progenitor cells expressing astrocytic and radial glial markers in the postnatal rat dentate gyrus. *Eur. J. Neurosci.*, 22, 1928-1941.

Nishiyama, A. (2007). Polydendrocytes: NG2 cells with many roles in development and repair of the CNS. *Neuroscientist* , 13, 62-76.

Nystul, T.G. and Spradling, A.C. (2006). Breaking out of the mold: diversity within adult stem cells and their niches. *Curr. Opin. Genet. Dev.*, 16, 463-468.

Okano, H.; Sakaguchi, M.; Ohki, K.; Suzuki, N.; and Sawamoto, K. (2007). Regeneration of the central nervous system using endogenous repair mechanisms. *J. Neurochem.*, 102, 1459-1465.

Ourednik, V.; Ourednik, J., Flax, J.D.; Zawada, W.M.; Hutt, C.; Yang, C.; Park, K.I.; Kim, S.U.; Sidman, R.L.; Freed, C.R.; and Snyder, E.Y. (2001). Segregation of human neural stem cells in the developing primate forebrain. *Science*, 293, 1820-1824.

Owens, D.F. and Kriegstein, A.R. (2002). Is there more to GABA than synaptic inhibition? *Nat. Rev. Neurosci.*, 3, 715-727.

Noctor, S.C.; Flint, A.C.; Weissman, T.A.; Dammerman, R.S.; and Kriegstain, A.R. (2001). Neurons derived from radial glial cells establish radial units in neocortex. *Nature*, 409, 714-720.

Nottebohm, F. (2002). Why are some neurons replaced in adult brain? *J. Neurosci.*, 22, 624-628.

Palmer, T.D.; Ray, J. and Gage, F.H. (1995). FGF-2-reponsive neuronal progenitors reside in proliferative and quiescent regions of the adult rodent brain. *Mol. Cell. Neurosci.*, 6, 474-486.

Palmer, T.D.; Markakis, E.A.; Willhoite, A.R.; Safar, F.; and Gage, F.H. (1999). Fibroblast growth factor-2 activates a latent neurogenic program in neural stem cells from diverse regions of the adult CNS. *J. Neurosci.*, 19, 8487-8497.

Palmer, T.D.; Schwartz, P.H.; Taupin, P.; Kaspar, B.; Stein, S.A.; and Gage, F.H. (2001). Cell culture. Progenitor cells from human brain after death. *Nature*, 411, 42-43.

Parent, J.M.; Yu, T.W.; Leibowitz, R.T.; Geshcwind, D.H.; Sloviter, R.S.; and Lowenstein, D.H. (1997). Dentate granule cell neurogenesis is increased by seizures and contributes to aberrant network reorganization in the adult rat hippocampus. *J. Neurosci.*, 17, 3727-3738.

Parent, J.M.; Vexler, Z.S.; Gong, C.; Derugin, N.; and Ferriero, D.M. (2002a). Rat forebrain neurogenesis and striatal neuron replacement after focal stroke. *Ann. Neurol.*, 52, 802-813.

Parent, J.M.; Valentin, V.V. and Lowenstein, D.H. (2002b). Prolonged seizures increase proliferating neuroblasts in the adult rat subventricular zone-olfactory bulb pathway. *J. Neurosci.*, 22, 3174-3188.

Parent, J.M.; von dem Bussche, N. and Lowenstein, D.H. (2006). Prolonged seizures recruit caudal subventricular zone glial progenitors into the injured hippocampus. *Hippocampus*, 16, 321-328.

Parish, C.L.; Beljajeva, A.; Arenas, E.; and Simon, A. (2007). Midbrain dopaminergic neurogenesis and behavioural recovery in a salamander lesion-induced regeneration model. *Development*, 134, 2881-2887.

Pencea, V.; Bingaman, K.D.; Wiegand, S.J.; and Luskin, M.B. (2001). Infusion of brain-derived neurotrophic factor into the lateral ventricle of the adult rat leads to new neurons in the parenchyma of the striatum, septum, thalamus, and hypothalamus. *J. Neurosci.*, 21, 6706-6717.

Peng, H. and Huard, J. (2004). Muscle-derived stem cells for musculoskeletal tissue regeneration and repair. *Transpl. Immunol.*, 12, 311-319.

Peretto, P.; Merighi, A.; Fasolo, A.; and Bonfanti, L. (1997). Glial tubes in the rostral migratory stream of the adult rat. *Brain Res. Bull.*, 42, 9-21.

Peretto, P.; Merighi, A.; Fasolo, A.; and Bonfanti, L. (1999). The subependymal layer in rodents: a site of structural plasticity and cell migration in the adult mammalian brain. *Brain Res. Bull.*, 49, 221-243.

Peretto, P.; Dati, C.; De Marchis, S.; Kim, H.H.; Ukhanova, M.; Fasolo, A.; and Margolis, F.L. (2004). Expression of the secreted factors noggin and bone morphogenetic proteins in the subependymal layer and olfactory bulb of the adult mouse brain. *Neuroscience*, 128, 685-696.

Peretto, P.; Giachino, C.; Aimar, P.; Fasolo, A.; and Bonfanti, L. (2005). Chain formation and glial tube assembly in the shift from neonatal to adult subventricular zone of the rodent forebrain. *J. Comp. Neurol.*, 487, 407-427.

Petreanu, L. and Alvarez-Buylla A. (2002). Maturation and death of adult-born olfactory bulb granule neurons: role of olfaction. *J. Neurosci.*, 22, 6106-6113.

Picard-Riera, N.; Decker, L.; Delarasse, C.; Goude, K.; Nait-Oumesmar, B.; Liblau, R.; Pham-Dinh, D.; and Van Evercooren, A.B. (2002). Experimental autoimmune encephalomyelitis mobilizes neural progenitors from the subventricular zone to undergo oligodendrogenesis in adult mice. *Proc. Natl. Acad. Sci. USA*, 99, 13211-13216.

Picard-Riera, N.; Nait-Oumesmar, B. and Van Evercooren, A.B. (2004). Endogenous adult neural stem cells: limits and potential to repair the injured central nervous system. *J. Neurosci. Res.*, 76, 223-231.

Pinto, L. and Gotz, M. (2007). Radial glial cell heterogeneity-The source of diverse progeny in the CNS. *Prog. Neurobiol.*, 83, 2-23.

Pixley, S.K.R. and De Vellis, J. (1984). Transition between immature radial glia and mature astrocytes studied with a monoclonal antibody to vimentin. *Dev. Brain Res.*, 15, 201-209.

Pluchino, S.; Quattrini, A.; Brambilla, E.; Gritti, A.; Salani, G.; Dina, G.; Galli, R.; Del Carro, U.; Bergami, A.; Furlan, R.; Comi, G.; Vescovi, A.L.; Martino, G., and Amadio, S. (2003). Injection of adult neurospheres induces recovery in a chronic model of multiple sclerosis. *Nature*, 422, 688-694.

Pluchino, S.; Zanotti, L.; Rossi, B.; Brambilla, E.; Ottoboni, L.; Salani, G.; Martinello, M.; Cattalini, A.; Bergami, A.; Furlan, R.; Comi, G.; Constantin, G.; and Martino, G. (2005). Neurosphere-derived multipotent precursors promote neuroprotection by an immunomodulatory mechanism. *Nature*, 436, 266-271.

Pollard, S.M. and Conti, L. (2007). Investigating radial glia in vitro. *Prog. Neurobiol.*, 83, 53-67.

Ponti, G. and Bonfanti, L. (2008). Postnatal and adult neurogenesis in the rabbit cerebellum. In: Bonfanti L, editor. *Postnatal and Adult Neurogenesis*. Research Signpost, Trivandrum, India.

Ponti, G.; Aimar, P. and Bonfanti, L. (2006a). Cellular composition and cytoarchitecture of the rabbit subventricular zone (SVZ) and its extensions in the forebrain. *J. Comp. Neurol.* 498, 491-507.

Ponti, G.; Peretto, P. and Bonfanti, L. (2006b).A subpial, transitory germinal zone forms chains of neuronal precursors in the rabbit cerebellum. *Dev. Biol.*, 294, 168-180.

Potten, C.S. and Loeffler, M. (1990). Stem cells: attributes, cycles, spirals, pitfalls and uncertainties. Lessons for and from the crypt. *Development*, 110, 1001-1020.

Quinones-Hinojosa, A.; Sanai, N.; Soriano-Navarro, M.; Gonzalez-Perez, O.; Mirzadeh, Z.; Gil-Perotin, S.; Romero-Rodriguez, R.; Berger, M.S.; Garcìa-Verdugo, J.M.; and Alvarez-Buylla, A. (2006). Cellular composition and cytoarchitecture of the adult human subventricular zone: a niche of neural stem cells. *J. Comp. Neurol.*, 494, 415-434.

Rakic P. (1990). Principles of neural cell migration. *Experientia*, 46, 882-891.

Rakic, P. (2002). Neurogenesis in adult primate neocortex: an evaluation of the evidence. *Nat. Rev. Neurosci.* 3, 65-71.

Rakic, P. (2003). Developmental and evolutionary adaptations of cortical radial glia. *Cereb. Cortex*, 13, 541-549.

Rakic, P.; Ang, E.S.B.C. and Breunig, J. (2004). Setting the stage for cognition: Genesis of the primate cerebral cortex. In: *The Cognitive Neurosciences III*. Gazzaniga, MS, (Ed.) The MIT Press, Cambridge.

Ramírez-Castillejo, C.; Sánchez-Sánchez, F.; Andreu-Agulló, C.; Ferrón, S.R.; Aroca-Aguilar, J.D.; Sánchez, P.; Mira, H.; Escribano, J.; and Fariñas, I. (2006). Pigment epithelium-derived factor is a niche signal for neural stem cell renewal. *Nat. Neurosci.*, 9, 331-339.

Represa, A. and Ben-Ari, Y. (2005). Trophic actions of GABA on neuronal development. *Trends Neurosci.*, 28, 278-283.

Reynolds, B.A. and Weiss, S. (1992). Generation of neurons and astrocytes from isolated cells of the adult mammalian central nervous system. *Science*, 255, 1707-1710.

Reynolds, B.A. and Rietze, R.L. (2005). Neural stem cells and neurospheres--re-evaluating the relationship. *Nat. Methods*, 2, 333-336.

Rochefort, C.; Gheusi, G.; Vincent, J.D.; and Lledo, P.M. (2002). Enriched odor exposure increases the number of newborn neurons in the adult olfactory bulb and improves odor memory. *J. Neurosci.*, 22, 2679-2689.

Rogers, J.; Mastroeni, D.; Leonard, B.; Joyce, J.; and Grover, A. (2007). Neuroinflammation in Alzheimer's disease and Parkinson's disease: are microglia pathogenic in either disorder? *Int. Rev. Neurobiol.*, 82, 235-246.

Romanko, M.J.; Rola, R.; Fike, J.R.; Szele, F.G.; Dizon, M.L.; Felling, R.J.; Brazel, C.Y.; and Levison, S.W. (2004). Roles of the mammalian subventricular zone in cell replacement after brain injury. *Prog. Neurobiol.* 74, 77-99.

Rossi, F.; Borsello, T. and Strata, P. (1992). Embryonic Purkinje cells grafted on the surface of the cerebellar cortex integrate in the adult unlesioned cerebellum. *Eur. J. Neurosci.*, 4, 589-593.

Rossi, F.; Borsello, T. and Strata, P. (1994). Embryonic Purkinje cells grafted on the surface of the adult uninjured rat cerebellum migrate in the host parenchyma and induce sprouting of intact climbing fibres. *Eur. J. Neurosci.*, 6, 121-136.

Rossi, F. and Cattaneo, E. (2002). Opinion: neural stem cell therapy for neurological diseases: dreams and reality. *Nat. Rev. Neurosci.*, 3, 401-409.

Sanai, N.; Tramontin, A.D.; Quinones-Hinojosa, A.; Barbaro, N.M.; Gupta, N.; Kunwar, S.; Lawton, M.T.; McDermott, M.W.; Parsa, A.T.; Garcìa-Verdugo, J.M.; Berger, M.S.; and Alvarez-Buylla, A. (2004). Unique astrocyte ribbon in adult human brain contains neural stem cells but lacks chain migration. *Nature* 2004, 427, 740-744.

Sawamoto, K.; Wichterle, H.; Gonzalez-Perez, O.; Cholfin, J.A.; Spassky, N.; Murcia, N.S.; Garcìa-Verdugo, J.M.; Marin, O.; Rubenstein, J.L.; Tessier-Lavigne, M.; Okano, H.; and Alvarez-Buylla, A. (2006). New neurons follow the flow of cerebrospinal fluid in the adult brain. *Science*, 311, 629-632.

Scadden, D.T. (2006). The stem-cell niche as an entity of action. *Nature*, 441, 1075-1079.

Seri, B.; Garcìa-Verdugo, J.M.; McEwen, B.S.; and Alvarez-Buylla, A. (2001). Astrocytes give rise to new neurons in the adult mammalian hippocampus. *J. Neurosci.*, 21, 7153-7160.

Seri, B.; Garcìa-Verdugo, J.M.; Collaudo-Morente, L.; McEwen, B.S.; and Alvarez-Buylla, A. (2004). Cell types, lineage, and architecture of the germinal zone in the adult dentate gyrus. *J. Comp. Neurol.*, 478, 359-378.

Seri, B.; Herrera, D.G.; Gritti, A.; Ferron, S.; Collado, L.; Garcìa-Verdugo, J.M.; and Alvarez-Buylla, A. (2006). Composition and Organization of the SCZ: A Large Germinal Layer Containing Neural Stem Cells in the Adult Mammalian Brain. *Cereb. Cortex*, 16 (Suppl 1), i103-i111.

Shihabuddin, L.S.; Horner, P.J.; Ray, J.; and Gage, F.H. (2000). Adult spinal cord stem cells generate neurons after transplantation in the adult dentate gyrus. *J. Neurosci.*, 20, 8727-8735.

Simard, A.R.; Soulet, D.; Gowing, G.; Julien, J.P.; and Rivest, S. (2006). Bone marrow-derived microglia play a critical role in restricting senile plaque formation in Alzheimer's disease. *Neuron.*, 49, 489-502.

Skogh, C.; Eriksson, C.; Kokaia, M.; Meijer, X.C.; Wahlberg, L.U.; Wictorin, K.; and Campbell, K. (2001). Generation of regionally specified neurons in expanded glial cultures derived from the mouse and human lateral ganglionic eminence. *Mol. Cell. Neurosci.*, 17, 811-820.

Smidt, M.P. and Burbach, J.P. (2007). How to make a mesodiencephalic dopaminergic neuron. *Nat. Rev. Neurosci.*, 8, 21-32.

Smith, A.G. (2001). Embryo-derived stem cells: of mice and men. *Annu. Rev. Cell. Dev. Biol.*, 17, 435-462.

Singec, I.; Knoth, R.; Meyer, R.P.; Maciaczyk, J.; Volk, B.; Nikkhah, G.; Frotscher, M.; and Snyder, E.Y. (2006). Defining the actual sensitivity and specificity of the neurosphere assay in stem cell biology. *Nat. Methods*, 3, 801-806.

Sotelo, C. and Alvarado-Mallart, R.M. (1991). The reconstruction of cerebellar circuits. *Trends Neurosci.*, 14, 350-355.

Sotelo, C.; Alvarado-Mallart, R.M.; Frain, M.; and Vernet, M. (1994). Molecular plasticity of adult Bergmann fibers is associated with radial migration of grafted Purkinje cells. *J. Neurosci.*, 14, 124-133.

Spassky, N.; Merkle, F.T.; Flames, N.; Tramontin, A.D.; Garcìa-Verdugo, J.M.; and Alvarez-Buylla, A. (2005). Adult ependymal cells are postmitotic and are derived from radial glial cells during embryogenesis. *J. Neurosci.*, 25, 10-18.

Spradling, A.S.; Drummond-Barbosa, D. and Kai, T. (2001). Stem cells find their niche. *Nature*, 414, 98-104.

Steiner, B.; Wolf, S. and Kempermann, G. (2006). Adult neurogenesis and neurodegenerative disease. *Regen. Med.*, 1, 15-28.

Tamura, Y.; Kataoka, Y.; Cui, Y.; and Yamada, H. (2004). Cellular proliferation in the cerebral cortex following neural excitation in rats. *Neurosci. Res.*, 50, 129-133.

Tamura, Y.; Kataoka, Y.; Cui, Y.; Takamori, Y.; Watanabe, Y.; and Yamada, H. (2007). Multi-directional differentiation of doublecortin- and NG2-immunopositive progenitor cells in the adult rat neocortex in vivo. *Eur. J. Neurosci*, 25, 3489-3498.

Tanapat, P.; Hastings, N.B.; Reeves, A.J.; and Gould, E. (1999). Estrogen stimulates a transient increase in the numeber of new neurons in the dentate gyrus of the adult female rat. *J. Neurosci.*, 19, 5792-5801.

Tramontin, A.D.; Garcìa-Verdugo, J.M.; Lim, D.A.; and Alvarez-Buylla, A. (2003). Postnatal development of radial glia and the ventricular zone (VZ): a continuum of the neural stem cell compartment. *Cereb. Cortex*, 13, 580-587.

Uchida, N.; Buck, D.W.; He, D.; Reitsma, M.J.; Masek, M.; Phan, T.V.; Tsukamoto, A.S.; Gage, F.H.; and Weissman, I.L. (2000). Direct isolation of human central nervous system stem cells. *Procl. Natl. Acad. Sci. USA*, 97, 10720-14725.

Van Praag, H.; Kempermann, G. and Gage, F.H. (1999). Running increases cell proliferation and neurogenesis, in the adult mouse dentate gyrus. *Nat. Neurosci.*, 2, 266-270.

Vescovi, A.L.; Reynolds, B.A.; Fraser, D.D.; and Weiss, S. (1993). bFGF regulates the proliferative fate of unipotent (neuronal) and bipotent (neuronal/astroglial) EGF-generated CNS progenitor cells. *Neuron*, 11, 951-966.

Vescovi, A.L.; Parati, E.A.; Gritti, A.; Poulin, P.; Ferrario, M.; Wanke, E.; Frolichsthal-Schoeller, P.; Cova, L.; Arcellana-Panlilio, M.; Colombo, A.; and Galli, R. (1999). Isolation and cloning of multipotential stem cells from the embryonic human CNS and establishment of transplantable human neural stem cell lines by epigenetic stimulation. *Exp. Neurol.*, 156, 71-83.

Voigt, T. (1989). Development of glial cells in the cerebral wall of ferrets: direct tracing of their transformation from radial glia into astrocytes. *J. Comp. Neurol.*, 289, 74-88.

Walker, B.E. and Leblond, C.P. (1958). Sites of nucleic acid synthesis in the mouse visualized by radioautography after administration of C14-labelled adenine and thymidine. *Exp. Cell Res.*, 14, 510-531.

Weiss, S.; Dunne, C.; Hewson, J.; Wohl, C.; Wheatley, M.; Peterson, A.C.; and Reynolds, B.A. (1996). Multipotent CNS stem cells are present in the adult mammalian spinal cord and ventricular neuroaxis. *J. Neurosci.*, 16, 7599-7609.

Wichterle, H.; Lieberam, I.; Porter, J.A.; and Jessell, T.M. (2002). Directed differentiation of embryonic stem cells into motor neurons. *Cell*, 110, 385-397.

Woodward, W.A.; Chen, M.S.; Behbod, F.; and Rosen, J.M. (2005). On mammary stem cells. *J. Cell Sci.*, 118, 3585-3594.

Yagita, Y.; Kitagawa, K.; Ohtsuki, T.; Takasawa, K.I.; Miyata, T.; Okano, H.; Hori, M.; and Matsumoto, M. (2001). Neurogenesis by progenitor cells in the ischemic adult rat hippocampus. *Stroke*, 32, 1890-1896.

Yamaguchi, M. and Mori, K. (2005). Critical period for sensory experience-dependent survival of newly generated granule cells in the adult mouse olfactory bulb. *Proc. Natl. Acad. Sci. USA*, 102, 9697-9702.

Yamashita, T.; Ninomiya, M.; Hernandez Acosta, P.; Garcia-Verdugo, J.M.; Sunabori, T.; Sakaguchi, M.; Adachi, K.; Kojima, .T; Hirota, Y.; Kawase, T.; Araki, N.; Abe, K.; Okano, H.; and Sawamoto, K. (2006). Subventricular zone-derived neuroblasts migrate and differentiate into mature neurons in the post-stroke adult striatum. *J. Neurosci.*, 26, 6627-6636.

Young, K.M.; Fogarty, M.; Kessaris, N.; and Richardson, W.D. (2007). Subventricular zone stem cells are heterogeneous with respect to their embryonic origins and neurogenic fates in the adult olfactory bulb. *J. Neurosci.*, 27, 8286-8296.

Zammit, P.S.; Partridge, T.A.; and Yablonka-Reuveni, Z. (2006). The skeletal muscle satellite cell: the stem cell that came in from the cold. *J. Histochem. Cytochem.*, 54, 1177-1191.

Zhang, R.L.; Zhang, Z.G.; Zhang, L.; and Chopp, M. (2001). Proliferation and differentiation of progenitor cells in the cortex and the subventricular zone in the adult rat after focal cerebral ischemia. *Neuroscience*, 105, 33-41.

Zhang, R.L.; Zhang, Z.G. and Chopp, M. (2007). Gene profiles within the adult subventricular zone niche: proliferation, differentiation and migration of neural progenitor cells in the ischemic brain. *Curr. Mol. Med.*, 7, 459-462.

Zhao, M.; Momma, S.; Delfani, K.; Carlen, M.; Cassidy, R.M.; Johansson, C.B.; Brismar, H.; Shupliakov, O.; Frisen, J.; and Janson, A.M. (2003). Evidence for neurogenesis in the adult mammalian substantia nigra. *Proc. Natl. Acad. Sci. USA*, 100, 7925-7930.

Zigova, T.; Pencea, V.; Wiegand, S.J.; and Luskin, M.B. (1998). Intraventricular administration of BDNF increases the number of newly generated neurons in the adult olfactory bulb. *Mol. Cell. Neurosci.*, 11, 234-245.

In: Stem Cell Applications in Disease and Health
Editors: W. Burnsides and R. Ellsley
ISBN: 978-1-60456-255-2

Chapter IV

In Vivo Regulation of Neural Stem Cells by Extrinsic Factors in Adult Mammals: Issues for Cellular Therapy in Clinics (a Review)

Emmanuel Moyse, Olivier Liard* and Stephanie Segura
Lab. Physiologie Neurovégétative, UMR 6153 CNRS, Université Aix-Marseille-3,
Av. Escadrille Normandie-Niemen, 13397 Marseille, France

Abstract

The discovery of neural stem cells in the central nervous system of adult mammals in 1992 raised great expectations for regenerative therapy after lesions or degenerative diseases. A tremendous accumulation of in vitro data about the dynamics of neural stem cells occurred in the last decade. However, these results do not yet allow successful cell therapy, neither in animal models nor in humans. The major issue is to identify the tissular factors that commit stem cell progeny to a given fate, which is still unresolved. It is well established indeed that neural stem cell function is primarily determined by extrinsic factors from the microenvironment. But the nature of these regulators mostly remains to unravel.

We review current knowledge of in vivo competent signals of the micro-environnement, i.e. both matricial and diffusible ones. Also reviewed are the published responses of adult neurogenesis systems to in vivo experimental models of neurogenic proliferation. These in vivo models are likely to provide original cues for identification of extracellular regulators, especially the post-lesional ones which elicit massive and synchronous reactions of the stem cells. Informative models to date are axotomies, apoptosis-mediated lesions, and moderate ischemic damages, but also physiological contexts like stress and various hormonal challenges. Emerging mediators of in vivo recruitment of intrinsic stem cells are some growth factors and cytokines, inflammatory signals, astrocytic and microglial messengers.

* Hopital Laveran, Service de Santé des Armées, 34 Bld Laveran, 13013 Marseille, France.

The current state-of-the-art in heterotopic graft of lineage-labelled stem cells point to the expected requirement for combination with pharmacological treatment using competent chemical signals. The source of primary stem cells for grafting is a critical issue: embryonic versus adult-derived, neural versus non-neural (hemetopoietic) stem cells, and human versus animal donors.

Altogether, many pertinent tracks are being opened to cell therapy progress, that all deserve basic research and animal model studies in the immediate future.

Introduction

The discovery of neural stem cells in the central nervous system of adult Mammals in 1992 (Reynolds & Weiss 1992) raised great expectations for regenerative therapy after lesions or degenerative diseases (Miller 2006). A tremendous accumulation of data about the dynamics of neural stem cells occurred in the last decade. However, these results do not allow yet successful cell therapy, neither in animal models nor in humans. The purpose of the review is to first summarize the present state of knowledge about basic cell biology of neural stem cells and tissular factors that restraint their in vivo regenerative potential, and then to present pertinent tracks for cell therapy progress in clinics.

I. Neural Stem Cells of Adult Mammals

I.1. Concept History

Indirect evidence for the existence of neural stem cells in adult mammals was initially brought by Altman's discovery of in situ neurogenesis in adult rat brain (Altman 1962, 1969; Altman & Das 1965, 1967). Following in vivo administration of the DNA synthesis precursor ^{3}H-thymidine in adult rodents, some restricted areas of the brain displayed autoradiographic labelling in some histologically stained neurons derived therefore from a mitotic cell division at the time of ^{3}H-thymidine injection. Since neurons are irreversibly post-mitotic cells, no mitotic division having ever been observed in histologically-identified neurons, this observation led Altman to suggest that adult mammalian brain should contain proliferative, stem-like cells. This foreseeing hypothesis was demonstrated three decades later, after discovery of molecular determinants of neural ontogeny, by the two neuro-embryologists B.A. Reynolds and S. Weiss who postulated that putative neural stem cells of adult brain should respond to the same mitogens (i.e., proliferation-stimulating factors) as embryonic neuroblasts. These authors sampled a tissue block from the most proliferative brain region of Altman's studies: the subventricular zone of dorso-medial striatum, and dissociated it into primary cell culture in the presence of Epidermal Growth Factor and basic Fibroblast Growth Factor (20 ng/mL each in Dubelcco's Modified Essential Medium); after several days in vitro, while most of initially seeded cells had died, some floating spherical masses of undifferentiated cells appeared and grew progressively as a sign of proliferation; removing the two growth factors at this stage (by changing the culture medium) resulted into sphere flattening and adhesion to the substratum; these differenciated spheres were shown to contain the three cell lineages of neural tissue (neurons, astrocytes, oligodendrocytes) by immunocytochemistry of cell-type-specific antigens and were called therefore "neurospheres" (Reynolds & Weiss 1992). This elegant proof-of-concept allowed to demonstrate in adult mammalian brain the presence of cells that fulfil the basic criteria of stem cells, as defined on the archetypical model of mammalian hematopoiesis and later generalized in numerous tissues (McKay 1997; table 1).

Table 1. Definition criteria for stem cells and progenitors

Stem cells	Progenitors
1- Unlimited capacity for mitotic division	Limited capacity to divide
2- Self-renewal capacity through symmetrical division	Limited self-renewal
3- Capacity to generate all cell types of the host tissue	3- Restriction of progenies to 1-2 cell lineages
4- Physiologically-triggered (in particular, lesion-induced) renewal of host tissue in vivo	4- Physiologically-triggered (in particular, lesion-induced) renewal of one given cell type

I.2. Isolation and Expansion of Native Neural Stem Cell by the "Neurosphere Assay"

Neurosphere production in primary cultures from adult neural tissue remains the basic assay for isolation of native neural stem cells and progenitors (Reynolds & Rietze 2005, Rietze & Reynolds 2006, Marshall et al. 2007; figure 1). It has been used as a colony-forming assay to demonstrate the presence of neural stem cells in various structures of central nervous system (Weiss et al. 1996, Markakis et al. 2005, Charrier et al. 2006), in retina (Tropepe et al. 2000), in olfactory mucosa (Carter et al. 2004, Murrell et al. 2005) and in enteric plexuses (Estrada-Mondaca et al. 2007). All the cardinal criteria of stem cells (table 1) can be fulfilled by using this in vitro method (figure 1).

The number of neurospheres produced from a given cell dilution has been used to assess and compare the stem cell content of different nervous structures and/or ages (Weiss et al. 1996, Shingo et al. 2003, Charrier et al. 2006). The diameter of neurosphere is commonly used to gauge the proliferation rate of neural stem cells; this application has been challenged though by time-lapse video microscopy-assisted demonstration, that spheres merge easily and absorb isolated cells (Singec et al. 2006). So "sphere formation must not be used as a metric tool", unless with initial plating dilution not higher than 500 cell/mL and sphere size lower than 100μm (Singec et al. 2006).

Self-renewal is demonstrated by obtention of novel spheres after dissociation of neurospheres and subsequent replatting. As in primary culture, since the culture conditions support only survival and growth of cells that are responsive to EGF and/or bFGF, i.e. both neural stem and progenitor cells, most of the replated cells die. Hence, upon passage, the generation of new secondary spheres represents a renewal of the previous population; passage also allows in vitro expansion of the neural stem cell population, with a 50 times magnification factor when using clonal density (Reynolds & Rietze 2005). However, 1-2 passages are not sufficient to establish the self-renewal criterion, since secondary spheres could be obtained from progenitor cells of hippocampus (Seaberg & van der Kooy 2002). Under the above culture conditions, growth-factor responsive cells can theoretically be passaged infinitely – in practice for 10 passages at least (Reynolds & Rietze 2005) up to 70 ones (Foroni et al. 2007).

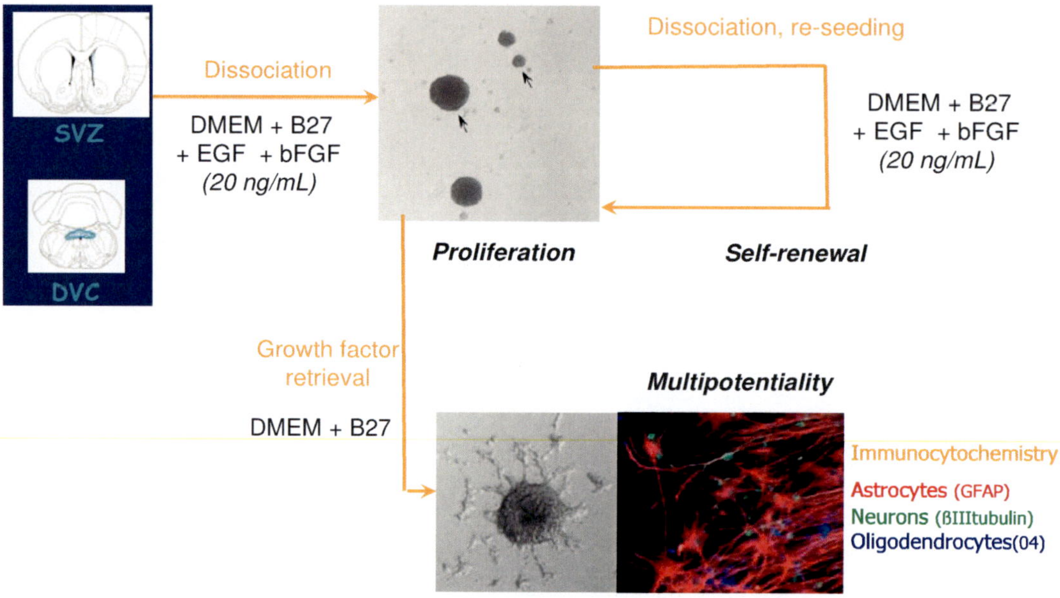

Figure 1. Neurosphere assay on microdissected tissues of adult rat.

Differentiation into the three cell types of nervous tissue is addressed by immunocytochemistry several days after mitogen removal (Reynolds & Weiss 1992). The lineage composition of differentiated spheres, as determined, is repetitively 75-80% GFAP-immunoreactive astrocytes, 12-20% βIIItubulin-immunoreactive neurons, 1-5% O4-immunoreactive oligodendrocytes (Figure 1; Reynolds & Rietze 2005). Some spheres contain exclusively 2 or even 1 of the neural lineages, which corresponds to clonal proliferation from one restricted progenitor. These lineage-restricted spheres do not usually display self-renewal (Seaberg & van der Kooy 2002). The neuronal differentiation can reach formation of functional synapses (Liebau et al. 2007).

I.3. Proliferation Mechanisms in Neural Stem Cells

The molecular mechanisms underlying mitotic activity of adult neural stem cells involve the general "cell cycle engine" of eukaryotic cells. In brief, entry and progression in the division cycle phases G1-S-G2-M are determined by sequential and transient induction of regulatory proteins of the cyclin family, that bind and activate respectively specific ligands of the cyclin-dependent-kinase (CDK) family of constitutive cytoplasmic enzymes; catalytic activities of the CDK-cyclin complexes are regulated by intrinsic inhibitors (CDKI) that were discovered as tumor suppressors (Nurse 2000, 2002). *In neural stem cells as in all cell types assessed so far, mitogen-driven division onset relies on transcriptional induction of a cyclin D that activates CDKs 4 and 6* (Ferguson et al 2000, Doetsch et al 2002).

However, *adult neural stem/progenitor cells display some original features*. Analysis of knockout mice for either one of the cyclin D family revealed indeed that cell cycle onset is driven through cyclin D2 exclusively in neural stem/progenitor cells of SVZ and hippocampus, versus cyclin D1 in the olfactory progenitors of olfactory epithelium (Kowalczyk et al. 2004). Cyclin D1 was shown to target inducible, instead of constitutive CDK4 in olfactory epithelium (Kastner et al. 2000), and to mediate proliferation trigger in both the neural stem cells of adult brainstem (Charrier et al 2006) and the neuronal progenitors of olfactory epithelium (Simpson et al. 2007). In addition, intrinsic modulations of the cell cycle engine can produce some cell dynamic variants. Thus, transgenic mice with constitutive activation of p53 and overexpression of p21 yield decreased number and size of neurospheres from SVZ which corresponded to decreased self-renewal and amplification capacities (Medrano et al. 2007). Also, aging-related decrease of neural stem cell proliferation was correlated with increased expression of the CDK-inhibitor p16INK4a (Molofsy et al. 2007).

I.4. Phenotype of Neural Stem / Progenitor Cells

Studies of neural stem / progenitor cells have been hampered by the lack of specific markers. Recent data accumulation has though began to reveal *stage-specific combinations of molecular markers* (table 2; Parker et al. 2005, Schwartz et al. 2005, Klassen et al. 2007, Ma et al. 2007, Raponi et al. 2007, Skalnikova et al. 2007), which display some analogies with maturation stages of embryonic neuroblasts and neuronal or glial progenitors (Kempermann 2006, Guillemot 2007).

The nature of neural stem cells has been elucidated much later than demonstration of their existence by Reynolds & Weiss in 1992. A diversity of indirect in vivo experimental evidences first pointed that neural stem cells of adult rodent SVZ were located in the nervous parenchyma just below the ependymal lining, probably belonging to astrocytic lineage (Doetsch et al. 1999). At the same time, another set of indirect in vivo data pointed that neural stem cells of adult rodent spinal cord were a subpopulation of the ependymal cells lining the medullary canal (Johansson et al. 1999). The astrocytic nature of neural stem cells has been directly demonstrated in SVZ by selective transgenesis-induced ablation of proliferating astrocytes in adult mouse brain, which blocked endogenous neurogenesis in forebrain, whereas reciprocally the whole forebrain neurogenesis system was labelled after transgenic labelling of proliferating astrocytes (Garcia et al. 2004). Neural stem cells of SVZ and hippocampus are thus established as a subpopulation of astrocytes, and more precisely a subpopulation of radial glia (see section III-1; Bonfanti & Peretto 2007, Ihrie & Alvarez-Buylla 2007, Laywell et al. 2007). In the spinal cord, neural stem cells have also been formally identified in situ with another transgenesis approach, allowing immunohistochemical labelling of either nestin-expressing cells or their neuronal progenies (Carlen et al. 2006); using this genetic tool, nestin-immunoreactive cells were seen in spinal cord ependymocytes, which after in vitro purification proved able to generate neurospheres (Carlen et al. 2006). Therefore, neural stem cells of the telencephalon and of spinal cord belong to two different glial lineages, i.e. astrocytes and ependymocytes respectively. The neural stem cells of mammalian retina have been formally identified to belong to a third glial lineage: the Müller cells (Klassen et al. 2007). The nature of other brain populations of neural stem cells (hypothalamus, brainstem) has not been elucidated yet. Altogether, at least *three anatomically distinct subpopulations of neural stem cells consist of three different glial cell lineages* respectively. Consistently, significant differences of either transcriptome or proteome have been reported between neurospheres deriving from different brain regions (Ma et al. 2007, Skalnikova et al. 2007). The impact of such phenotype heterogeneity on the functional properties of neural stem cells remains to be determined.

Table 2. Molecular markers of neural stem / progenitor cells and their progenies.

	Non-neural stem cell	Neural stem cell	Transit-amplifying cell	Progenitor / Migrating neuroblast	Immature cell	Mature, differentiated cell
nestin		+				
Vimentin		+				
GFAP		+				+ (astrocyte)
Ki-67	+	+	+	+		
Sox2	+	+				
Sox1		+	+			
BMI1		+				
CD133	+	+				
Musashi1	+	+				
BLBP		+				
NF1		+				
LIF-R	+	+				
Notch - delta		+				
ALDH		+	+			
EGF-R		+	+	+		
HSA			+	+		
DCX				+	+	
Pax6				+		
PSA-NCAM				+	+	
Dlx				+	+	
Neuro D				+	+	+
Calretinin					+	
MASH-1				+ (neuronal)	+ (neuron)	
Olig1,2					+ (oligodendr)	
PDGF-β					+ (oligodendr)	
β-III-tubulin						
TOAD					+ (neuron)	
NeuN						+ (neuron)
Hu C/D						+ (neuron)
calbindin						+ (neuron)
NSE						+ (neuron)
S100β						+ (astrocyte)
O4						+ (oligodend)

I.5. Resistance to Tumorigenesis

Since cancer is well-established to arise mainly from tissues that harbour stem cells in adult (Sherr 1994), the existence of neural stem cells provides a candidate mechanism for brain tumor etiology (Sanai et al. 2005). Consistently, diverse brain tumors contain transformed neural precursors that respond to the same mitogens as neural stem cells of adult healthy brain and fulfil stem cell criteria and possess some of the molecular features of neural stem cells: Sox2, BMI1, CD133, Musashi1 (Vescovi et al. 2006, Poppleton & Gilbertson 2007). CD133, being a cell-surface protein, can be used for FACS purification of tumor stem-like cells, that were labelled and found to initiate new tumours after in vivo transplantation. Gliomas in human occur indeed more frequently in the vicinity of SVZ than in non-proliferative brain parenchyma, and exogenous carcinogens are much more efficient when injected into SVZ than in perfused cortex (Sanai et al. 2005). Tumorigenesis from neural stem cells must though rely on transforming mutations, which are rare and involve more than one among the following: i) overexpression of PDGFα, EGF-receptor, NF1, Notch receptors and/or their ligands Delta (DLK1) and Jagged (JAG1), BMI1; ii) mutation or loss of Trp53, CDKN2A (encoding two tumor suppressors), Sonic-Hedgehog (SHH) and its downstream effectors (glioma-associated oncogenes GLI 1, 2, 3), PTEN and the canonical Wnt-β-catenin pathway (reviewed in Vescovi et al 2006).

However, the putative propensity of neural stem cells to transform into tumor cells has been experimentally challenged by a recent detailed analysis, using the neurosphere assay (Foroni et al 2007). Adult neural stem cells, after in vitro propagation over one year through 70 passages, were found indeed to maintain a strikingly stable profile with regard to self-renewal, differentiation, growth-factor dependence, karyotype and molecular profiling (Foroni et al. 2007). More importantly, such long-term culturing of adult neural stem cells did not result in the formation of tumors in vivo upon grafting, even when neural stem cells were transfected with Myc and Ras oncogenes (Foroni et al. 2007). *This cancer resistance indicates that adult neural stem cells must possess specific anti-tumorigenic mechanisms, which are presently unknown.* Investigation of these putative mechanisms would bring about, not only progress regarding neural stem cells, but also new cues for cancer and cell cycle control research.

I.6. Lineage Labelling for Progeny Tracking after Graft

In order to know the role and fate of neural stem cells in vivo, in basal state as well as under patho-physiological adaptations, it is necessary to perform a "lineage labelling" i.e. to introduce in neural stem cells a marker that will be expressed permanently, by the labelled cells themselves and by their progenies across successive division cycles. The key experimental approach for this purpose consists in incorporating into the target cell genome an exogenous gene sequence coding for a molecule that can be visualized, either directly (fluorescent proteins like GFP) or by histochemistry (bacterial galactosidase LacZ). The gene sequence to be used as a lineage marker must be first linked by molecular biology to the virus genome region that drives integration into the host cell DNA. This category of genomic manipulation has been extensively developed for gene therapy purposes (for review, see

Waehler et al. 2007). For in vivo tracking of neural stem cell progeny, two experimental strategies are available. First, the viral-assisted labeling of neural stem cells can be performed in vitro, and the labeled cells are later transplanted into the brain by using stereotaxy. Second, in vivo labeling of endogenous neural stem cells can be performed by stereotaxic injection of the viral construct. In all cases, the molecular construct contains only non-pathogenic regions of viral genomes that allow entry into the cell and integration into its genome. The labeling strategy to be used and the experimental goal determine the type of viral construct tools. Adenoviruses allow efficient transfer of large gene sequences into both dividing and quiescent cells at relatively low efficiency; in vivo, adenovirus-based DNA injections are potent inducers of inflammation, which is now known to modulate adult neurogenesis (Cayre et al. 2006, Danilov et al. 2006). Retroviruses infect only dividing cells and allow robust transfer of small genomic sequences; this is the most dedicated tool for lineage labelling of neural stem cells, both in vitro and in vivo. Lentiviruses infect dividing and quiescent cells, including stem cells and surrounding ones, which did a priori preclude in vivo applications. However, it has been empirically found that lentiviral gene transfer into SVZ can provide specific and efficient labelling of neural stem cells and their progeny in vivo (Consiglio et al. 2004). It must be recalled indeed, that virology knowledge is based on extensive analysis of the few cell types that are either immune reaction effectors (lymphocytes and dendritic cells) or gene therapy targets. Neural stem / progenitor cells of adults can be efficiently targeted with lentiviral vectors that have been optimized for cells of the immune system displaying similar cellular dynamics, i.e. slow mitotic rate: the antigen-presenting dendritic cells (Mangeot et al. 2002, Negre et al. 2002). Incidently therefore, *adult neurogenesis analysis will provide progresses in basic virology.*

BrdU incorporation cannot provide a tool for this purpose, because of the semi-conservative mode of DNA replication; incorporated BrdU is thus diluted by half at each mitotic division happening after labelling, which was experimentally found to make it undetectable after 1-2 divisions (Wojtowicz & Kee 2006). In addition, adult nervous tissue contains some non-neuronal cell types that keep the potential for mitotic proliferation: astrocytes, ependymocytes, microglial cells.

Altogether, it is clear by now that *adult neural stem cells of mammals are basically different from embryonic stem cells*, which are the common precursors of all types of adult tissue-specific stem cells, *and from the neuroblasts of embryonic neural tube*. Sensitivity to extracellular signals and resistance to tumorigenesis are indeed drastically different among these three types of proliferating cells. Although all of them are endowed with neurogenic potential, the more readily available embryonic stem cells and neuroblasts can more and more hardly be considered as relevant model-systems for adult neural stem cells. This distinction is specially important for therapeutical attempts of cellular therapy in the coming future (see section VIII).

II. In Vivo Restriction of Multipotentiality

II.1. Anatomical Restriction of Stem-Cell-Derived Neurogenesis

The production of new neurons in adult is strikingly restricted to a few discrete structures of nervous system. The most classical and studied ones are the olfactory bulb, which is the migratory target of the subventricular zone (SVZ) progenies, and the hippocampus (reviewed in Kempermann 2006). A few additional neurogenesis foci have been recently characterized: hypothalamus (Kokoeva et al. 2005, 2007), dorsal vagal complex (Bauer et al 2005, Pecchi et al 2007) and vestibular nuclei (Tighilet et al. 2007) of the brainstem, retina (Klassen et al. 2007), olfactory mucosa (ref), enteric wall (Estrada-Mondaca et al. 2007).

These neurogenesis-endowed areas display quantitative differences in vivo with regard to neurogenic activity (Figure 2). The proliferation rate is highest in SVZ, amounting 20 times that of hippocampus or dorsal vagal complex. Interestingly, the proliferation rate of the non-neurogenic lumbar spinal cord is as high as in SVZ (Weiss et al. 1996). The neurogenic yield, i.e. the proportion of newborn neurons among stem cell progenies, is highest in hippocampus (90-100% versus 70-80% in olfactory bulb and 10-30% in brainstem foci).

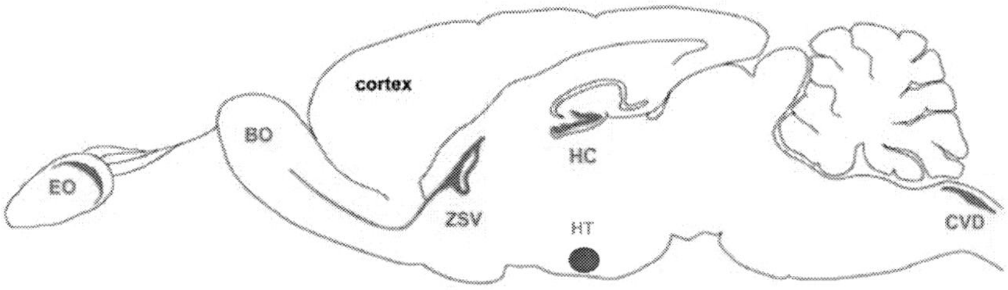

neurogenic system In vivo	Proportion of neurons among new cells per day	Number of new neurons	proliferation rate
SVZ – olfactory bulb(OB)	70 -80 %	80 000	> 20 X
Hippocampus (HC)	90 - 100 %	8 000	X
Dorsal vagal complex	35 %	N.D.	X

Figure 2. Adult neurogenesis in vivo: quantitative comparison.

It is striking to note that the neurogenic yields of SVZ neural stem cells are conspicuously much higher in vivo (70-80% of SVZ progenies in olfactory bulb) than in vitro (12-20% of neurosphere cells). This indicates that in vivo tissular microenvironment should be a determinant stimulator for intrinsic neurogenic potential of neural stem cells; which will be discussed in details below.

II.2. Non-neurogenic Populations of Neural Stem Cells In Vivo

Some populations of neural stem cells have been detected by using the neurosphere assay in diverse areas of the central nervous system where no neurogenesis occurs in vivo (Weiss et al. 1996, Emsley et al. 2005, Reynolds & Rietze 2005). Therefore, adult neural parenchyma contains widely distributed potential for neurogenesis that is not expressed in standard conditions, most likely because of tissue-induced repression, but may provide a candidate mechanism to be recruited for brain repair (Buffo 2007). It is therefore crucial to identify the tissular signals that inhibit neuronal differentiation of neural stem cell progeny. On the other hand, accurate histological analysis after in vivo mitogenic challenge indicated that proliferative cell populations display anatomically localized distributions (Doetsch et al. 1997, Martens & Van der Kooy 2002, Pecchi et al. 2007). More precisely, in vivo intracerebroventricular injections of the neural stem cell mitogens were found to stimulate neurogenic proliferation in nervous parenchyma thickness exclusively within the neurogenic niches, whereas BrdU incorporation in neighbouring areas was restricted to a thin ventricular lining (Martens & Van der Kooy 2002, Pecchi et al. 2007).

II.3. Plasticity of Adult Neural Stem Cells

A conceptual breakthrough was made by demonstrating that transplantation of primary neural stem cells from adult mammalian brain could restore long-term functional hematopoiesis in irradiated recipient mice that would otherwise have died shortly (Bjornson et al. 1999). Reciprocally, primary hematopoietic stem cells from adult mammalian bone marrow or umbilical cord blood can generate neural cells, including neurons, upon transplantation into adult brain (Mezey et al. 2000, Chen et al. 2005, Hermann et al. 2006). Following the same experimental design, hematopoietic stem cell transplantation was also found to rescue infarcted myocardium (Orlic et al. 2001). Interestingly, neural potential could be induced in the stem cells from adult mammalian skin epithelium (Fernandes et al. 2006) and hair follicle (Mignone et al. 2007). Within a few years, these discoveries have been generalized with a bulk of experimental evidence in diverse tissues, supporting the concept that *adult mammalian stem cells from one given tissue can be reprogrammed under heterotopic grafting to ensure functional renewal of the host tissue cells.* This has important value for clinical repair applications.

III. The "Neurogenic Niche" Concept

III.1. Cytoarchitecture of Tissular Microenvironnements Harboring Neural Stem Cells

In general, adult stem cells are embedded within their host tissue in a very definite topography that can be visualized by conventional histology. At the molecular level, neighbouring cells and/or extracellular matrix can make a scaffold that harbors the stem cells.

In the last decade, the cellular functions of stem cells (proliferation, self-renewal, lineage commitment) progressively appeared to receive in vivo various modulatory signals from their tissular microenvironnement: diffusible signals and adhesiotopes from neighbouring cells, extracellular matrix components, systemic hormones that are trapped and accumulated around the stems. This modulatory interplay between stem cells and their host tissue has been summarized by the concept of "*stem cell niche*" (Scadden 2006).

In the case of neural stem cells, the cytoarchitecture of their microenvironnement, or *neurogenic niche*, has been thoroughly described in SVZ and hippocampal dentate gyrus before identification of the stem cells per se, which was actually helped by ultrastructural and immunohistochemical investigations of these two proliferative nests with or without antimitotic drug treatments (Doetsch et al. 1997, 1999). Common cytoarchitectural features were detected among the two telencephalic niches (Doetsch 2003), among which the morphological distinction between multipotent stem cells (astrocyte-like "B" cells) and transit-amplifying progenitor cells ("C" and "A" cells in SVZ, "2a", "2b", "D" in dentate gyrus) has been updated more accurately with recent molecular profiling data (Kempermann 2006). The proliferative cells and migrating progenies are surrounded by abundant astrocytes (Doetsch 2003) including a subpopulation of radial glia (Ihrie & Alvarez-Buylla 2007). Like in embryonic neural tube, the radial astrocytic profiles express specific markers (vimentin, nestin, BLBP) along with GFAP and provide physical support for migration of stem cell progenies (Sundholm-Peters et al. 2004, Pecchi et al. 2007). These radial astrocytes provide an obvious candidate for the "niche" concept in adult neurogenesis (Scadden 2006, Conover & Notti 2007). A similar niche cytoarchitecture has been found in adult human SVZ (Lim et al. 2007, Quinones-Hinojosa et al. 2007). *Three neurogenic niches, containing both neural stem cells and radial glia, are known in the adult central nervous system* so far: SVZ, hippocampus, dorsal vagal complex of the brainstem (Doetsch et al. 1997, Doetsch 2003, Sundholm-Peters 2004, Nern & Momma 2006, Lim et al. 2007, Pecchi et al. 2007). A fourth brain structure is endowed with both intrinsic neural stem cells and neurogenesis: hypothalamus (Weiss et al. 1996, Markakis et al. 2004, Kokoeva et al. 2005); however the stem cells have not yet been localized with anatomical accuracy (Xu et al. 2005), and the occurrence of a radial glia-like niche is not established, although probably related to the tanycyte glial population of medio-basal hypothalamus. In addition, olfactory epithelium and retina can also be considered as neurogenic niches, the radial guidance of stem cell progenies being provided respectively by sustentacular cells and Müller glial cells.

Another recently recognized component of neural stem cell niches is the vascular compartment. Neural stem cells are systematically found in the vicinity of blood capillaries, at least in SVZ and hippocampus (Palmer et al. 2000), but also along the migratory stream from SVZ to olfactory bulb (Bovetti et al. 2007). Proliferation of neural stem cells and neurogenesis are potently stimulated by endothelial cells (Shen et al.2004), which is mediated by Vascular Endothelial Growth Factor (VEGF, Sun et al. 2006). Blood vessel endothelium thus makes part of the neurogenic niche (Tonchev et al. 2007).

A semantic detail must be noted: the "neurogenic niche" must not be assimilated a priori with "neurogenesis site or focus". For instance, the archetypical model of neurogenic niche, SVZ, does not harbour any neurogenesis (i.e. genesis of new neurons) since the progenies of

SVZ neural stem cells migrate all to the olfactory bulb, i.e. the exclusive and remote site of new neuron differentiation and integration in this neurogenic system.

III.2. Adult Tissue Competence for Neurogenesis: Assay with Heterologous Grafts

Neural stem cells were shown to be determinantly regulated by their tissular microenvironnement in vivo, which can be viewed as a particular case of the recently demonstrated trans-tissular plasticity of stem cells (see II.3). The landmark experiment consisted in grafting primary neural stem cells, after in vitro expansion and virally-assisted lineage labelling (see I.6) from one non-neurogenic structure like the cervical spinal cord into the highly neurogenic hippocampus; grafted cells proliferated and produced typically hippocampal neurons, whereas the same cells in their original location generated only astrocytes (Shihabuddin et al. 2000). Similarly, after transplantation into the adult olfactory bulb, pre-labelled primary hippocampal progenitor cells generated interneurons that displayed characteristic phenotypes of the host structure (Suhonen et al. 1996).

These data led to define a novel criterion that should be fulfilled before any structure of adult nervous system can be accepted as a neurogenic focus (like hippocampus and olfactory bulb). Lineage-labelled neural stem cells from SVZ must be transplanted into the studied structure and, after a few weeks survival, generate neurons that display phenotypes of the host structure (Kokoeva et al. 2005, Klassen et al. 2007). The archetypical neural stem cells of SVZ, after in vitro culture and lineage labelling, should thus be used as probes to test whether a neural structure is indeed endowed with neurogenesis.

More recently however, by specifically targeting stem cells and following their progeny in vivo, postnatal stem cells in different regions were shown to produce different types of neurons, even when heterotopically grafted or grown in culture (Merkle et al. 2007). This suggests that rather than being plastic and homogeneous, neural stem cells would include a restricted and diverse population of progenitors. These new data are in keeping with the highly reproducible in vitro behaviour of neural stem/progenitor cells from different structures of adult nevraxis in the neurosphere assay conditions (Weiss et al. 1996, Charrier et al. 2006). It is though highly probable that in vitro culture conditions of neural stem/progenitor cells deserve specific optimization for different structures and/or species (ref Stem Cell company), whereas in most studies the "neurosphere assay" uses the protocol that had been initially optimized for mouse SVZ (Reynolds & Weiss 1992).

The question remains thus open, whether the major determinants of neural stem cell function and cellular yield are intrinsic (intracellular) or derived from micro-environnement (extracellular). It is obviously a critical issue for stem cell-assisted nervous repair in clinics, i.e., "cellular therapy", which requires additional cell biology studies in vitro and in vivo.

III.3. Regulatory Role of Astrocytes in Neurogenic Niches

Co-culture experiments between astrocytes from one adult nervous structure with neurospheres from another, have revealed that proliferation and/or neuronal differentiation of neural stem cells are stimulated by astrocytes from neurogenic structures and inhibited by astrocytes from non-neurogenic ones (Song et al. 2002, Agasse et al. 2005, Barkho et al. 2006). The same effects can be reproduced with "conditioned media" from primary astrocyte cultures instead of co-culture, indicating that astrocytes exert their modulatory actions via secretion of diffusible factors, which remain largely unknown. Consistently, transcriptomal analysis of astrocyte cultures from different brain regions, i.e. extensive comparison of respective mRNA extracts by DNA array methods, has revealed region-specific variations of gene expression patterns of astrocytes (Lafon-Cazals et al. 2003, Emsley & Macklis 2006).

Astrocytic cells are involved in some injury-induced reactions of neurogenic niches (Zhu & Dahlström 2007).

Altogether, the tissular niche that surrounds neural stem cells in vivo provides a bulk of neurogenesis-modulating signals, either of local origin or diffusing from general circulation (Scadden 2006). The elucidation of niche-characterizing components, that would be lacking in non-neurogenic regions, is therefore a crucial challenge for future cell therapy strategies (Lim et al. 2007). These niche components are likely to include signalling molecules (paracrine diffusible messengers, cell-influencing epitopes of extracellular matrix and membrane-linked adhesiotopes) and extracellular components that can adsorb circulating messengers and concentrate them at the vicinity of neural stem cells and early progenies (Scaden 2006, Riquelme et al. 2007).

IV. Physiological Regulation and Functions of Neural Stem Cells in Adults

It is now well-established that the major role of adult neurogenesis in mammals is to provide an effector for experience-dependent plasticity and neuroadaptations (Abrous et al. 2005, Emsley et al. 2005, Kemperman 2006, Sohur et al. 2006).

IV.1. Adult Neurogenesis Involvement in Experience-Dependent Plasticity

IV.1.1. Behavioural Plasticity-Related Alterations of Neurogenesis

The first phenomenon of this category was actually reported in birds: neurogenesis in the song center undergoes an important rise at the breeding season which is correlated with enhanced capacity to learn new congener songs and with testosterone surge (Rasika et al. 1994).

In adult rodents, two-month-exposure to enriched environnement triggers significant increase of hippocampal neurogenesis which is correlated with improvements of short-term memory performances like the water maze, as compared to rats kept in standard laboratory housing conditions; this phenomenon occurs in 18-month-aged rats as well as in young ones, although absolute neurogenesis baseline is lower in aged rats than in young ones (Kemperman et al. 1997, 1998). Hippocampal neurogenesis was also significantly improved by a hippocampus-dependent learning paradigm; the rise in neuron production was maximal 10 days following the end of the training period (Bruel-Jungerman et al. 2007). Hippocampal neurogenesis is enhanced in adult rats by chronic food restriction (Lee et al. 2002) which is considered as a "successful aging" model with regard to somatic, weight gain and also cognitive performances (Martin et al. 2006); its impact on hippocampal neurogenesis is more probably related to the cognitive incidence than the metabolic outcome of this regimen. Physical exercise has been extensively documented to stimulate hippocampal neurogenesis as well as cognitive performances (Van Praag et al. 1999 a,b; Pereira et al. 2007). Lineage-labeled, adult-generated interneurons of hippocampus were directly visualized being integrated into anatomical circuits that support spatial memory, i.e. one key specific function of hippocampus (Kee et al. 2007). Conversely, hippocampal neurogenesis was decreased in depression models (Paizanis et al. 2007) and after various paradigms of chronic stress (Heine et al 2004, 2005, Hitoshi et al. 2007, Veenema et al. 2007), which is blocked by antidepressant treatments (Hitoshi et al. 2007). Hippocampal neurogenesis is stimulated in rat by antidepressants (Newton & Duman 2007), but this appears to relate to behavioural effects of antidepressors and not to a causal role of adult neurogenesis in depression per se (Sahay & Hen 2007). Adult neurogenesis is stimulated by the mood-normalizing drug lithium (Su et al. 2007); it is also significantly altered by addictive drugs of abuse (Abrous et al. 2005, Eisch & Harburg 2006). SVZ-derived neurogenesis of olfactory bulb is strongly stimulated by olfactory experience, especially in odor discrimination learning context (Petreanu & Alvarez-Buylla 2002, Lledo et al. 2006, Mandairon et al. 2006, Lledo & Lazarini 2007); these sensory-triggered stimulations are correlated with the spatial coding of repetitively exposed odor (Mandairon et al. 2006). In line with these results, olfactory bulb neurogenesis is transiently enhanced in female rodents at the very end of gestation, during the key period of maternal imprinting; this effect being correlated with a perinatal surge of the hormone prolactin (Shingo et al. 2003).

Altogether, a diversity of well-known models of experience-dependent behavioural plasticity in adult mammals was found to trigger significant modulations of neurogenesis in *vivo*. In the present state of knowledge, neural stem cells are recruited much more by such physiological adaptation-eliciting challenges in healthy adults, than by lesions (see section V). This is a striking difference with all other types of adult stem cells.

Interestingly, in vivo neurogenesis modulation by one given physiological challenge displays neuroanatomical selectivity, i.e. one neurogenic system is modulated whereas others are not affected. For instance, neurogenesis stimulation by late gestation concerns specifically the SVZ-olfactory bulb system, hippocampus being not modulated in the same animals (Shingo et al. 2003).

All these observations raise the question to know whether neurogenesis modulation is a cause or a consequence of experience-dependent plasticity. This is discussed in the next section.

IV.1.2. Causal Role of Neurogenesis in Experience-Induced Adaptations

The best approach to test whether adult neurogenesis plays a causal role in experience-induced plasticity, is to block neurogenesis in vivo with an anti-mitotic drug and to assess the effect of this experimental manipulation on the physiological-behavioral adaptation that has been correlated with neurogenesis enhancement. Although the principle seems simple, this approach encounters practical difficulties: i) systemic administration of anti-mitotic agents like those used in human chimiotherapy has important side-effects that are likely to mask the specific effects of neurogenesis blockade; ii) local administration relies upon stereotaxic surgery, which also produces morbidity. However, this approach has been used successfully to demonstrate that hippocampal neurogenesis is causally involved in hippocampus-mediated learning; inhibition of behavioural performances was though partial, which indicated that adult neurogenesis is only one among several mechanisms underlying the learning processes (Shors et al. 2001). Similarly, CNTF-induced stimulation of adult neurogenesis in rodent hypothalamus was shown to be the causal mechanism of CNTF-induced anorexia, by local blockade of hypothalamic neurogenesis by stereotaxic injection of the anti-mitotic drug AraC in the third ventricle (Kokoeva et al. 2005).

The causal role of adult neurogenesis in some physiological adaptations could be demonstrated by using alternative, less direct strategies. Thus, the causal role of prolactin in pregnancy-induced stimulations of both olfactory learning and olfactory bulb neurogenesis, was demonstrated by using genetic deletion of prolactin receptor in mouse, which blocked these two effects of late pregnancy (Shingo et al. 2003).

In several cases therefore, the causal role of adult neurogenesis in experience-dependent plasticity has been formally demonstrated.

IV.2. In Vivo Mechanisms of Neural Stem Cell / Neurogenesis Regulation

Neuro-glio-genesis from adult neural stem cells involve several functions at the cellular level, which could a priori provide candidate effector mechanisms for in vivo regulation: proliferation, cell death and survival, differentiation commitment switch (es).

In the course of aging, the proliferation rate of neural stem cells is strongly decreased as for all types of proliferative cells in Mammals (Tropepe et al. 1997, Maslov et al. 2004). However, neural stem cells seem to retain their other functional characteristics intact through aging, bth in vivo and in vitro (Kemperman et al. 1997, Foroni et al. 2007).

V. Post-Lesional Reactivity of Endogenous Neural Stem Cells

Adult nerve tissue, especially in the central nervous system, displays very poor repairing reactivity to lesion in physiological conditions, both in animal models and in human. This is a striking originality of neural stem cells as compared with all other types of adult mammalian stem cells. The general criteria of stem cells include indeed stimulation by lesions of the host tissue, and the major role of stem cells is tissue renewal and repair. Reciprocally, the diversified causal roles of neural stem cells in physiological adaptations and in experience-dependent plasticity are unique among stem cells of adult mammals. However, in the last few years, more and more examples of neural stem cell recruitment by lesions are being unravelled. These processes were detected essentially in mild, discrete models of neural lesioning.

V.1. Apoptosis-Induced Stimulation of Neurogenic Proliferation

The first characterized lesion-induced activation of neural stem cells and reparative neurogenesis was detected in a very mild and anatomically restricted lesion model: neuronal death induction in a small cerebral cortex area by stereotaxic injection of a neurotoxin-coupled retrograde tracer in the terminal arborisation field of a discrete cortico-thalamic pathway (Magavi et al. 2000); the neuronal loss is partial (20% of the projection neurons) and proceeds via apoptosis, i.e. programmed cell death avoiding any tissular inflammation (Krantic et al. 2005), in a cortical area that is normally devoid of any neurogenesis in adult. This mild lesion being combined with in situ viral-assisted lineage labelling of SVZ neural stem cells, some SVZ progenies were found to migrate to the lesioned cortical area and to differentiate into cortico-thalamic neurons that integrated morphologically and anatomically in the partially lesioned projection (Magavi et al. 2000). The same group applied the same approach to the pyramidal cortico-spinal tract, which is a much larger and longer projection than the previous one; similarly, some SVZ progenies migrated to the partially lesioned motor cortex and differentiated into typical cortico-spinal premotor neurons, which ensured full recovery of lesion-altered motor function (Chen et al. 2004). Adult SVZ-derived neurogenesis can therefore be recruited by a lesion via progeny migration and neuronal differentiation into sites that are normally devoid of any neurogenesis. Cortical tissue has even been shown to secrete diffusible factors that inhibit neurogenesis in vitro (Agasse et al. 2005). These data suggest that migration and neuronal differentiation of SVZ progenies to the site of lesion could be triggered by diffusible signals arising from apoptotic neurons.

The latter hypothesis is strenghthened by our own data in another lesional model: the selective apoptotic death that is induced in the olfactory epithelium neurons of adult rodents by ablation of their common synaptic target, olfactory bulb (Michel et al. 1994, 1997, Moyse & Michel 1997) and is followed by mitotic stimulation of intrinsic neuronal progenitors ensuring neuroepithelium regeneration (Kastner et al. 2000, Giannetti et al. 2004). In this model, the axotomized neurons that are committed to apoptosis were shown by in situ hybridization to enhance by 10 times their secretion of Leukemia Inhibitory Factor (LIF), which is necessary and sufficient to trigger lesion-induced proliferation of local progenitors (Bauer et al. 2003a). Therefore, in vivo apoptosis induction in these neurons triggers their secretion of a mitogenic factor that stimulates subsequent lesion-induced proliferation of neurogenic progenitors. These results have wider relevance than tissular homeostasis of olfactory epithelium, since LIF was also found to modulate SVZ-derived neurogenesis in adult rodents (Bauer & Patterson 2006, Bauer et al. 2007), and since an extensive correlation between apoptosis and neurogenesis has been documented in the brain (Biebl 2000, Bauer & Patterson 2005). Still wider, this novel phenomenon also has key relevance for cancer therapy and research, since an important proportion of novel anticancer therapeutic strategies rely upon apoptosis induction in tumoral cells (Fleischer et al. 2006, Fernandez-Luna 2007). By analogy with olfactory neurons (above, Bauer et al. 2003a), tumoral cells might conceivably secrete mitogenic factors at the same time as entering apoptosis; this issue should be checked in the various tumor types that are thus targeted.

V.2. Well-Documented Stem Cell Reactions to Neurological Pathologies

We will review shortly pathologies that are firmly established to modulate adult neurogenesis. The status of Alzheimer's disease with regard to neural stem cells remains highly controversial, although much effort has been directed to this major disease since a few years (see for instance Uchida et al 2007).

V.2.1. Demyelinating Lesions

Demyelinating lesions are dramatically hampering pathologies that result from oligodendrocytic glia dysfunction and are frequently worsened by the immune system. Remyelination by oligodendrocytes genesis from neural stem cells has lately begun to be investigated, maybe because of the low proportion of this glial lineage in stem cell progenies in vivo as well as in vitro (Butovski et al. 2006, Cayre et al. 2006, Ben-Hur 2007). Interestingly, exposure of adult mice to enriched environnement was shown to improve neural progenitor cell mobilization to demyelinated areas in vivo (Magalon et al. 2007). So one given lifestyle context can stimulate neurogenesis in relation with cognitive performances in one neurogenic structure (hippocampus, Kemperman et al 1997, see section IV) and repair potential in another (SVZ, Magalon et al. 2007). Finally, an experimental model of multiple sclerosis in adult rodent has been able to promote neurogenesis in the non-neurogenic spinal cord of adult rodent (Danilov et al. 2006). These recent findings bring more accurate hopes for repair-targeted recruitment of endogenous neural stem cells.

V.2.2. Proinflammatory Conditions

In vivo models of neuroinflammation induction, like experimental autoimmune encephalitis (EAE), were recently found to stimulate endogenous neurogenesis in hippocampus (Battista et al. 2006), in SVZ-derived rostral migratory stream (Cayre et al. 2006) and in spinal cord (Danilov et al. 2006). Interestingly, neuroinflammation is systematically associated with proliferation and activation of microglia, which is well-known as a source for numerous proinflammatory and plasticity-promoting messengers. Underlying mechanisms deserve to be addressed in the near future.

V.2.3. Ischemia and Vascular Infarct

In vivo ischemia models are currently among the most investigated lesion paradigms because of their direct relevance to cerebral vascular infarct. Transplanted human fetal NSC were first shown to be stimulated by ischemia (Kelly et al. 2004). More lately, endogenous neurogenesis was found to be recruited by ischemia in adult hippocampus (Yamashima et al. 2007).

V.3. Repair Reactions to Mechanical Lesions of the Spinal Cord

Post-sectional repair of spinal cord is an important challenge in human clinics. Complete trans-section of spinal cord failed to induce neurogenesis in adult mammals in vivo (Horky et al. 2006). However, milder paradigms of spinal injury, like partial section or compression of spinal cord, were found to recruit neurogenesis with reparative outcome. This was first achieved with transplantation of embryonic stem cells, which are much more multipotent and tumorigenesis-prone than adult neural stem cells but did promote functional recovery (McDonald et al. 2002), then with intravenously injected adult neural stem cells (Vroemen et al. 2004). More recently, by using transgenic labelling of nestin-expressing cells, the compression spinal cord injury model was found in mice to recruit endogenous neural stem cells of spinal cord, by enhanced proliferation in the ependymal canal lining, rapid progeny migration towards the injury site and genesis of new neurons in this place (Ke et al. 2006). The latter study proved that *endogenous neurogenesis potential of spinal cord can be unmasked in certain post-lesional conditions*, which confirms the hope for a clinical use of adult neurogenesis in repair of acute lesion-induced damages.

VI.4. Mismatch Between In Vivo and In Vitro Reactions of Neural Stem Cells to Lesion

When neurogenic niche tissue was sampled at the peak of in vivo injury-induced proliferation of local stem cells, neurosphere assay failed to reveal any change in neural stem cell dynamics. This has been observed in SVZ (Dizon et al. 2006) and in dorsal vagal complex (Charrier et al. 2006). This mismatch cannot be a technical bias due to the highly stimulating culture conditions of neurosphere assay, since some significant modulations of neurosphere-forming activity have been detected in the same conditions under in vitro hormonal challenge (Shingo et al. 2003, Kim et al. 2004). Rather, this indicates that in vivo recruitment of neural stem cells by injuries does not involve instructive, persistent alteration of proliferating cells.

However, at the molecular level, transcriptomal analyses has revealed ischemia-induced significant differences of gene expression pattern in neural stem/progenitor cells of SVZ in vitro as well as in vivo (Liu et al. 2007). The neurosphere assay thus deserves to be more extensively applied to well-characterized lesional models, since this should indicate how to select neural stem cells in the most suitable state for transplantation.

VI. Neural Stem Cell Regulation by Extracellular Signals

VI.1. Biological Responses of Neural Stem Cells to Regulatory Signals

The previous sections (IV, V) have shown that neural stem cell functions can be modulated by a large variety of adult life experiences and pathological challenges. In several cases, the demonstration of causal involvement of neural stem cells has allowed to reveal specific cellular mechanisms underlying behaviour, and thus represented a highly integrated approach "from cell to behaviour". Such modulation of precise cellular populations by integrated functions or events, at the whole organism level, must be mediated by intercellular communication systems of the organism, i.e. in adult the nervous, endocrine and/or immune systems. Intercellular communications all ultimately rely on chemically coded messages, from cells emitting chemical messengers to target cells that express specific receptors for these messengers and display specific biological reactions to messenger reception. The biological reactions of cells to extracellular signals belong to true intercellular communication, only if they represent specific receptor-triggered modulations that do not consist in mere consequence of signal metabolism (like glucose-induced increase in muscle cell energy production). In the case of neural stem cells, potential effectors of physiological regulators have been identified at the cellular and molecular levels.

At the cellular level, effector processes of physiological regulations include mitotic division, programmed cell death (the most frequent one being apoptosis), cell survival, cell differentiation into one of the three neural lineages, secretion of intercellular messengers (diffusible ones, membrane-bound or extracellular matrix components). The widely documented "proliferation" control actually encompasses two possible cellular functions, i.e. mitotic division rate and programmed cell death triggering. For instance, in vivo regulation of SVZ-olfactory bulb neurogenesis by late gestation involves exclusively mitotic rate enhancement by prolactin with unaltered apoptosis frequency (Shingo et al. 2003), whereas in vivo enhancement of SVZ-olfactory bulb neurogenesis by noradrenergic innervation relies upon apoptosis inhibition in olfactory bulb with no change of mitotic rate in SVZ (Bauer et al. 2003b). Overall, on the basis of all the physiological regulations of neural stem cell "proliferation" that have been studied at the cellular level, *the actual effector mechanism of neurogenesis regulation is either mitosis or apoptosis, but never both simultaneously.* Therefore, any case of neurogenesis modulation deserves quantitative assays of mitotic rate and of programmed cell death.

At the molecular level, intracellular effector mechanisms are commonly transcription factors, i.e. the inducible proteins that regulate gene expression at the DNA level in a combinatorial manner (Guillemot 2007). Physiologically induced *alterations of transcription factor combinations in neural stem cells will in turn determine genetically-coded cellular programs* like mitotic division, differentiation, secretion of growth factors, and expression of receptors for extracellular signals. Expression of receptors for bFGF (i.e., one of the two key mitogens for neural stem cells, cf section I) was thus shown to be up-regulated by exposure to the other mitogen EGF (Ciccolini & Svendssen 1998). Programmed cell deaths like apoptosis are the only cellular effectors of neural stem cell regulations which involve strictly post-transcriptional mechanisms (Krantic et al. 2005).

Identification of underlying effector mechanisms has prompted extensive investigation of competent intercellular signals, in vitro and at a lesser extent in vivo (see next section).

VI.2. Diversity of Extracellular Regulators

This topic has been extensively reviewed in recent publications (Hagg 2005, 2007, Grote & Hannan 2007, Ninkovic & Gotz 2007, Riquelme et al. 2007). We will summarize briefly the categories of extracellular messengers which have been shown to modulate neural stem cell function, in vitro and/or in vivo.

VI.2.1. Growth Factors

Growth factor are a homogeneous category of intercellular mediators, which have been defined by their capacity to modulate proliferation, cell death, cell survival, migration of immature cells, and/or lineage commitment and differentiation. This operational definition has matched with the common chemical nature of these factors (large monomeric peptides) and their common transduction mechanism (intracellular phosphorylation triggering at the receptor level, either by presence of a kinase site in the receptor molecule, or by cytoplasmic kinase activation via "adaptator proteins").

The two specific mitogens of neural stem cells are the epidermal- and basic-fibroblast-growth factors (see I). Whereas EGF primarily stimulated mitotic division of neural stem/progenitor cells, bFGF has slightly complementary role, in self-renewing and in commitment to differentiation, the latter looking like an "instructive" signal in the embryologist sense (Maric et al. 2007). It is tempting to relate this emerging complementarity of EGF and bFGF, to the cross-talk between their respective receptors that had been evidenced in vitro (Ciccolini & Svendssen 1998).

Several other growth factors were found to modulate neural stem cell function. VEGF mediates stimulatory impact of blood vessels on neurogenesis, by stimulating neural stem cell proliferation (Sun et al. 2006). Transforming growth factor-alpha stimulates neurogenic proliferation (Tropepe et al. 1997). The insulin-growth-factor-I stimulates proliferation of neural stem cells in vivo (Kalluri et al. 2007) and the pigment epithelium-derived factor (PEDF) stimulate their self-renewal (Ramirez-Castillejo et al. 2006).

Brain-derived neurotrophic factor (BDNF) was elegantly shown to stimulate neurogenesis in vivo even in brain structures that are devoid of any neurogenesis (Benraiss et al. 2001, Pencea et al. 2001); this growth factor actually stimulates differentiation and survival of neuronal progenies, which is correlated with inhibition of mitotic activity. Interestingly, a recent accumulation of data has revealed in the last years that BDNF mediates numerous processes of experience-dependent plasticity (Brahman & Messaoudi 2005, Soule et al. 2006); this growth factor is a good candidate for physiologically-triggered recruitments of neural stem cells that have been reviewed above (see section IV). Consistently, BDNF was demonstrated to mediate enhancement of hippocampal neurogenesis by chronic caloric restriction in adult rodents (Lee et al. 2002). Modulation of BDNF signalling has also been involved in recovery enhancement following experimental stroke in mouse (Nygren et al. 2006).

VI.2.2. Embryonic Inductors

Some of the chemical messengers that mediate embryonic inductions were found to modulate adult neural stem/progenitor cells: the one morphogenetic proteins (BMPs) that are expressed in mature neurons of olfactory epithelium and tonically repress neuronal progenitor proliferation (Shou et al.1999), Wnt-Frizzled (Malaterre et al. 2007), Delta-Notch (Campos et al. 2006, Basak & Taylor 2007).

VI.2.3. Neurotransmitters

The neurotransmitters of diffuse projection systems of the brain have been shown to modulate neurogenesis: serotonin, dysfunction of which causes depression and anxiety (Brezun & Daszuta 2000); norepinephrin, that displays deficits in early stages of neurodegenerative diseases (Bauer et al. 2003b); acetylcholine, i.e. the key mediator of memory formation in hippocampus and a contrast sharpener in cerebral cortex (Mechawar et al. 2005); dopamine, dysfunctions of which cause Parkinson's disease, addiction, schizophrenia (Coronas et al. 1997, 2004; Hoglinger et al. 2005). These mediators are specially relevant since diffuse projection systems are potent modulators of plasticity effectors in long-term adaptation contexts. Their functional effectors are diverse: dopamine stimulate mitotic activity of neural stem cells (Coronas et al. 1997, 2004; Höglinger et al. 2005), norepinephrine and acetylcholine increase the survival by inhibiting apoptosis of SVZ cell progenies in olfactory bulb (Bauer et al. 2003b, Mechawar et al. 2004).

The fast neurotransmitters glutamate and GABA stimulate adult hippocampal neurogenesis (Cameron et al. 1995, Tashiro et al. 2006, Jagasia et al. 2006, Platel et al. 2007, Earnheart et al. 2007), which has potential relevance to epilepsy. The gaseous neurotransmitter nitric oxide is a potent inhibitor of adult neurogenesis in hippocampus, SVZ and olfactory epithelium (Moreno-Lopez et al. 2004, Calof 2006). Endocannabinoids stimulate neurogenesis in hippocampus (Jiang et al. 2007).

Several neuropeptides modulate adult neurogenesis, in particular Neuropeptide Y (Horwell et al. 2003), PACAP (Mercer et al. 2004), opioid peptides (Abrous et al. 2005).

On a conceptual viewpoint, by these biological actions on adult neural stem cells and neurogenesis, *neurotransmitters exert a range of "growth-factor-like activity"* that fall in the same semantic category as their previously reviewed morphogenetic actions on neuronal differentiation (Weiss et al. 1998).

VI.2.4. Hormones

Some hormones exert direct modulatory actions on neural stem cells: stimulation by prolactin (Shingo et al. 2003, see above), estrogens (Suzuki et al. 2007).

Glucocorticoids exert direct inhibitory action on neural progenitors of adult hippocampus (Cameron et al. 1995, Kim et al. 2007) as on embryonic neural stem cells (Sundberg et al. 2006), which provides putative mechanism for the well-documented effects of stress on adult neurogenesis in hippocampus, i.e. the target of glucocorticoid feed-back on the hypothalamo-pituitary axis of stress response.

VI.2.5. Cytokines

Cytokines are immune messengers that mediate numerous aspects of inflammation impact on nervous system. Adult neurogenesis and neural stem cells are influenced by Leukemia Inhibitory Factor (LIF), interleukins 1β and 6 (reviewed in Bauer et al. 2007).

VI.2.6. Extracellular Matrix Components and Adhesiotopes

The polysialylated form of the neural cell adhesion molecule (PSA-NCAM), which is ubiquitous in embryonic nervous system and restricted to a few plasticity-prone structures in adulthood, is systematically associated with neural stem and progenitor cells (Seki 2002, 2003).

Determinant regulatory roles of the extracellular matrix are currently emerging, which are mediated by the matrix ligands of the cell surface receptors integrins. Potent effects of this type of communication were demonstrated in the SVZ-olfactory system, on both stem cell proliferation (Chenn et al. 2002) and progeny migration (Belvindrah et al. 2007). In cancer research indeed, extracellular matrix has been well established to exert decisive influence in tumor progression to malignancy and metastasis (Nelson & Bissell 2006). By analogy, extracellular matrix composition is likely to provide key elements of the niche and neurogenic properties of adult nerve tissues. Indirect matrix influence can also consist in capturing and concentrating competent growth factors in the extracellular vicinity of neural stem cells (Kerever et al. 2007). This pertinent track has a drastic importance for future applications of stem cells in repararative medicine; it deserves comparative analysis of molecular matrix composition among healthy neurogenic versus non-neurogenic tissues in laboratory animals, which can be easily performed nowadays at the transcriptomal level by DNA arrays, and at the protein level by proteomic approaches.

The major matrix component collagen, has been reported to stimulate neuronal commitment in vitro (Branvall et al. 2007); in vivo relevance of this phenomenon has not been determined, but it is clearly an interesting cue for practical progenitor transplantation.

Another pertinent cue resides in the endogenous regulators of extracellular matrix composition that were recently recognized to have determinant roles in oncogeny. The most documented is a category of cell-surface-bound enzymes that display specific and multifaceted adhesive and protease activities: the A-Disintegrin and Metalloprotease transmembrane proteins of the ADAM family (Yang et al. 2006). The metalloprotease family is also a good candidate to investigate, since it has already been related to numerous cases of lesion-induced axonal plasticity (Jourquin et al. 2005).

Neurogenesis-modulating roles of some glycosylating chains or epitopes are currently emerging, especially at the level of neurogenic niches (Von Holst et al. 2006).

The above listed regulating factors are emitted or diffused by the tissular microenvironment of neural stem cells, which provides *a molecular substratum to the "neurogenic niche" concept* (see section III). In addition, neural stem/progenitor cells can secrete themselves neurogenesis-modulating messengers, like growth factors (BDNF: ref), cytokines (Klassen et al. 2003) and extracellular ATP (Mishra et al. 2006).

VII. Experimental Grafting of Lineage-Labeled Stem Cells into Animal and Human Nervous Systems

It has been long believed that the damaged central nervous system does not regenerate upon injury (Cajal). Owing to Altman (1965) and Reynolds & Weiss (1992), there is an emerging hope for regeneration-based therapy of the damaged central nervous system (CNS) due to the progress of developmental biology and regenerative medicine including stem cell biology. This hope is the same for spine cord (SC) and peripheral nerve system (PNS). But neurological disorders are variable with stroke, degenerative diseases, injuries, therefore experimental grafting of lineage-lebeled stem cells are testing on different animal models of pathology. For in situ detection, whether 5-bromo-2-deoxyuridine (BrdU) is usually used to label these cells, it makes problem for chronic models. Lentivirus- assisted labelling of cells seem to be the best solution in that case (see section I.6). Recent progress in imaging improve possibilities, for example to test the ability of magnetic resonance imaging (MRI) in detecting transplanted NSCs, NSC line can be labeled in vitro with superparamagnetic iron oxide (SPIO) or PKH26, a fluorescent dye. Quantum dots (QDs) are a class of Fluorescent semiconductor nanocrystal which can be effectively used to label mammalian NSPCs in vivo and will be useful for studies of in vivo fate mapping, cellular migration, and NSPC differentiation during mammalian development.

VII.1. Animals

Mouse, rat, pig and non-human primate models have been established and may increase in importance as a phase before clinical trials. These studies on animals allow to define models for each pathology and form of cell therapy, cell types, combination, administration mode : intravenous (Chen J et al. 2001) or on injury site; timing : acute-subacute model (day 9 after injury) or chronic model (starting 28 days after injury). But it makes problems of extrapolation pertinence between the human pathology and the chosen model of lesion. To assess functional results it is necessary to use the same evaluation scale for the same animal. Significant studies on animal have led to human therapy.

VII.1.1. Animal CNS

Parkinson's disease (PD) is a chronic progressive neurodegenerative movement disorder characterized by a profound and selective loss of nigrostriatal dopaminergic neurons. Clinical manifestations of this complex disease include motor impairments involving resting tremor, bradykinesia, postural instability, gait difficulty and rigidity. There are at least three ways to develop animal models of PD (Hirsch 2006). One of these is to induce a nigrostriatal lesion. In mice, the most common is injection of 1-methyl-4-phenyl-1,2,3,6-tetrahydropyridine (MPTP) ; cell therapy studies using chronic mouse model of Parkinson's disease can reproduce progressive degeneration of dopamine (DA) cells with a small daily dose (4 mg/kg/day for 20 days). Many therapeutic strategies have focused on restoring striatal dopamine levels, including stem cells graft, direct gene transfer to striatal cells, using viral vectors that express specific dopamine biosynthetic enzymes. One of the problems is the optimal time course, cell preparation and surgical targets. Cell therapy needs systematic examinations of both potential benefits and side effects of dopamine neuron cell transplantation as shown recently in primate models of PD (Redmond et al. 2007).

Huntington's disease: Time of transplantation and cell preparation determine neural stem cell survival in a mouse model of Huntington's disease (Johann et al. 2007). Cell transplants were often studied into the striatum of quinolinic acid (QA)-lesioned rats. Possibilities to study human neural stem cell transplants are another aspect before clinical trials. The fact that Human neural stem cell transplants improved motor function in a rat model of Huntington's disease (McBride et al. 2004) was decisive.

Stroke (Bliss et al. 2007, Kelly et al. 2007), multiple sclerosis, Alzheimer's disease are also major challenges. Animal models are absolutely necessary to understand the pathology and to study cell-based therapy.

VII.1.2. Animal Spinal Cord Injuries (SCI)

SCI results in loss of oligodendrocytes demyelination of surviving axons and severe functional impairment. Spontaneous remyelination is limited in rat model of spinal cord contusion injury. It is why transplantation of exogenous cells is one approach to spinal cord repair that could potentially enhance the growth and myelination of endogenous axons. Three kinds of stem/progenitor cells have been used in cell therapy in animal models of SCI: embryonic stem cells, bone marrow mesenchymal stem cells, and neural stem cells. Cell-based therapies following spinal cord injury are based on three possibilities: (1) replacement of degenerated spinal cord parenchyma by an axon growth supporting scaffold; (2) remyelination of regenerating axons; and (3), local delivery of growth promoting molecules. It had been shown that embryonic stem cell-derived motor neurons implanted into the spinal cord could innervate muscle targets and improved functional recovery. Many models of SCI can be used, complete or partial transections, injury levels, compression injury. Motor function of rats' hind limbs have been observed and HE and X-gal immunocytochemical staining, in situ hybridization, and retrograde HRP tracing. It is necessary to perform several histological and functional methods.

To assess functional results, the Basso, Beattie and Bresnahan (BBB) locomotor rating scale have been usually used in rat models of acute motor neuron injury.

It appears that cell-based therapy and neuroprotective therapy must be joined up. For example, combined treatment of neural stem cells and neurotrophin-3 (NT-3) gene can facilitate functional recovery (Zhang et al. 2007). Another aspect is that animals require external stimulation (tail pinch) to elicit the behavior. Lastly, the timing of neural stem cell application therefore is critical to successful therapy in the injured spinal cord: rostral and caudal injections must be compared to direct injections into the site of injury.

VII.1.3. Animal PNS

Generally SCs have been transplanted to a peripheral nerve gap. Murakami et al. (2003) studied a 15-mm defect prepared in the sciatic nerve of mature rats, and the nerve was bridged with a silicone tube filled with neuronal progenitor cells (1 x 10(5)) embedded in collagen gel. The transplanted neuronal progenitor cells were labeled in advance with (BrdU). Generally, studies have been based on axonal regeneration and evaluated by the sciatic functional index (SFI) and by histological analysis. These results show that when NSCs are transplanted to a peripheral nerve defect they differentiate into supportive cells that contribute to the promotion of axonal regeneration. In fact, PNS microenvironment seems to exert an influence over labeled transplanted stem cells. Including adipose-derived stem cells which differentiate into a Schwann cell phenotype and promote neurite outgrowth in vitro (Kingham et al. 2007).

VII.2. Human

Stem cells have been used in human CNS for Parkinson's disease (1991), Hungtinton's disease (2000) and Batten's disease (2005). It makes problem of integration, differentiation, clinical improvment evaluation after graft by fonctional scale and post mortem analysis. Difficulties such as the low survival rate in vivo and tumor formation must be solved. At least, one must be able to distinguish between continued neurological declines as a result of the ongoing pathobiology versus a potential negative effect of the graft (Reier et al. 2004).

A better understanding of interactions between neural stem cells and the neurodegenerative environment is necessary. Cell-based therapy strategies can be classified into two approaches: (i) the replenishment of lost neural cells and influence and modulate endogenous repair (ii) the induction of axonal regeneration.

What are the signals controlling proliferation and differentiation of neural stem cells (see section VI) and can these be utilized to allow repair of the CNS? The ability to establish and maintain polarized excitatory synaptic contacts would be one of the basic requirements for intercellular communication and functional integration into existing neuronal networks (Lieubau et al. 2007)

VII.2.1.Human CNS (Ourednik 2007)

Parkinson's disease (PD) is a common neurological disorder of the brain which has as a part of its core pathology the progressive degeneration of the dopaminergic nigrostriatal pathway. Therefore, cell therapies that aim to restore this degenerated dopaminergic network represent a promising strategy in helping to cure PD. Together with the solid animal experimental data the clinical observations support the idea that neural transplantation can be developed into an effective therapy in Parkinson's disease. Surgeries involving transplantation of fetal dopamine (DA) neurons into the caudate-putamen of patients with Parkinson's disease (PD) have been performed in various clinical trials to examine a potential restoration of motor function (Lindvall 1991). The findings indicate that fetal DA neurons can survive in the human parkinsonian brain and produce therapeutically valuable functional effects. (Mendes et al. 2005) report the first post-mortem analysis of two patients with Parkinson's disease who received fetal midbrain transplants as a cell suspension in the striatum, and in one case also in the substantia nigra. Then its adoption in the clinic is discussed, including the ethical and practical problems with its use, the varied composition of VM tissue that is implanted with the graft. This may account for some of the problems seen in the clinical trials using this tissue, especially graft-induced dyskinesia. Finally other promising sources of tissue for PD cell therapy must be studied, including mesenchymal and embryonic stem cells, before concluding on what is the best approach to the cellular repair of the parkinsonian brain.

Stroke is the most common cause of disability in the United States, and one of the leading causes of mortality and disability in the world. Regenerative cell-based therapies can lead to functional recovery in stroke patients. Stem cells can differentiate into neural lineages to replace lost neurons. Moreover, they provide trophic support to tissue at risk in the penumbra surrounding the infarct area, enhance vasculogenesis, and help promote survival, migration, and differentiation of the endogenous precursor cells after stroke. Stem cells are highly migratory and seem to be attracted to areas of brain pathology such as ischemic regions. For maximal functional recovery, however, regenerative therapy may need to follow combinatorial approaches, which may include cell replacement, trophic support, protection from oxidative stress, and the neutralization of the growth-inhibitory components for endogenous neuronal stem cells (Okano et al. 2007).

For Hungtington disease, a recent study (Keene et al. 2007) with two autopsies and pathologic examination, histochemistry and immunohistochemistry, confirmed previous findings of neuronal differentiation and survival of transplanted fetal tissue from the ganglionic eminence and also demonstrated viability of neurons from fetal transplants in human neostriatum for more than 6 years. Despite prolonged survival, these grafts displayed poor integration within host striatum that is likely responsible for lack of clear clinical improvement in these patients.

VII.2.2. Human Spinal Cord

The aim is here to target such the direct injury and the demyelination process. Regrowth of axons is impeded by the loss of astrocytic pathways caused at the time of injury. Substitutive therapy by cell therapy and gene therapy must be joined up with neuroprotective therapie. Recent clinical trials respectful from the ethical aspect are interesting, which were based on bone marrow stem cells (Syková et al. 2006) through either autologous bone marrow cell transplantation or bone marrow stimulation with granulocyte macrophage-colony stimulating factor: Phase I/II clinical trial (Yoon et al. 2007).

At least, transplantation of olfactory ensheathing cells (Raisman et al. 2007) is another promising issue (see section VIII).

VII.2.3. Human PNS

Peripheral nerves possess the capacity of self-regeneration after traumatic injury when the gap is less than 3 cm. For short and medium gaps, transected peripheral nerves can be bridged by direct surgical coaptation of the two nerve stumps or by interposing autografts or biological (veins) or synthetic nerve conduits (NC). For long nerve gaps, find an alternative to nerve autograft remains a challenge.

The stepping stone concept, which associate cell-based therapy with extracellular matrix is still the aim. Multi-component gels of collagen-1, fibronectin, laminin and hyaluronic acid shown improve of neurite extension (Deister 2007). Hodde and coll (2007) conclude that a biologic scaffold can be prepared for human use and still retain significant bioactivity.

VIII. Available Sources of Stem Cells for Neural Repair

The suitability of various human stem cell sources for such treatments of both slowly progressing conditions, such as Parkinson's disease, Huntington's disease and multiple sclerosis, and injuries, such as stroke and spinal cord injury (Miller 2006).

1). Embryonic stem cells (ESC) many problems remain to be solved in order for ES cells to be used for clinical applications (Freed 2001)
2). Fœtal stem cells (FSC) are presently used in Parkinson's disease
3). The use of adult stem cells (ASC) in tissue regeneration appears to be a powerful research tool, due to the intrinsic characteristics of these cells, i.e., self-renewal and capacity for proliferation. They seem less tumorogenic than ESCs (Foroni et al 2007). These neural precursors are found in the subventricular zone and dentate gyrus of the adult brain and they have demonstrated capacity to become neurons or glial cells after transplantation, at least into neurogenesis-prone areas (see previous sections). Adult neural stem cells (NSCs) are potential sources for cell therapy of neurodegenerative diseases and for drug screening.

4). Despite their potential benefits, ethical and practical considerations limit the application of NSCs derived from human embryonic stem cells (ES) or adult brain tissue Neural stem cells from other origin could be more easier ; this requires to find a novel method for providing limitless numbers of neuronal precursors from a readily accessible autologous adult human source.

5). Olfactory ensheathing cells (OEs) are a non-myelinating glial cell type which is found exclusively in the olfactory nerve and peripheral layer of olfactory bulb of adult mammals, and displays some phenotypic features of both astrocytes and oligodendrocytes (Franssen et al. 2007). These cells were extensively shown to acquire unequalled capacities for myelination and axon regrowth when they are transplanted in vivo at the site of a central or peripheral pathway section in adult; therefore there is a hope with these cells for spinal cord injury (Raisman & Li 2007). Olfactory ensheathing cells, as all glial cell types, have kept ability to divide and can easily be purified and expanded in primary culture. OEs do not derive from known neural stem cells, neither from SVZ, neither from olfactory mucosa.

6). Ectomesenchymal stem cells (EMSCs) originate from the cranial neural crest are a potential source of neuronal and Schwann cells (SCs) of the peripheral nervous system (PNS). In particular, mesenchymal stem cells (MSCs) obtained from bone marrow (Hermann 2006) or peripheral blood can be easily isolated, cultivated, propagated and can be differentiated into several specialized cell types thanks to their plasticity. Among these MSCs can evolve into neural cell lineages. For example, MSCs transplantation exerted a positive influence upon the brain tissue reparation after stroke.

7). Human umbilical cord blood progenitors have the potential to become neural, like hematopoietic cells (Chen 2005)

8). HSC: Xu (2007) is suggesting that hMSCs retain the ability to differentiate into neuroectodermal cell types .

9). Skin-derived precursors (SKPs), a neural crest-like precursor that can be isolated and expanded from mammalian skin. Likewise, cells in the bulge region of the hair follicle have stem cell-like properties, are multipotent, and can effectively generate cells of neural lineage in vitro and in vivo (Mignone 2007).

10). Human somatic stem cells, isolated by minimally invasive periodontal surgery (Windera 2007) could be another source for cell therapy.

11). Finally, transduced cells (Golden 2007) should improve the possibilities of cell therapy.

References

Abrous DN, Koehl M, LeMoal M (2005) Adult neurogenesis: from precursors to network and physiology. *Physiol. Rev.* 85:523-69.

Agasse F, Roger M, Coronas V (2004) Neurogenic and intact or apoptotic non-neurogenic areas of adult brain release diffusible molecules that differentially modulate the development of subventricular zone cell cultures. *Eur. J. Neurosci.* 19(6):1459-68.

Altman J (1962) Are new neurons formed in the brains of adult Mammals? *Science* 135:1127-8.

Altman J (1969) Autoradiographic and histological studies of postnatal neurogenesis. IV. Cell proliferation and migration in the anterior forebrain, with special reference to persisting neurogenesis in the olfactory bulb. *J. Comp. Neurol.* 137(4):433-57.

Altman J, Das GD (1965) Autoradiographic and histological evidence of post-natal hippocampal neurogenesis in rats. *J. Comp. Neurol.* 124(3):319-35.

Altman J, Das GD (1967) Post-natal neurogenesis in the guinea-pig. *Nature* 214 1098-101.

Bariohay B, Lebrun B, Moyse E, Jean A. 2005. Brain-derived neurotrophic factor plays a role as an anorexigenic factor in the dorsal vagal complex. *Endocrinology.* 146(12):5612-20.

Barkho BZ, Song H, Aimone JB, Smrt RD, Kuwabara J, Nakashima K, Gage FH, Zhao X (2006) Identification of astrocyte-expressed factors that modulate neural stem/progenitor cell differentiation. *Stem Cells Dev.* 15(3):407-21.

Basak O, Taylor V (2007) Identification of self-replicating multipotent progenitors in the embryonic nervous system by high Notch activity and Hes5 expression.*Eur. J. Neurosci.* 25(4) :1006-22.

Battista D, Ferrari CC, Gage FH, Pitossi FJ (2006) Neurogenic niche modulation by activated microglia: transforming growth factor beta increases neurogenesis in the adult dentate gyrus. *Eur. J. Neurosci.* 23(1):83-93.

Bauer S, Patterson PH (2005) The cell cycle-apoptosis connection revisited in the adult brain. *J. Cell Biol.* 171(4):641-50.

Bauer S., Patterson PH (2006) Leukemia inhibitory factor promotes neural stem cell self-renewal in the adult brain. *J. Neurosci.* 26(46):12089-99.

Bauer S, Rasika S, Mauduit C, Raccurt M, Morel G, Jourdan F, Benahmed M, Moyse E, Patterson PH (2003a) Leukemia Inhibitory Factor induction is a key component of lesion-triggered renewal of olfactory sensory neurons in adult mouse. *J. Neurosci.* 23(5):1792-1803.

Bauer S, Moyse E, Jourdan F, Colpaert F, Martel JC, Marien M (2003b) Effects of the alpha2-adrenoceptor antagonist, dexefaroxan, on neurogenesis in the olfactory bulb of the adult rat in vivo : selective protection against apoptosis. Neuroscience 117:281-291.

Bauer S, Hay M, Amilhon B, Jean A, Moyse E (2005) *In vivo* neurogenesis in the dorsal vagal complex of adult rat brainstem. *Neuroscience* 130(1):75-90.

Bauer S, Kerr BJ, Patterson PH (2007) The neuropoietic cytokine family in development, plasticity, disease and injury. *Nat Rev Neurosci.* 8(3):221-32.

Belvindrah R, Hankel S, Walker J, Patton BL, Muller U (2007) Beta-1 integrins control the formation of cell chains in the adult rostral migratory stream. *J. Neurosci.* 27(10):2704-17.

Ben-Hur T (2007) Immunomodulation by neural stem cells. *J Neurol Sci.*, online.

Benraiss A, Chmielnicki E, Lerner K, Roh D, Goldman SA. 2001. Adenoviral brain-derived neurotrophic factor induces both neostriatal and olfactory neuronal recruitment from endogenous progenitor cells in the adult forebrain. *J. Neurosci.* 21(17):6718-31.

Biebl M, Cooper CM, Winkler J, Kuhn HG (2000) Analysis of neurogenesis and programmed cell death reveals a self-renewing capacity in the adult rat brain. *Neurosci Lett* 291, 17-20.

Bjornson CR, Rietze RL, Reynolds BA, Magli MC, Vescovi AL (1999) Turning brain into blood: a hematopoietic fate adopted by adult neural stem cells in vivo. *Science* 283:534-7.

Bliss T, Guzman R, Daadi M, Steinberg GK (2007) Cell transplantation therapy for stroke. *Stroke* 38:817.

Bonfanti L, Peretto P. (2007) Radial glia origin of the adult neural stem cells in the subventricular zone. *Prog. Neurobiol.* 83(1):24-36.

Bovetti S, Hsieh YC, Bovolin P, Perroteau I, Kazunori T, Puche AC (2007) Blood vessels form a scaffold for neuroblast migration in the adult olfactory bulb. *J. Neurosci.* 27(22):5976-80.

Bramham CR, Messaoudi E (2005) BDNF function in adult synaptic plasticity: the synaptic consolidation hypothesis. *Prog. Neurobiol.* 76(2):99-125.

Branvell K, Bergman K, Wallenquist U, Svahn S, Bowden T, Hilborn J, Forsberg-Nilsson K (2007) Enhanced neuronal differentiation in a three-dimensional collagen-hyaluronan matrix. *J. Neurosci. Res.* 85(10):2138-46.

Brezun JM, Daszuta A (2000) Serotonergic reinnervation reverses lesion-induced decreases in PSA-NCAM labeling and proliferation of hippocampal cells in adult rats. *Hippocampus* 10(1):37-46.

Bruel-Jungerman E, Rampon C, Laroche S (2007) Adult hippocampal neurogenesis, synaptic plasticity and memory : facts and hypotheses. *Rev. Neurosci.* 18(2):93-114.

Buffo A (2007) Fate determinant expression in the lesioned brain : Olig2 induction an dits implication for neural repair. *Neurodegener. Dis.* 4(4):328-32.

Butovski O, Landa G, Kunis G, Ziv Y, Avidan H, Greenberg N, Schwartz A, Smirnov I, Pollack A, Jung S, Schwartz M (2006) Induction and blockage of oligodendrogenesis by differently activated microglia in an animal model of multiple sclerosis. *J. Clin. Invest.* 116(4):905-15.

Cameron HA, McEwen BS, Gould E (1995) Regulation of adult neurogenesis by excitatory input and NMDA receptor activation in the dentate gyrus. *J. Neurosci.* 15:4687-92.

Campos LS, Decker L, Taylor V, Skarnes J (2006) Notch, Epithelial Growth Factor-receptor and β1-integrin pathways are coordinated in neural stem cells. *J. Biol. Chem.* 281(8):5300-9.

Carlen M, Meletis K, Barnabe-Heider F, Frisen J (2006) Genetic visualization of neurogenesis. *Exp. Cell Res.* 312(15):2851-9.

Carter LA, MacDonald JL, Roskams AJ (2004) Olfactory horizontal basal cells demonstrate a conserved multipotent progenitor phenotype. *J. Neurosci.* 24(25) 5670-83.

Cayre M, Bancila M, Virard I, Borges A, Durbec P (2006) Migrating and myelinating potential of subventricular zone neural progenitor cells in white matter tracts of the adult rodent brain. *Mol. Cell. Neurosci.* 31(4):748-58.

Charrier C, Coronas V, Fombonne J, Roger M, Jean A, Krantic S, Moyse E (2006) Characterization of neural stem cells in the dorsal vagal complex of adult rat brainstem by in vivo proliferation labelling and in vitro neurosphere assay. *Neuroscience* 138(1):5-16.

Chen J, Sanberg PR, Li Y, Wang L, Lu M, Willing AE, Sanchez-Ramos J, Chopp M (2001) Intravenous administration of human umbilical cord blood reduces behavioral deficits after stroke in rats. *Stroke* 32:2682.

Chen J, Magavi SS, Macklis JD (2004) Neurogenesis of corticospinal motor neurons extending spinal projections in adult mice. *Proc. Natl. Acad. Sci. U S A.* 101(46) 16357-62.

Chen N, Hudson JE, Walczak P, Misiuta I, Garbuzova-Davis S, Jiang L, Sanchez-Ramos J, Sanberg PR, Zigova T, Willing AE (2005) *Human umbilical cord blood progenitors: the potential of these hematopoietic cells to become neural stem cells.* 23(10):1560-1570.

Chenn A, Walsh CA (2002) Regulation of cerebral cortical size by control of cell cycle exit in neural precursors. *Science* 297, 365-369.

Ciccolini F, Svendsen CN (1998) Fibroblast growth factor 2 (FGF-2) promotes acquisition of epidermal growth factor (EGF) responsiveness in mouse striatal precursor cells: identification of neural precursors responding to both EGF and FGF-2. *J. Neurosci.* 18, 7869-80.

Conover JC, Notti RQ (2007) The neural stem cell niche. Cell Tissue Res., online.

Consiglio A, Gritti A, Dolcetta D, Foillenzi A, Bordignon C, Gage FH, Vescovi AL, Naldini L (2004) Robust in vivo gene transfer into adult mammalian neural stem cells by lentiviral vectors. Pr*oc. Natl. Acad. Sci. U.S.A.* 101(41):14835-14840.

Coronas V, Bantubungi K, Fombonne J, Krantic S, Schiffmann SN, Roger M (2004) Dopamine D3 receptor stimulation promotes the proliferation of cells derived from the post-natal subventricular zone. *J. Neurochem.* 91:1292-1301

Corti S, Nizzardo M, Nardini M, Donadoni C, Locatelli F, Papadimitrion D, Salami S, Del Bo R, Ghezzi S, Strazza S, Bresolin N, Comi GP (2007) Isolation and characterization of murine neural stem / progenitor cells based on prominin-1 expression. *Exp. Neurol.* 205:547-62.

Danilov AI, Covacu R, Moe MC, Langmoen IA, Johansson CB, Olsson R, Brundin L (2006) Neurogenesis in the adult spinal cord in an experimental model of multiple sclerosis. *Eur. J. Neurosci.* 23(2):394-400.

Deister C, Aljabari S, Schmidt CE (2007) Effects of collagen 1, fibronectin, laminin and hyaluronic acid concentration in multi-component gels on neurite extension. *J. Biomater. Sci. Polym. Ed.* 18(8):983-97.

Dhar S, Yoon ES, Kachgal S, Evans GR (2007) Long-term maintenance of neuronally differentiated human adipose tissue-derived stem cells. Tissue Eng. Online.

Dizon ML, Shin L, Sundholm-Peters NL, Kang E, Szele FG (2006) Subventricular zone cells remain stable in vitro after brain injury. *Neuroscience* 142(3):717-25.

Doetsch F, Garcia-Verdugo JM, Alvarez-Buylla A (1997) Cellular composition and three-dimensional organization of the subventricular germinal zone in the adult mammalian brain. *J. Neurosci.* 17:5046-61.

Doetsch F, Caille I, Lim DA, Garcia-Verdugo JM, Alvarez-Buylla A (1999) Subventricular zone astrocytes are neural stem cells in the adult mammalian brain. *Cell* 97:703-16.

Doetsch F, Verdugo JM, Caille I, Alvarez-Buylla A, Chao MV, Casaccia-Bonnefil P (2002) Lack of the cell-cycle inhibitor p27Kip1 results in selective increase of transit-amplifying cells for adult neurogenesis. *J. Neurosci.* 22:2255-64.

Earnheart JC, Schweitzer C, Crestani F, Iwasato T, Itohara S, Mohler H, Luscher B (2007) GABAergic control of adult hippocampal neurogenesis in relation to behavior indicative of trait anxiety and depression states. *J. Neurosci.* 27(14):3845-54.

Eisch AJ, Harburg GC (2006) Opiates, psychostimulants, and adult hippocampal neurogenesis: Insights for addiction and stem cell biology. *Hippocampus* 16(3):271-86.

Emsley JG, Mitchell BD, Kempermann G and Macklis JD (2005) Adult neurogenesis and repair of the adult CNS with neural progenitors, precursors, and stem cells. *Prog. Neurobiol.* 75:321-341.

Emsley JG, Macklis JD (2006) Astroglial heterogeneity closely reflects the neuronal-defined anatomy of the adult murine CNS. *Neuro Glia Biol.* 2(3):175-86.

Estrada-Mondaca S, Carreon-Rodriguez A, Belkind-Gerson J (2007) Biology of the adult enteric neural stem cell. *Dev. Dyn.* 236(1):20-32.

Ferguson KL, Callaghan SM, O'Hare MJ, Park DS, Slack RS (2000) The Rb/CDK4/6 signalling pathway is critical in neural precursor cell cycle regulation. *J. Biol. Chem.* 275(43):33593-600.

Fernandes KJ, Kobayashi NR, Gallagher CJ, Barnabe-Heider F, Aumont A, Kaplan DR, Miller FD (2006) Analysis of the neurogenic potential of multipotent skin-derived precursors. *Exp. Neurol.* 201(1):32-48.

Fernandez-Luna JL (2007) Apoptosis regulators as targets for cancer therapy. *Clin. Transl. Oncol.* 9(9):555-62.

Fleischer A, Ghadiri A, Dessaug F, Duhamel M, Rebollo MP, Alvarez-Franco F, Rebollo A (2006) Modulating apoptosis as a target for effective therapy. *Mol. Immunol.* 43(8):1065-79.

Foroni C, Galli R, Cipelletti B, Caumo A, Alberti S, Fiocco R, Vescovi A (2007) Resilience to transformation and inherent genetic and functional stability of adult neural stem cells ex vivo. *Cancer Res.* 67(8):3725-33.

Franssen EH, de Bree FM, Verhaagen J (2007) Olfactory ensheathing glia: Their contribution to primary olfactory nervous system regeneration and their regenerative potential following transplantation into the injured spinal cord. Brain Res. Rev. online.

Freed CR, Greene P, Breeze R, Yanntsai W, Dumouchel W, Kao R, Dillon S, Winfield H, Culver S, Trojanovski JQ, Eidelberg D, Fahn S (2001) Transplantation of embryonic dopamine neurons for severe parkinson's disease. *N. Engl. J. Med.* 344(10):710-9.

Garcia ADR, Doan NB, Imura T, Bush TG, Sofroniew MV (2004) GFAP-expressing progenitors are the principal source of constitutive neurogenesis in adult mouse forebrain. *Nat. Neurosci.* 7(11):1233-41.

Giannetti N, Moyse E, Ducray A, Bondier JR, Jourdan F, Kastner A (2004) Accumulation of Ym1/2 protein in the mouse olfactory epithelium during regeneration and aging. *Neuroscience* 123(4):907-917.

Golden KL, Pearse DD, Blits B, Garg MS, Oudega M, Wood PM, Bunge MB (2007) Transduced Schwann cells promote axon growth and myelination after spinal cord injury. *Exp Neurol.* 207(2):203-17.

Grote HE, Hannan AJ (2007) Regulators of adult neurogenesis in the healthy and diseased brain. *Clin. Exp. Pharmacol. Physiol.* 34(5-6):533-45.

Guillemot F (2007) Spatial and temporal specification of neural fates by transcription factor codes. *Development* 134:3771-80.

Hagg T (2005) Molecular regulation of adult CNS neurogenesis: an integrated view. *Trends Neurosci* 28(11):589-595.

Hagg T (2007) Endogenous regulators of adult CNS neurogenesis. *Curr. Pharm. Des.* 13(18):1829-40.

Hall PE, Lathia JD, Miller NG, Caldwell MA, Ffrench-Constant C (2006) Integrins are markers of human neural stem cells. *Stem Cells* 24(9):2078-84.

Heine VM, Maslam S, Joels M, Lucassen PJ (2004) Prominent decline of newborn cell proliferation, differentiation and apoptosis in the aging dentate gyrus, in absence of an age-related hypothalamus-pituitary-adrenal axis activation. *Neurobiol. Aging* 25(3):361-75.

Heine VM, Zareno J, Maslam S, Joels M, Lucassen PJ (2005) Chronic stress reduces cell proliferation in the adult dentate gyrus near the vasculature and VEGF and Flk-1 protein expression. *Eur. J. Neurosci.* 21(5):1304-14.

Hermann A, Liebau S, Gastl R, Fickert S, Habisch HJ, Fiedler J, Schwarz J, Brenner R, Storch A (2006) Comparative analysis of neuroectodermal differentiation capacity of human bone marrow stromal cells using various conversion protocols. *J. Neurosci. Res.* 83(8):1502-14.

Hirsch EC (2006) How to judge animal models of Parkinson's disease in terms of neuroprotection. *J. Neural Transm. Suppl.* 2006(70):255-60.

Hitoshi S, Maruta N, Higashi M, Kumar A, Kato N, Ikenaka K (2007) Antidepressant drugs reverse the loss of adult neural stem cells following chronic stress. *J. Neurosci.* Res. Online.

Hodde J, Janis A, Hiles M (2007) Effects of sterilization on an extracellular matrix scaffold: part II. Bioactivity and matrix interaction. *J. Mater. Sci. Mater. Med.* 18(4):545-50.

Höglinger GU, Rizk P, Muriel MP, Duyckaerts C, Oertel WH, Caille I, Hirsch EC (2004) Dopamine depletion impairs precursor cell proliferation in Parkinson disease. *Nat. Neurosci.* 7:726-35.

Horky LL, Galimi F, Gage FH, Horner PJ (2006) Fate of endogenous stem/progenitor cells following spinal cord injury. *J. Comp. Neurol.* 498(4):525-38.

Howell OW, Scharfman HE, Herzog H, Sundstrom LE, Beck-Sickinger A, Gray WP (2003) Neuropeptide Y is neuroproliferative for post-natal hippocampal precursor cells. *J. Neurochem.* 86, 646-659.

Hwang IK, Yoo KY, Jung BK, Cho JH, Kim DH, Kang TC, Kwon YG, Kim YS, Won MH (2006) Correlations between neuronal loss, decrease of memory, and decrease expression of brain-derived neurotrophic factor in the gerbil hippocampus during normal aging. *Exp. Neurol.* 201(1):75-83.

Ihrie RA, Alvarez-Buylla A (2007) Cells in the astroglial lineage are neural stem cells. Cell Tissue Res., online.

Jagasia R, Song H, Gage FH, Lie DC (2006) New regulators on adult neurogenesis and their potential role for repair. *Trends Mol. Med.* 12(9):400-5.

Jiang W, Zhang Y, Xiao L, Van Cleemput J, Ji SP, Bai G, Zhang X (2007) Cannabinoids promote embryonic and adult hippocampus neurogenesis and produce anxiolytic- and antidepressant-like effects. *J. Clin. Invest. Doi:*10.1172/JCI25509.

Johann V, Schiefer J, Sass C, Mey J, Brook G, Krüttgen A, Schlangen C, Bernreuther C, Schachner M, Dihné M, Kosinski CM (2007) Time of transplantation and cell preparation determine neural stem cell survival in a mouse model of Huntington's disease. *Exp. Brain Res.* 177(4):458-70.

Johansson CB, Momma S, Clarke DL, Risling M, Lendahl U, Frisen J (1999) Identification of a neural stem cell in the adult mammalian central nervous system. *Cell* 96(1):25-34.

Jourquin J, Trembaly E, Bernard A, Charton G, Chaillan FA, Marchetti E, Roman FS, Soloway PD, Dive V, Yiotakis A, Khrestchatisky M, Rivera S (2005) Tissue inhibitor of metalloproteinases-1 (TIMP-1) modulates neuronal death, axonal plasticity, and learning and memory. *Eur. J. Neurosci.* 22(10):2569-78.

Kalluri HSG, Vemuganti R, Dempsey RJ (2007) Mechanism of insulin-growth-factor-I-mediated proliferation of adult neural progenitor cells: role of Akt. *Eur. J. Neurosci.* 25:1041-8.

Karussis D, Kassis I (2007) Use of stem cells for the treatment of multiple sclerosis. *Expert Rev. Neurother.* 7(9):1189-201.

Kastner A, Moyse E, Bauer S, Jourdan F, Brun G (2000) Unusual regulation of cyclin-dependent kinases cdk2 and cdk4 during in vivo mitotic stimulation of olfactory neuron progenitors in adult mouse. *J. Neurochem.* 74(6): 2343-2349.

Ke Y, Chi L, Xu R, Luo C, Gozal D, Liu R (2006) Early response of endogenous adult neural progenitor cells to acute spinal cord injury in mice. *Stem Cells* 24:1011-19.

Kee N, Teixeira CM, Wang AH, Frankland PW (2007) Preferential incorporation of adult-generated granule cells into spatial memory networks in the dentate gyrus. *Nat. Neurosci.* 10(3):355-62.

Kelly S, Bliss TM, Shah AK, Sun GH, Ma M, Foo WC, Masel J, Yenari MA, Weissman LL, Uchida N, Palmer T, Steinberg GK (2004) .Transplanted human fetal neural stem cells survive, migrate, and differentiate in ischemic rat cerebral cortex. *Proc. Natl. Acad. Sci. U.S.A.* 101(32):11839-44.

Kemperman G (2006) Adult neurogenesis. Oxford University Press.

Kempermann G, Kuhn HG, Gage FH (1997) More hippocampal neurons in adult mice living in an enriched environment. *Nature* 386:493-5.

Kempermann G; Kuhn HG, Gage FH (1998) Experience-induced neurogenesis in the senescent dentate gyrus. *J. Neurosci.* 18:3206-12.

Kerever A, Schnack J, Vellinga D, Ichikawa N, Moon C, Arikawa-Hirasawa E, Efird JT, Mercier F (2007) Novel extracellular matrix structures in the neural stem cell niche capture the neurogenic factor fibroblast growth factor-2 from the extracellular milieu. *Stem Cells* 25(9):2146-57.

Kim JB, Ju JY, Kim JH, Kim TY, Yang BH, Lee YS, Son H (2004) Dexamethasone inhibits proliferation of adult hippocampal neurogenesis in vivo and in vitro. *Brain Res.* 1027(1-2):1-10.

Kingham PJ, Kalbermatten DF, Mahay D, Armstrong SJ, Wiberg M, Terenghi G (2007) Adipose-derived stem cells differentiate into a Schwann cell phenotype and promote neurite outgrowth in vitro. *Exp. Neurol.* 207(2):267-74.

Klassen HJ, Imfeld KL, Kirov II, Tai L, Gage FH, Young MJ, Berman MA (2003) Expression of cytokines by multipotent neural progenitor cells. *Cytokine* 22:101-6.

Klassen H, Kiilgaard JF, Zahir T, Ziaeian B, Kirov II, Schefig E, Warfvinge K, Young MJ (2007) Progenitor cells from the porcine neural retina express photoreceptor markers after transplantation to the subretinal space of allorecipients. *Stem Cells* 25(5):1222-30.

Kokoeva MV, Yin H, Flier JS (2005) Neurogenesis in the hypothalamus of adult mice: potential role in energy balance. *Science* 310:679-83.

Kowalczyk A, Filipowski RK, Rylski M, Wilczynski GM, Konopacki FA, Jaworski J, Ciemeryck MA, Sicinski P, Kaczmarek L (2004) The critical role of cyclin D2 in adult neurogenesis. *J. Cell Biol.* 167(2):209-13.

Krantic S, Reix S, Mechawar N, Quirion R (2005) Molecular basis of programmed cell death involved in neurodegeneration. *Trends Neurosci.* 28(12):670-6.

Lafon-Cazal M, Adjali O, Galeotti N, Poncet J, Jouin P, Homburger V, Bockaert J, Marin P (2003) Proteomic analysis of astrocytic secretion in the mouse. Comparison with the cerebrospinal fluid proteome. *J. Biol. Chem.* 278:24438-48.

Lauder (1993) Morphogenetic actions of neurotransmitters. Trends Neurosci.

Laywell ED, Steindler DA, Silver DJ (2007) Astrocytic stem cells in the adult brain. N*eurosurg. Clin. N. Am.* 18(1):21-30.

Lee J, Duan W, Mattson MP (2002) Evidence that brain-derived neurotrophic factor is required for basal neurogenesis and mediates, in part, the enhancement of neurogenesis by dietary restriction in the hippocampus of adult mice. *J. Neurochem.* 82:1367-75.

Liebau S, Vaida B, Storch A, Boeckers TM (2007) Maturation of synaptic contacts in differentiating neural stem cells. *Stem Cells.* 25(7):1720-9.

Lim DA, Huang YC, Alvarez-Buylla A (2007) The adult neural stem cell niche: lessons for future neural cell replacement strategies. *Neurosurg. Clin. N. Am.* 18(1) 81-92.

Lindvall O (1991) Transplants in Parkinson's disease. *Eur. Neurol.* 31 Suppl 1:17-27.

Liu XS, Zhang ZG, Zhang RL, Gregg SR, Meng H, Chopp M (2007) Comparison of in vivo and in vitro gene expression profiles in subventricular zone neural progenitor cells from the adult mouse after middle cerebral artery occlusion. *Neuroscience* 146(3) 1053-61.

Lledo PM, Alonso M, Grubb MS (2006) Adult neurogenesis and functional plasticity in neuronal circuits. *Nat. Rev. Neurosci.* 7(3):179-193.

Lledo PM, Lazarini F (2007) Neuronal replacement in microcircuits of the adult olfactory system. *C. R. Biol.* 330(6-7):510-20.

Ma L, Sun B, Hood L, Tian Q (2007) Molecular profiling of stem cells. *Clin. Chim. Acta* 378(1-2):24-32.

Magalon K, Cantarella C, Monti G, Cayre M, Durbec P (2007) Enriched environment promotes adult neural progenitor cell mobilization in mouse demyelination models. *Eur. J. Neurosci.* 25(3) :761-71.

Magavi SS, Leavitt BR, Macklis JD (2000) Induction of neurogenesis in the neocortex of adult mice. *Nature* 405:951-5.

Malaterre J, Ramsay RG, Mantamadiotis T (2007) Wnt-Frizzled signalling and the many paths to neural development and adult brain homeostasis. *Front. Biosci.* 12 492-505.

Mandairon N, Jourdan F, Didier A (2003) Deprivation of sensory inputs to the olfactory bulb up-regulates cell death and proliferation in the subventricular zone of adult mice. *Neuroscience* 119(2):505-16.

Mandairon N, Sacquet J, Jourdan F, Didier A (2004) Long-term fate and distribution of newborn cells in the adult mouse olfactory bulb: Influences of olfactory deprivation. *Neuroscience* 141(1):443-51.

Mandairon N, Sacquet J, Garcia S, Ravel N, Jourdan F, Didier A (2007) Neurogenic correlates of an olfactory discrimination task in the adult olfactory bulb. *Eur. J. Neurosci.* 24(12):3578-88.

Mangeot PE, Duperier K, Negre D, Basson B, Rigal D, Cosset FL, Darlix JL (2002) High levels of transduction of human dendritic cells with optimized SIV vectors. *Mol. Ther.* 5(3):283-90.

Maric D, Fiorio Pla A, Chang YH, Barker JL (2007) Self-renewing and differentiating properties of cortical neural stem cells are selectively regulated by basic fibroblast growth factor (FGF) signalling via specific FGF receptors. *J. Neurosci.* 27(8):1836-52.

Markakis EA, Palmer TD, Randolph-Moore L, Rakic P, Gage FH (2004) Novel neuronal phenotypes from neural progenitor cells. *J. Neurosci.* 24(12):2886-97.

Marshall GP 2nd, Reynolds BA, Laywell ED (2007) Using the neurosphere assay to quantify neural stem cells in vivo. *Curr Pharm Biotechnol.* 8(3):141-5.

Martin B, Mattson MP, Maudsley S (2006) Caloric restriction and intermittent fasting: two potential diets for successful brain aging. *Ageing Research Reviews* 5:332-53.

Maslov AY, Barone TA, Plukett RJ, Pruitt SC (2004) Neural stem cell detection, characterization, and age-related changes in the subventricular zone of mice. *J. Neurosci.* 24(7):1726-33.

Mc Kay R (1997) Stem cells in the central nervous system. *Science* 276:66-71.

Mechawar N, Saghatelyan A, Grailhe R, Scoriels L, Gheusi G, Gabellec MM, Lledo PM, Changeux P (2004) Nicotine receptors regulate the survival of newborn neurons in the adult olfactory bulb. *Proc. Natl. Acad. Sci. U.S.A.* 101(26):9822-6.

Medrano S, Burns-Cusato M, Atienza MB, Rahimi D, Scrable H (2007) Regenerative capacity of neural precursors in the adult mammalian brain is under the control of p53. Neurobiol. Aging, doi:10.1016/j.neurobiolaging.2007.07.016.

Mendez I, Sanchez-Pernaute R, Cooper O, Viñuela A, Ferrari D, Björklund L, Dagher A, Isacson O (2005) Cell type analysis of functional fetal dopamine cell suspension transplants in the striatum and substantia nigra of patients with Parkinson's disease. *Brain* 128(7):1478-9.

Mercer A, Ronnholm H, Holmberg J, Lundh H, Heidrich J, Zachrisson O, Ossoinak A, Frisen J, Patrone C (2004) PACAP promotes neural stem cell proliferation in adult mouse brain. *J. Neurosci. Res.* 76:205-215.

Merkle FT, Mirzadeh Z, Alvarez-Buylla A (2007) Mosaic organization of neural stem cells in the adult rat brain. *Science* 317:381-4.

Mezey E, Chandross KJ, Harta G, Maki RA, McKercher SR (2000) Turning blood into brain: cells bearing neuronal antigens generated in vivo from bone marrow. *Science* 290:1779-82.

Michel D, D, Moyse E, Brun G, Jourdan F (1994) Induction of apoptosis in mouse olfactory epithelium by synaptic target ablation. *Neuroreport* 5:1329-32.

Michel D, Moyse E, Trembleau A, Jourdan F, Brun G (1997) Clusterin/ApoJ expression is associated with neuronal apoptosis in the olfactory mucosa of the adult mouse. *J. Cell Sci.* 110:1635-43.

Mignone JL, Roig-Lopez JL, Fedtsova N, Schones DE, Manganas LN, Maletic-Savatic M, Keyes WM, Mills AA, Gleiberman A, Zhang MQ, Enikolopov G (2007) Neural potential of a stem cell population in the hair follicle. *Cell Cycle*.6(17):2161-70.

Miller RH (2006) The promise of stem cells for neural repair. *Brain Res.* 1091(1) 258-64.

Mishra SK, Braun N, Shutka V, Fullgrabe M, Schomerus C, Korf HW, Gachet C, Ikehara Y, Sevigny J, Robson SC, Zimmermann H (2006) Extracellular nucleotide signalling in adult neural stem cells: synergism with growth factor-mediated cellular proliferation. *Development* 133:675-84.

Molofsky AV, Slutsky SG, Joseph NM, He S, Pardal R, Krishnamurthy J, Sharpless NE, Morrison SJ (2006). Increasing p16INK4a expression decreases forebrain progenitors and neurogenesis during ageing. *Nature* 443:448-52.

Moreno-Lopez B, Romero-Grimaldi C, Noval J, Murillo-Carretero M, Matarredona E, Estrada C (2004) Nitric oxide is a physiological inhibitor of neurogeneisi in the adult mouse subventricular zone and olfactory bulb. *J. Neurosci.* 24, 85-95.

Moyse E, Michel D (1997) Analysis of apoptosis-associated DNA fragmentation in vivo during neurodegeneration of the peripheral olfactory system in adult mammals. Neuromethods, vol. 29, ed. J. Poirier, pp. 135-161.

Murakami T, Fujimoto Y, Yasunaga Y, Ishida O, Tanaka N, Ikuta Y, Ochi M (2003) Transplanted neuronal progenitor cells in a peripheral nerve gap promote nerve repair. *Brain Res.* 974(1-2):17-24.

Murrell W, Feron F, Wetzig A, Cameron N, Splatt K, Bellette B, Bianco J, Perry C, Lee G, Mackay-Sim A (2005) Multipotent stem cells from olfactory mucosa. *Dev. Dyn.* 233(2):496-515.

Negre D, Duisit G, Mangeot PE, Moullier P, Darlix JL, Cosset FL (2002) Lentiviral vectors derived from simian immunodeficiency virus. Curr. Top. *Microbiol. Immunol.* 261:53-74.

Nelson CM, Bissell MJ (2006) Of extracellular matrix, scaffolds and signalling: tissue architecture regulates development, homeostasis and cancer. *Annu. Rev. Cell Dev. Biol.* 22:287-309.

Nern C, Momma S (2006) The realized niche of adult neural stem cells. *Stem Cell Rev.* 2(3):233-40.

Newton SS, Duman RS (2007) Neurogenic actions of atypical antipsychotic drugs and therapeutic implications. *CNS Drugs* 21(9):715-25.

Ninkovic J, Götz M (2007) Signaling in adult neurogenesis : from stem cell niche to neuronal networks. *Curr. Opin. Neurobiol.* 17(3):338-44.

Nurse P (2000) A long twentieth century of the cell cycle and beyond. *Cell* 100(1) 71-8.

Nurse P (2002) Cyclin dependent kinases and cell cycle control (nobel lecture). *Chembiochem*. 3(7):596-603.

Nygren J, Kokaia M, Wieloch T (2006) Decreased expression of brain-derived neurotrophic factor in BDNF +/- mice is associated with enhanced recovery of motor performance and increased neuroblast number following experimental stroke. *J. Neurosci. Res*. 84:626-31.

Okano H, Sakaguchi M, Ohki K, Suzuki N, Sawamoto K (2007) Regeneration of the central nervous system using endogenous repair mechanisms. *J Neurochem*. 102(5):1459-65.

Orlic D, Kajstura J, Chimenti S, Jakoniuk I, Anderson SM, Li B, Pickel J, McKay R, Nadal-Ginard B, Bodine DM, Leri A, Anversa P (2001) Bone marrow cells regenerate infarcted myocardium. *Nature* 410:701–705.

Ourednik V, Ourednik J (2007) Plasticity of the central nervous system and formation of "auxiliary niches" after stem cell grafting: an essay. *Cell Transplant*. 16(3):263-71.

Paizanis E, Hamon M, Lanfumey L (2007) Hippocampal neurogenesis, depressive disorders and antidepressant therapy. *Neural Plast*. 73754.

Parker MA, Anderson JK, Corliss DA, Abraria VE, Sidman RL, Park KI, Teng YD, Cotanche DA, Snyder EY (2005) Expression profile of an operationally-defined neural stem cell clone. *Exp. Neurol*. 194(2):320-32.

Pecchi E, Dallaporta M, Charrier C, Pio J, Jean A, Moyse E, Troadec JD (2007). Glial fibrillary acidic protein (GFAP)-expressing cells exhibiting a radial morphology are present in the nucleus of the solitary tract of adult rat. *J. Comp. Neurol*. 501(3):353-368.

Pencea V, Bingaman KD, Wiegand SJ, Luskin MB (2001) Infusion of brain-derived neurotrophic factor into the lateral ventricle of the adult rat leads to new neurons in the parenchyma of the striatum, septum, thalamus and hypothalamus. *J. Neurosci*. 21(17):6706-17.

Pereira AC, Huddleston DE, Brickman AM, Sosunov AA, Hen R, McKhann GM, Sloan R, Gage FH, Brown TR, Small SA (2007) An in vivo correlate of exercise-induced neurogenesis in the adult dentate gyrus. *Proc. Natl. Acad. Sci. USA* 104(13):5638-43.

Petreanu L, Alvarez-Buylla A (2002) Maturation and death of adult-born olfactory bulb granule neurons: role of olfaction. *J. Neurosci.* 22(14):6106-13.

Platel JC, Lacar B, Bordey A (2007) GABA and glutamate signaling: homeostatic control of adult forebrain neurogenesis. *J. Mol. Histol*. 38(4):303-11.

Poppleton H, Gilbertson RJ (2007) Stem cells of ependymoma. *Br. J. Cancer* 96(1) 6-10.

Quinones-Hinojosa A, Sanai A, Gonzalez-Perez O, Garcia-Verdugo JM (2007) The human brain subventricular zone: stem cells in this niche and its organization. *Neurosurg. Clin. N. Am*. 18(1):15-20.

Raisman G, Li Y (2007) Repair of neural pathways by olfactory ensheathing cells. *Nat. Rev. Neurosci.* 8(4):312-9.

Ramirez-Castillejo C, Sanchez-Sanchez F, Andreu-Agullo C, Ferron SR, Aroca-Aguilar JD, Sanchez P, Mira H, Escribans J, Farinas I (2006) Pigment epithelium-derived factor is a niche signal for neural stem cell renewal. *Nat. Neurosci.* 9(3):331-9.

Raponi E, Agenes F, Delphin C, Assard N, Baudier J, Legraverend C, Deloulme JC (2007) S100B expression defines a state in which GFAP-expressing cells lose their neural stem cell potential and acquire a more mature developmental stage. *Glia* 55(2):165-77.

Rasika S, Nottebohm F, Alvarez-Buylla A (1994) Testosterone increases the recruitment and/or survival of new high vocal center neurons in adult female canaries. *Proc. Natl. Acad. Sci. U S A.* 91(17):7854-8.

Redmond DE Jr, Vinuela A, Kordower JH, Isacson O (2007) Influence of cell preparation and target location on the behavioral recovery after striatal transplantation of fetal dopaminergic neurons in a primate model of Parkinson's disease. *Neurobiol Dis.* Online.

Reier PJ (2004) Cellular transplantation strategies for spinal cord injury and translational neurobiology. *NeuroRx.* 1(4): 424–451.

Reynolds BA, Rietze RL (2005) Neural stem cells and neurospheres--re-evaluating the relationship. *Nat. Methods* 2(5):333-6.

Reynolds BA, Weiss S (1992) Generation of neurons and astrocytes from isolated cells of the adult mammalian central nervous system. *Science* 255:1707-10.

Rietze RL, Reynolds BA (2006) Neural stem cell isolation and characterization. *Methods Enzymol.* 419:3-23.

Riquelme PA, Drapeau E, Doetsch F (2007) Brain micro-ecologies: neural stem cell niches in the adult mammalian brain. Philos. Trans. R. Soc. Lond. B. Biol. Sci. online.

Sahay A, Hen R (2007) Adult hippocampal neurogenesis in depression. *Nat. Neurosci.* 10(9):1110-5.

Sanai N, Alvarez-Buylla A, Berger MS (2005) Neural stem cells and the origin of gliomas. *N. Engl. J. Med.* 353:811-22.

Scadden DT (2006) The stem cell niche as an entity of action. *Nature* 441:1075-9.

Schwartz PH, Nethercott H, Kirov II, Ziaeian B, Young MJ, Klassen H (2005) Expression of neurodevelopmental markers by cultured porcine neural precursor cells. *Stem Cells* 23(9):1286-94.

Seaberg RM, van der Kooy D (2002) Adult rodent neurogenic regions: the ventricular subependyma contains neural stem cells, but the dentate gyrus contains restricted progenitors. *J. Neurosci.* 22:1784-93.

Seki T (2002) Hippocampal adult neurogenesis occurs in a microenvironment provided by PSA-NCAM-expressing immature neurons. *J. Neurosci. Res.* 69(6):772-83.

Seki T (2003) Microenvironmental elements supporting adult hippocampal neurogenesis. *Anat. Sci. Int.* 78(2):69-78.

Shen Q, Goderie SK, Jin L, Karanth N, Sun Y, Abramova N, Vincent P, Pumiglia K, Temple S (2004) Endothelial cells stimulate self-renewal and expand neurogenesis of neural stem cells. *Science* 304:1338-40.

Sherr C (1994) G1 phase progression : cycling on cue. *Cell* 79:551-5.

Shihabuddin LS, Horner PJ, Ray J, Gage FH (2000) Adult spinal cord stem cells generate neurons after transplantation in the adult dentate gyrus. *J. Neurosci.* 20:8727-8735.

Shingo T, Gregg C, Enwere E, Fujikawa H, Hassam R, Geary C, Cross JC, Weiss S (2003) Pregnancy-stimulated neurogenesis in the adult female forebrain mediated by prolactin. *Science* 299:117-20.

Shou J, Rim PC, Calof AL (1999) BMPs inhibit neurogenesis by a mechanism involving degradation of a transcription factor. *Nat. Neurosci.* 2(4):339-45.

Simpson PJ, Moon C, Kleman AM, Connolly E, Ronnett GV (2007) Progressive and inhibitory cell cycle proteins act simultaneously to regulate neurotrophin-mediated proliferation and maturation of neuronal precursors. *Cell Cycle* 6(9):1077-89.

Singec I, Knoth R, Meyer RP, Maciaczyk J, Volk B, Nikklah G, Frotscher M, Snyder EY (2006) Defining the actual sensitivity of the neurosphere assay in stem cell biology. *Nat. Meth.* 3(10):801-6.

Skalnikova H, Halada P, Vodicka P, Motlik J, Rehulka P, Horning O, Chmelik J, Norregaard Jensen O, Kovarova H (2007) A proteomic approach to studying the differentiation of neural stem cells. *Proteomics* 7(11):1825-38.

Song H, Stevens CF, Gage FH (2002) Astroglia induce neurogenesis from adult neural stem cells. *Nature* 417:39-44.

Soule J, Messaoudi E, Bramham CR (2006) Brain-derived neurotrophic factor and control of synaptic consolidation in the adult brain. *Biochem. Soc. Trans.* 34(Pt.4) 600-4.

Su H, Chu TH, Wu W (2007) Lithium enhances proliferation and neuronal differentiation of neural progenitor cells in vitro and after transplantation into the adult rat spinal cord. *Exp. Neurol.* 206:296-307.

Suhonen JO, Peterson DA, Ray J, Gage FH (1996) Differentiation of adult hippocampus-derived progenitors into olfactory neurons in vivo. *Nature* 383:624-627.

Sun Y, Jin K, Childs JT, Xie L, Mao XO, Greenberg DA (2006) Vascular endothelial growth factor-B (VEGFB) stimulates neurogenesis: evidence from knockout mice and growth factor administration. *Dev. Biol.* 289(2):329-335.

Sun D, McGinn MJ, Zhou Z, Harvey HB, Bullock MR, Colello RJ (2007) Anatomical integration of newly generated dentate granule neurons following traumatic brain injury in adult rats and its association to cognitive recovery. *Exp. Neurol.* 204(1):264-72.

Sundberg M, Savola S, Hienola A, Korhonen L, Lindholm D (2006) Glucocorticoid hormones decrease proliferation of embryonic neural stem cells through ubiquitin-mediated degradation of cyclin D1. *J. Neurosci.* 26(20):5402-10.

Sundholm-Peters NL, Yang HK, Goings GE, Walker AS, Szele FG (2004) Radial glia-like cells at the base of the lateral centricles in adult mice. *J. Neurocytol.* 33:153-64.

Suzuki S, Gerhold LM, Böttner M, Rau SW, Dela Cruz C, Yang E, Zhu H, Yu J, Cashion AB, Kindy MS, Merchenthaler I, Gage FH, Wise PM (2007) Estradiol enhances neurogenesis following ischemic stroke through estrogen receptors alpha and beta. *J. Comp. Neurol.* 500(6):1064-75.

Tashiro A, Sandler VM, Toni M, Zhao C, Gage FH (2006) NMDA-receptor-mediated, cell-specific integration of new neurons in the adult dentate gyrus. *Nature* 442:929-33.

Teixeira AI, Duckworth JK, Hermanson O (2007) Getting the right stuff: controlling neural stem cell state and fate in vivo and in vitro with biomaterials. *Cell Res.* 17(1) 56-61.

Tighilet B, Brezun JM, Sylvie GD, Gaubert C, Lacour M (2007) New neurons in the vestibular nuclei complex after unilateral vestibular neurectomy in the adult cat. *Eur. J. Neurosci.* 25(1):47-58.

Tonchev AB, Yamashima T, Guo J, Chaldakov GN, Takakura N (2007) Expression of angiogenic and neurotrophic factors in the progenitor cell niche of adult monkey subventricular zone. *Neuroscience* 144(4):1425-35.

Tropepe V, Craig CG, Morshead CM, van der Kooy D (1997) Transforming growth factor-alpha null and senescent mice show decreased neural progenitor cell proliferation in the forebrain subependyma. *J. Neurosci.* 17:7850-9.

Tropepe V, Coles BL, Chiasson BJ, Horsford DJ, Elia AJ, McInnes RR, Van der Kooy D (2000) Retinal stem cells in the adult mammalian eye. *Science* 287:2032-6.

Uchida Y, Nakano S, Gomi F, Takahashi H (2007) Differential regulation of basic helix-loop-helix factors Mash1 and Olig2 by β-amyloid accelerates both differentiation and death of cultured neural stem/progenitor cells. *J. Biol. Chem.* 282(27):19700-9.

Van Praag H, Christie BR, Sejnowski TJ, Gage FH (1999a) Running enhances neurogenesis, learning and long-term potentiation in mice. *Proc. Natl. Acad. Sci. U.S.A.* 96:13427-31.

Van Praag H, Kempermann G, Gage FH (1999b) Running increases cell proliferation and neurogenesis in the adult mouse dentate gyrus. *Nat. Neurosci.* 2:266-70.

Veenema AH, de Kloet ER, de Wilde MC, Roelofs AJ, Kawata M, Buwalda B, Neumann ID, Koolhaas JM, Lucassen PJ (2007) Differential effects of stress on adult hippocampal cell proliferation in low and high aggressive mice. *J. Neuroendocrinol.* 19(7):489-98.

Vescovi AL, Galli R, Reynolds BA (2006) Brain tumour stem cells. *Nat. Rev. Cancer* 6:425-36.

Von Holst A, Sirko S, Faissner A (2006) The unique 473 HD-chodroitin sulphate epitope is expressed by radial glia and involved in neural precursor cell proliferation. *J. Neurosci.* 26:4082-94.

Waehler R, Russell SJ, Curiel DT (2007) Engineering targeted viral vectors for gene therapy. *Nat. Rev. Genet.* 8(8):573-87.

Weiss ER, Maness P, Lauder JM (1998) Why do neurotransmitters act like growth factors? *Perspect. Dev. Neurobiol.* 5(4):323-35.

Weiss S, Dunne C, Hewson J, Wohl C, Wheatley M, Peterson AC, Reynolds BA (1996) Multipotent CNS stem cells are present in the adult mammalian spinal cord and ventricular neuroaxis. *J Neurosci.* 16(23):7599-609.

Widera D, Grimm WD, Moebius JM, Mikenberg I, Piechaczek C, Gassmann G, Wolff NA, Thévenod F, Kaltschmidt C, Kaltschmidt B (2007) Highly efficient neural differentiation of human somatic stem cells, isolated by minimally invasive periodontal surgery. *Stem Cells Dev.* 16(3):447-60.

Wojtowicz JM, Kee N (2006) BrdU assy for neurogenesis in rodents. *Nat. Protoc.* 1(3):1399-405.

Xu Y, Tamaki N, Noda T, Kimura K, Itokazu Y, Matsumoto N, Dezawa M, Ide C (2005) Neurogenesis in the ependymal layer of the adult rat third ventricle. *Exp. Neurol.* 192(2):251-64.

Yamashima T, Tonchev AB, Borlongan CV (2007) Differential response to ischemia in adjacent hippocampal sectors : neuronal death in CA1 versus neurogenesis in dentate gyrus. Bi*otechnol. J.* 2(5):596-607.

Yang P, Baker KA, Hagg T (2006) The ADAMs family: coordinators of nervous system development, plasticity and repair. *Prog. Neurobiol.* 79(2):73-94.

Zhu H, Dahlström A (2007) Glial-fibrillary-acidic-protein-expressing cells in the neurogenic regions in normal and injured adult brains. *J. Neurosci. Res.* 85(12): 2783-92.

In: Stem Cell Applications in Disease and Health
Editors: W. Burnsides and R. Ellsley
ISBN: 978-1-60456-255-2

Chapter V

Normal Stem Cells and Cancer Stem Cells in the Gastrointestinal Tract

Gang-Ming Zou*
Department of Pathology, Johns Hopkins University School of Medicine
1550 Orleans Street, Baltimore, MD 21231

Abstract

Cancer stem cell population within the solid tumor with indefinite proliferation potential drives the growth and metastasis of cancer. Under the model of cancer stem cells, it is suggested that present therapy strategy may not be efficient at targeting cancer stem cells and, consequently, could not really cure these patients. Cancer stem cells exhibit a low rate of division and proliferation in their niche that help them to withstand chemotherapy and radiation. Epithelial cancers are believed to originate from transformation of tissue stem cells. Bone marrow-derived cells, which are frequently recruited to sites of tissue injury and inflammation, might also represent a potential source of malignancy in the gastrointestinal tract. Pancreatic cancer is one of most common causes of cancer-related death. Pancreatic cancer stem cells have been characterized recently through serial transplantation of human pancreatic cancer cells. The phenotype of pancreatic cancer stem cells has been defined as $CD24^{+}CD44^{+}CD326(ESA)^{+}$ cells within the cancer. More recently, other cancer stem cells in gastrointestinal tract, such as colon cancer stem cells and liver cancer stem cells, have been also characterized in their phenotype. These advances clearly will lead to a new strategy in cancer treatment and control in the gastrointestinal tract. In this chapter, the author reviews the current status and progress of normal stem cells and cancer stem cell research in the gastrointestinal tract and discusses the implications of these studies in new cancer treatment strategy.

Key words: Stem cells, esophageal stem cells, pancreatic stem cells, pancreatic cancer stem cells, colon cancer stem cells, liver cancer stem cells.

---

* E-mail address: gzou1@jhmi.edu, Tel: 1-410-955-3511, Fax: 1-410-614-0671

Introduction

Stem cells are a term to describe a cell that can differentiate into multiple cell types and maintain a self-renewal activity (Kondo et al., 2003; McCulloch and Till, 2005). Self-renewal is a process by which a daughter cell maintains the same properties of its parent cell. It indicates cell proliferation without any differentiation. Stem cells are immature, unspecialized cells that are capable of perpetuating themselves as stem cells and of undergoing differentiation into one or more specialized types of cells. They are most active during embryonic development, and give rise to all the tissues in the body. Adult stem cells are found in differentiated tissues and can differentiate to form all the cell types in the tissue from which they originated. They are most commonly found in the bone marrow, where they give rise to the different blood cells. But it has been described recently that stem cells also reside within the liver, stomach, and colon in the gastrointestinal tract

Though the existence of cancer stem cells had been hypothesized for many decades, it was not until 1997 that malignant stem cells were isolated from patients with acute myeloid leukemia. Subsequently, cancer stem cells have been isolated from breast and brain cancers. Cancer stem cells may derive from mutations in normal stem cells. Alternatively, differentiated tumor cells may acquire the characteristics of stem cells. Genomic instability cause spontaneously transformation of human nontumorigenic stem cells to cancer stem cells (Shiras et al., 2007). Currently, it is believed that cancer stem cells are present in small numbers in tumors. Cancers are composed of a heterogeneous mixture of cancer cells at various levels of differentiation, very similar to the structure of an organ. It is suggested that all cells within a tumor are unequal at any give time. There is a predetermined cell population with the "stem cells" phenotype, enabling this cell to perpetuate the cancer, while other cells of the same cancer are incapable of self-renewal. The standard analysis to examine the existence of a subpopulation of cancer stem cells is the demonstration of these cells to transfer the tumor and recapitulate the phenotypic heterogeneity present within the parental tumor. The same subpopulation should again fulfill these criteria when re-isolated from a secondary tumor (Lynch et al., 2006). In this review, the author focuses on the introduction and discussion of the current status in the study of normal stem cells and cancer stem cells in the gastrointestinal tract.

Esophagus

Stem Cells in the Esophagus

The identification and characterization of esophageal stem cells are critical to our understanding of the biology of the esophageal epithelium in health and disease. The basal cells of the mouse esophagus can be separated into three phenotypically and functionally distinct subpopulations based on the expression of $\alpha6$ integrin and transferrin receptors (CD71). Cells that express high levels of $\alpha6$ integrin and low levels of CD71 ($\alpha6^{bri}CD71^{dim}$) are a minor subpopulation of undifferentiated cells that might represent a putative mouse esophageal stem cell population. These cells are quiescent. They are long-lived with high proliferative potential in vitro. They have the ability to form colonies in *in vitro* cell culture assessed by using a limiting dilution culture assay, and have *in vivo* tissue reconstitution ability (Croagh et al., 2007). At the present, the characterization of similar stem cells from normal human esophageal is lacking.

Cancer Stem Cells in Esophagus Cancer

Currently, the knowledge about esophagus cancer stem cells is limited. Cancer stem cells are also named side population (SP) cells in the literature. These SP cells exhibit both self-renewal and differential potential, so they can be recognized as stem cells. Cancer cell line from the esophagus, such as TE1, TE2, and TE13, contained 0.3 ~1.4 % SP cells (Haraguchi et al., 2006). However, the question of whether these SP cells have self-renewal potential is unclear since serial transplantation assay is lacking for these cells. So, the question of whether those SP cells are cancer stem cells needs to be further characterized.

Stomach

Stem Cells in the Stomach

Animal studies demonstrate that multipotential stem cells play a role in the development of the gastric epithelium. Bjeknes and Chen (2002) recently used chemical mutagenesis to label random epithelial cells by loss of transgene function in adult hemizygous ROSA26 mice, a mouse strain expressing the transgene lacZ in all tissues. Many clones derived from such cells contained all the major epithelial cell types. These results indicate existence of functional multipotential stem cells in adult mouse gastric epithelium. Interesting, they also obtained similar results in the duodenum and colon, showing that this mouse model is suitable for lineage tracing in all regions of the gastrointestinal tract. Turnover of the epithelial cell lineages within the gastrointestinal tract is a constant process, occurring every 2-7 days under normal homeostasis and increasing after damage. This process is regulated by multipotent stem cells, which give rise to all gastrointestinal epithelial cell lineages and can regenerate whole intestinal crypts and gastric glands (Brittan et al., 2002). The growth factor response pathways are prominent in mouse gastric epithelial progenitor, with insulin-like growth factor appearing to play a key role in these cells (Mills et al., 2002). In addition, Wnt and Notch/Delta signaling pathways have been suggested to play major role in regulation of stem cell proliferation and differentiation in the gastrointestinal tract. Retinoids have long been known to modulate proliferation and differentiation of various renewing epithelia, and the expression of their receptors has been demonstrated in the gastric mucosa. Karam et al. (2005) recently examined the effects of retinoic acid on progenitor cell proliferation and cell lineage formation in the mouse stomach. They found that retinoic acid enhances the proliferation of gastric epithelial progenitors. The emerging evidence suggests that a single stem cell in gastric glands indirectly gives rise to a clone of all differentiated cells, by production of committed progenitor cells. These multipotential stem cells can give rise to the ulcer-associated cell lineage and gastric carcinomas. These stem cells occupies a niche in the isthmus composed of mesenchymal cells and extracellular matrix factors, which regulates the function of the cell via mesenchymal-epithelial cross talk (Modlin et al., 2003).

Cancer Stem Cells in the Stomach

Epithelial cancers are believed to originate from transformation of tissue stem cells. However, bone marrow-derived cells (BMDCs), which are frequently recruited to sites of tissue injury and inflammation, might also represent a potential source of malignancy in the stomach. Chronic infection of C57BL/6 mice with Helicobacter, a known carcinogen, induces repopulation of the stomach with BMDCs. These cells can progress through metaplasia and dysplasia to intraepithelial cancer. These observations suggest that epithelial cancers can originate from marrow-derived sources (Houghton et al., 2004). Cancer stem cells may exist in some gastric cancer cell lines. In gastric cell line NUGC3, MKN1, MKN7 and MKN28, the cancer stem cells (i.e., SP cells) are about 0.6 ~ 2.2% (Haraguchi et al., 2006). But the growth characteristics and signaling in regulation of these stem cells currently remain unclear.

Liver

Stem Cells in the Normal Liver

Normal hepatic stem cells have been suggested to play a major role in liver development and regeneration. Mouse primitive hepatic progenitor cells have been isolated from the mouse liver. These cells express both CD49f and CD29 ($\alpha6$ and $\beta1$ integrin subunits), but do not express hematopoietic cell antigens such as CD45, TER119, and c-Kit. When these cells are transplanted into the spleen, they migrate to the recipient liver and differentiate into liver parenchymal cells. These hepatic progenitor cells can be enriched by FACS and can they form colonies in *in vitro* culture (Suzuki et al., 2000). These hepatic stem/progenitor cells exhibit multilineage differentiation potential and self-renewing capability. They differentiate into hepatocytes and cholangiocytes with reconstitution of hepatocyte and bile duct structures, when they are transplanted into recipient animals. Interestingly, these cells differentiated into pancreatic ductal and acinar cells or intestinal epithelial cells when transplanted into pancreas or duodenal wall. Using transplantation assay, the existence of a bipotent epithelial liver stem cell in nonhuman primates has been suggested (Allain et al., 2002). More recently, it has been demonstrated that human hepatic stem cells are located in ductal plates in fetal and neonatal livers, an in pediatric and adult livers with the frequency of liver stem cells remaining relatively constant throughout life. The phenotype of these hepatic stem cells is $CD133.1^{+}ESA^{+}$ (Schmelzer et al., 2007).

Liver Cancer Stem Cells

Recently, there is emerging evidence to show the existence of liver cancer stem cells within liver cancer (Suetsugu et al., 2006; Ma et al., 2007; Chiba et al., 2007). The CD133 antigen, identified originally as a hematopoietic stem cell marker, appears in various human embryonic epithelia including the neural tube, gut, and kidney. Mouse prominin-1 is homologue to human CD133 (Ma et al., 2007). CD133^{+} cells isolated from human hepatocellular carcinoma cell lines possess cancer stem/progenitor cell-like properties. The CD133 antigen was found to be expressed on the surface of Huh-7 liver cancer cells line. CD133^{+} cells isolated from Huh-7 liver cancer cell line existed a higher in vitro proliferative potential and lower mRNA expressions of mature hepatocyte markers (glutamine synthetase and cytochrome P450), than CD133^{-} population of Huh-7 cells. In vivo study revealed that CD133^{+} liver cancer cells formed tumors. In contrast, CD133^{-} liver cancer had a very limited effect in this regard, when either CD133^{+} or CD133^{-} cells were subcutaneously injected into SCID mice (Suetsugu et al., 2006). This phenomenon regarding CD133^{+} cell population in human liver cancer cell line has been further confirmed by a research group in University of Hong Kong (Ma et al., 2007). They demonstrated the identification and isolation of a population of cancer stem cells expressing a CD133 surface phenotype from human liver cell lines. These CD133^{+} liver cancer cells exhibit increased potential for self-renewal, an increased colony-forming efficiency, higher proliferative potential in vitro, and greater ability to form tumor in vivo. Interestingly, CD133 antigen is also found to represent only a minority of the tumor cell population in human primary HCC specimens.

It is a well accepted concept currently that in the hematopoietic system, transformed hematopoietic stem/progenitor cells with an enhanced or acquired self-renewal capability function as leukemic stem cells (Zou, 2007). However, it remains unclear whether disruption of normal stem cell function directly contributes to cancer initiation in solid tumor, such as liver cancer. To address this question, a collaborative research group in Japan did functional analyses of polycomb-group protein Bmi1 and Wnt/β-catenin, the molecules that are responsible for the self-renewal capability of many types of stem cells, in liver stem/progenitor cells. These Bmi1 and Wnt genes were transduced in c-Kit^{-} CD29^{+}CD49f$^{+/low}$CD45^{-}Ter-119^{-} hepatic stem/progenitor cells using retrovirus- or lentivirus-mediated gene transfer. They found that forced expression of Bmi1 and constitutively active β-catenin mutant similarly promoted the self-renewal of hepatic stem/progenitor cells. The transplantation of Bmi1- or β-catenin-transduced cells clonally expanded from single hepatic stem/progenitor cells produced tumors. These observations indicate that the enhanced self-renewal level of normal hepatic stem/progenitor cells serves as an early event in hepatocarcinogenesis, and this study also highlights the roles of Bmi1 and the Wnt/β-catenin in regulating the self-renewal of normal or cancer stem cells in liver (Chiba et al., 2007).

Pancreas

Stem/Progenitor Cells in the Normal Pancreas

The pancreas is a gland organ in the digestive and endocrine systems of vertebrates. It has both exocrine (secreting pancreatic juice containing digestive enzymes) and endocrine (producing several important hormones, including glucagons and insulin) function. In anatomy, the human pancreas is a 15-25 cm (6-10 inches) elongated organ in the abdomen. Its weight is between 65 and 75 g. the pancreas organ has four regions: a head, neck, body and tail. The pancreatic neck springs from the right upper portion of the front of the head. The tail of the pancreas extends towards the spleen. In addition, the pancreatic duct runs the length of the pancreas and empties into the second part of the duodenum at the ampulla of Vater. In regard to organ development, the progenitor cell number, independently of regulation by growth factors, is suggested as a key determinant of organ size in the pancreas (Stanger et al., 2007). Recently, there is emerging evidence showing the existing progenitor cells in the duct of the pancreas, named pancreatic ductal progenitor cells (Noguchi et al., 2006; Oshima et al., 2007). Mouse pancreatic ductal epithelial progenitor cells express CD133 and c-Met (Oshima et al., 2007), indicating hepatic growth factor might be a growth regulator in this mouse pancreatic progenitor cell. Ductal progenitor cells in the pancreas may be useful for therapies that target β-cell replacement in diabetic patients. Overexpression of embryonic transcription factors NeuroD facilitates pancreatic stem/progenitor cell differentiation into insulin-producing cells (Noguchi et al., 2006). Notch signaling is suggested as a modulator of differentiation of this pancreatic progenitor cell (Murtaugh et al., 2003). Tissue turnover can occur by the differentiation of stem cells, as is well documented for blood, skin and intestine, or by the duplication of existing differentiated cells. Stem cell differentiation may not be the way in regeneration pancreatic β-cells, since pre-existing β-cells, rather than pluripotent stem cells, are the major source of new β-cells during adult life and after pancreatectomy in mice (Dor et al., 2004). These results indicate that terminally differentiated β-cells exhibit a high proliferative capacity in vivo and consequently raise a question whether adult stem cells have a significant role in β-cell regeneration. More recently, it has been demonstrated that bone marrow stem/progenitor cells cannot spontaneously transdifferentiate into insulin-producing cells in vivo, even after β-cell lesion and mobilization induced by cytokines (Lavazais et al., 2007). The $CD133^{+}c\text{-}Met^{+}$ pancreatic ductal progenitor cells might be the source of pancreatic cancer stem cells (Figure 1). However, since these progenitor cells found in mice species, it is unclear whether pancreatic ductal progenitor cells exist in the human pancreas.

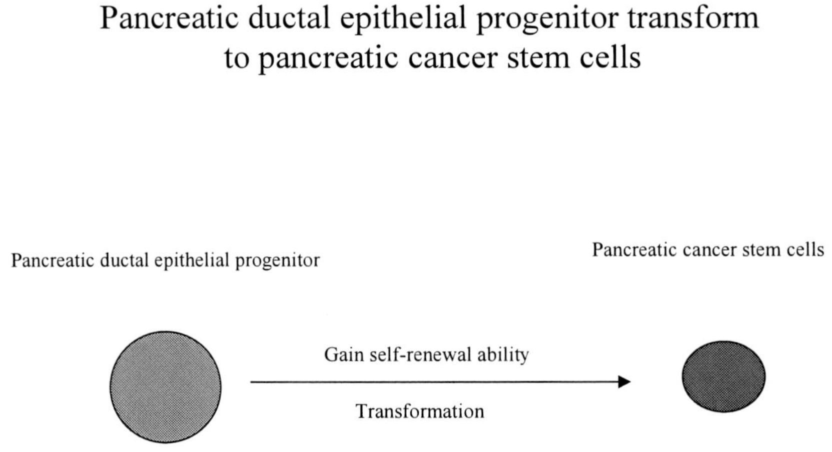

Figure 1. Pancreatic ductal progenitor cells and pancreatic cancer stem cells.

Pancreatic Cancer Stem Cells

Pancreatic cancer is the fourth most common cause of cancer-related death, accounting for approximately 32,000 lives each year in the United States. An overall five-year survival rate of <5% reflects the ineffectiveness of current therapies for this disease. Most patients in the United States undergoing surgical resection for pancreatic cancer will receive adjuvant chemo-radiation therapy, which in several studies has been shown to provide an incremental survival benefit over surgery alone. However, most patients usually develop loco-regional relapse in the pancreatic bed. Due to a high likelihood of vascular invasion and persistent locoregional disease following surgery, surgical resection is almost ineffective for these patients. Pancreatic intraepithelial neoplasia (PanIN) is a precursor to invasive pancreatic cancer (Kuang et al., 2006). PanIN lesions are premalignant lesions classified into PanIN-1A, PanIN-1B, Pan-IN2, and PanIN3 based on the degree of cytological and architectural atypia of pancreatic sections (Kern et al., 2001). The Nestin progenitor lineage is the compartment of origin for pancreatic intraepithelial neoplasia. Nestin cell lineage is highly responsive to Kras oncogenic activation and may represent the elusive progenitor population in which PDAC arises (Carrere et al., 2007). Currently, the cellular origin of pancreatic ductal carcinoma is controversial (Kuang et al., 2006). Grippo et al (2003) suggested that pancreatic cancer cells are from acinar-to-ductal metaplasia. But this hypothesis has been challenged, because the study with Smad7 transgene revealed PanIN lesion, without obvious acinar dysplasia surrounding the ductal change, indicating the premalignant changes of the ductal epithelium are not likely derived from acinar cells. In contrast, it may be possible that the premalignant ductal cells originate from islet cells or multipotent precursors due to transdifferentiation (Kuang et al., 2006).

Several kinds of multipotent stem/progenitor cells can be the pancreatic cancer precursors. They are summarized as follow:

1). *Pancreatic ductal epithelial progenitor cells:* These cells may gain self-renewal ability through mutation or other mechanisms. Consequently, they commit to transform to malignant stem cells in the pancreas. But this is only a hypothesis and currently lacks experimental evidence to support it.

2). *Pancreatic non-ductal exocrine cells:* Recently, Guerra and colleagues (2007) provided important evidence about the ability of specific pancreatic cell types to generate invasive pancreatic cancer. In their observation, the classical pancreatic "ductal" neoplasia can be induced by activation of oncogenic Kras in nonductal exocrine cells. Kras activation in immature acinar and centroacinar cells is readily able to induce ductal neoplasia. The results provide a new hypothesis on the "cell of origin" of pancreatic ductal cancer (Murtaugh et al., 2007).

3). *Pancreatic mesenchymal stem cells:* Mesenchymal stem cells are distributed within the pancreas (Seeberger et al., 2006). Despite recent emerging evidence suggesting that cancer stem cells reside in a variety of tumors, it is not yet fully elucidated whether postnatal stem cells are directly involved in tumorigenesis. Murine bone marrow-derived mesenchymal stem cells (BMMSCs) were used as a model to test a hypothesis that tumorigenesis may originate from spontaneous mutation of stem cells. Murine BMMSCs, after numerous passages, obtained unlimited population doublings and proceeded to a malignant transformation state, resulting in fibrosarcoma formation in vivo. Fibrosarcoma cells formed by transformed BMMSCs contained cancer progenitors, which were capable of generating colony clusters in vitro and fibrosarcoma in vivo by the second administration. The mechanism by which BMMSCs transformed to malignant cells was associated with accumulated chromosomal abnormalities, gradual elevation in telomerase activity, and increased c-myc expression. (Miura et al., 2006). So we hypothesize here that bone marrow MSC migrate to the pancreatic organ through blood flow. Then local inflammation stimulation promotes them for transformation, consequently leading to carcinogenesis in the pancreas (Figure 2).

4). *Bone marrow hematopoietic stem cells:* Transdifferentiation of pancreatic duct epithelium from bone marrow stem cells: Does pancreatic cancer originate from bone marrow multipotent stem cells? Recent reports suggest that bone marrow-derived cells engraft and differentiate into pancreatic tissue after pancreatic injury, though this frequency is low. All such studies have used adult recipients. Pluripotent stem cells or epithelial precursors in the bone marrow can migrate to the pancreas and differentiate into complex organ-specific structures during the neonatal period (Wang et al., 2006). Moreover, it has been suggested that gastric cancer can originate from bone marrow derived cells (BMDC) (Houghton et al., 2004). Therefore, we hypothesize that pancreatic cancers (at least some of them) might be caused by hematopoietic stem cell migration to the pancreas and transformation by mutation induced by local environment.

Cancer stem cells have been identified in human blood, brain, and breast cancers. Cancer stem cells in pancreatic cancer have been identified recently (Li et al., 2007, Hermann et al., 2007). Using a xenograft model in which primary human pancreatic adenocarcinomas were grown in immunocompromised mice, tumorigenic subpopulation of pancreatic cancer cells expressing the cell surface markers CD44, CD24, and epithelial-specific antigen (ESA) have been identified. Pancreatic cancer stem cells with the $CD44^+CD24^+ESA^+$ phenotype had a 100-fold increased tumorigenic potential compared with nontumorigenic cancer cells. The $CD44^+CD24^+ESA^+$ pancreatic cancer cells showed the stem cell properties of self-renewal, the ability to produce differentiated progeny, and increased expression of the developmental signaling molecule sonic hedgehog. (Li et al., 2007). CD133 has been indicated as a marker for cancer stem cells in solid tumors recently. In brain tumors, CD133 is a marker for cancer stem cells expressed by 6 ~ 29 % of the total tumor populations (Singh et al., 2004). Cancer stem cells in colorectal cancer are a small number of undifferentiated tumorigenic $CD133^+$ cells (Ricci-Vitiani et al., 2007). Human pancreatic cancer tissue contains cancer stem cells defined by CD133 expression that are exclusively tumorigenic and highly resistant to standard chemotherapy. The subpopulation of $CD133^+$ $CXCR4^+$ cancer stem cells related to tumor metastasis in pancreatic cancer (Hermann et al., 2007). The strategies of modulations the SDF-1/CXCR4 axis may have important clinical applications to inhibit metastasis of cancer stem cells, consequently improve the therapeutic effect in pancreatic cancer. Interestingly, pancreatic cancer cell lines, such as PK-9, may also contain cancer stem cells

(Haraguchi et al., 2006). Olempska et al (2007) recently reported that they examined CD133 expression in 5 pancreatic cancer cell lines: Panc1, Panc89, Colo357, PancTu1, and A818-6, only 2 cell line (PancTu1, A818-6) are positive in CD133 expression in Real-time RT-PCR assay. Another recent report shows that a subpopulation of PANC-1 cells can propagate to form spheres with properties of stem cells (Gou et al., 2007).

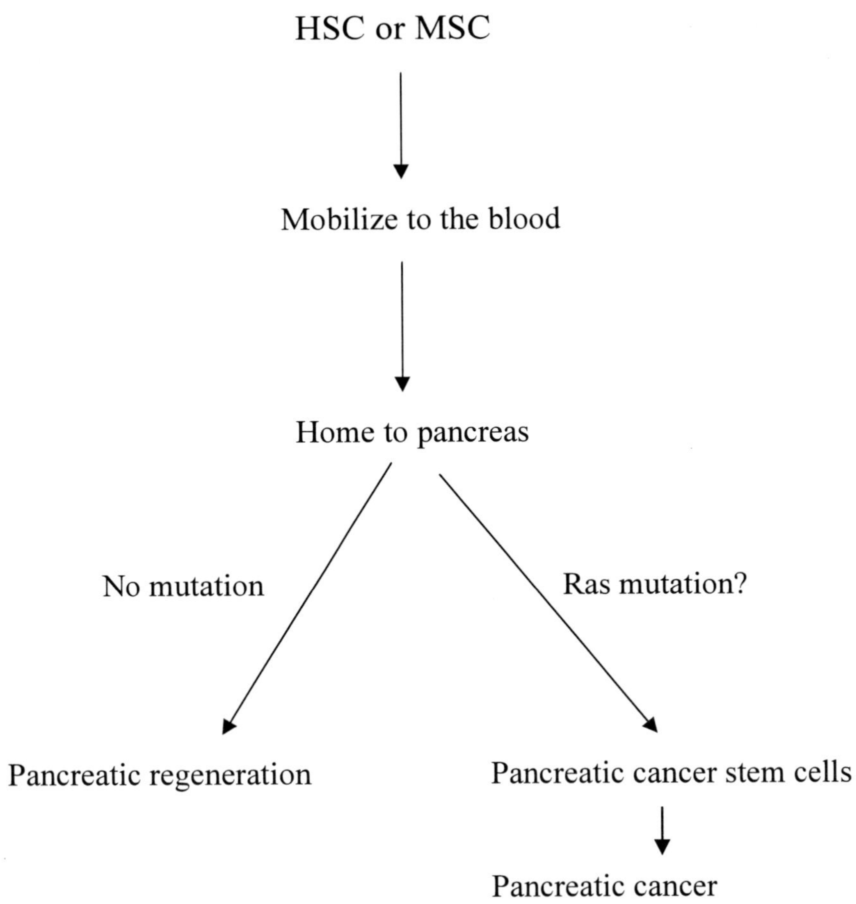

Figure 2. HSC, MSC and pancreatic cancer stem cells. HSC: hematopoietic stem cells. MSC: mesenchymal stem cells.

Intestine

Normal Intestinal Stem Cells

In mice, conditional inactivation of the *Bmpr1a* gene disturbs homeostasis of intestinal epithelial regeneration with an expansion of the stem and progenitor cell populations in the intestine, eventually leading to intestinal polyposis resembling human juvenile polyposis syndrome. It is suggested that BMP signaling may control the duplication of intestinal stem cells (He et al., 2004). The Notch signaling pathway is a highly conserved cell signaling system which is present in most multicellular organisms (Artavanis-Tsakonas et al., 1999). Notch possesses four different notch receptors, referred to as Notch1 to Notch4. In Drosophila melanogaster there are two ligands named Delta and Serrate. In mammals, the corresponding names are Delta-like and Jagged for the Notch. Multiple intestinal stem cells in Drosophila contain a high level of cytoplasmic Delta-rich vesicles, and Notch signaling is involved in the regulation of differentiation of these stem cells (Ohlstein and Spradling, 2007). Intestinal polyposis, a precancerous neoplasia, results primarily from an abnormal increase in the number of crypts, which contain intestinal stem cells. In mice, deletion of the tumor suppressor PTEN results in hamartomatous intestinal polyps with epithelial and stromal involvement. PTEN helps govern the proliferation rate and number of intestinal stem cells, and loss of PTEN results in an excess of intestinal stem cells. The PTEN-Akt pathway is suggested to govern intestinal stem cell activation by helping control nuclear localization of the Wnt pathway effector β-catenin (He et al., 2007). Recently, it is suggested that Wip1 phosphatase regulates homeostasis of intestinal stem cells (Demidov et al., 2007)

Colon Cancer Stem Cells

In the United States, colon carcinoma is the second most common cause of death from cancer. The existence of cancer stem cells in colon cancer has been reported in 2007 by different research groups (Ricci-Vitiani et al., 2007; O'Brian et al., 2007; Dalerbra et al, 2007). Tumorigenic cells in colon cancer are included in the high-density $CD133^+$ population, which accounts for about 2.5% of the tumor cells. Subcutaneous injection of colon cancer $CD133^+$ cells could reproduce the original tumor in immunodeficient mice, whereas $CD133^-$ cells did not have this ability. Such tumors were serially transplanted for several generations without significant phenotypic alterations. The phenotype of colon cancer stem cells also can be $ESA^+CD44^+CD166^+$. Further, it is also suggested that ALDH activity was useful for isolation of cancer stem cells in some, but not all colon cancer xenografts (Dalerba et al., 2007). $CD133^+$ colon cancer stem cells produce and utilize IL-4 to protect themselves from apoptosis. Treatment with anti-IL-4 neutralizing antibody significantly enhances the antitumor efficacy of chemotherapeutic drugs through selective sensitization of $CD133^+$ cells (Todaro et al., 2007).

Self-renewal Pathway in Cancer Stem Cells in the Gastrointestinal Tract

In pancreatic cancer stem cells, real-time PCR analysis reveals that Hedgehog expression was increased over 40 fold compared with normal pancreatic epithelial cells (Li et al., 2007). However, the information in SHH expression between pancreatic cancer stem cells and normal human pancreatic epithelial stem cells is lacking. Furthermore, it is undefined whether these high levels of SHH expression is only a constitutive expression in these malignant stem cells or they are essential in control of self-renewal of these stem cells. Bmi1 is a critical factor in control self-renewal capacity of HSCs based on gain-of-function and loss-of-function analyses (Iwama et al., Park et al., 2003). Recently, it has been reported that it also regulates self-renewal of normal murine hepatic stem/progenitor cells, and enhancement of self-renewal capacity of these normal hepatic stem cells through transduction of exogenous Bmi1 gene into these cells causes malignant transformation and hepaticellular formation when they are transplanted into mice (Chiba et al., 2007). Currently, it is unclear whether Bmi1 plays any role in self-renewal regulation of pancreatic cancer stem cells.

Conclusion

In malignancy, actively proliferating cells may be effectively targeted and killed by anti-cancer therapies, but stem cells may survive and support re-growth of the tumor. Thus, new strategies for the treatment of cancer obviously need to target cancer stem cells or their niche cells. Cancer stem cells may secret cytokines to sustain their survival, proliferation, etc. In another regard, niche cells around the cancer stem cells might also secret some cytokines to regulate cancer stem cells. So cytokines in the regulation of cancer stem cells in the gastrointestinal tract will be one important project in future study. Understanding the interaction between cancer stem cells and niche cells will be helpful for us to design more efficient strategies to target these malignant stem cells and cure cancer.

References

Allain JE, Dagher I, Mahieu-Caputo D, Loux N, Andreoletti M, Westerman K, Briand P, Franco D, Leboulch P, Weber A. Immortalization of a primate bipotent epithelial liver stem cell. *Proc Natl Acad Sci U S A*. 2002. 99:3639-3644.

Artavanis-Tsakonas S, Rand MD, Lake RJ. Notch signaling: cell fate control and signal integration in development. *Science*. 1999. 284:770-776.

Bjerknes M, Cheng H. Multipotential stem cells in adult mouse gastric epithelium. Am J Physiol Gastrointest Liver Physiol. 2002. 283:G767-777.

Brittan M, Wright NA. Gastrointestinal stem cells. *J Pathol.* 2002. 197:492-509.

Carriere C, Seeley ES, Goetze T, Longnecker DS, Korc M. The Nestin progenitor lineage is the compartment of origin for pancreatic intraepithelial neoplasia. *Proc Natl Acad Sci U S A*. 2007. 104:4437-4442.

Chiba T, Zheng YW, Kita K, Yokosuka O, Saisho H, Onodera M, Miyoshi H, Nakano M, Zen Y, Nakanuma Y, Nakauchi H, Iwama A, Taniguchi H. Enhanced Self-Renewal Capability in Hepatic Stem/Progenitor Cells Drives Cancer Initiation. *Gastroenterology*. 2007. 133:937-950.

Croagh D, Phillips WA, Redvers R, Thomas RJ, Kaur P. Identification of candidate murine esophageal stem cells using a combination of cell kinetic studies and cell surface markers. *Stem Cells*. 2007. 25:313-318.

Dalerba P, Dylla SJ, Park IK, Liu R, Wang X, Cho RW, Hoey T, Gurney A, Huang EH, Simeone DM, Shelton AA, Parmiani G, Castelli C, Clarke MF. Phenotypic characterization of human colorectal cancer stem cells. *Proc Natl Acad Sci U S A*. 2007. 104:10158-10163.

Demidov ON, Timofeev O, Lwin HNY, Kek C, Appella E, and Bulavin DV. Wip1 Phosphatase Regulates p53-Dependent Apoptosis of Stem Cells and Tumorigenesis in the Mouse Intestine. *Cell Stem Cell*. 2007. 1:180-190

Dor Y, Brown J, Martinez OI, Melton DA. Adult pancreatic beta-cells are formed by self-duplication rather than stem-cell differentiation. *Nature*. 2004. 429:41-46.

Gou S, Liu T, Wang C, Yin T, Li K, Yang M, Zhou J. Establishment of clonal colony-forming assay for propagation of pancreatic cancer cells with stem cell properties. *Pancreas*. 2007. 34:429-435.

Grippo PJ, Nowlin PS, Demeure MJ, Longnecker DS, Sandgren EP. Preinvasive pancreatic neoplasia of ductal phenotype induced by acinar cell targeting of mutant Kras in transgenic mice. *Cancer Res*. 2003. 63:2016-2019.

Guerra C, Schuhmacher AJ, Canamero M, Grippo PJ, Verdaguer L, Perez-Gallego L, Dubus P, Sandgren EP, Barbacid M.. Chronic pancreatitis is essential for induction of pancreatic ductal adenocarcinoma by K-Ras oncogenes in adult mice. *Cancer Cell*. 2007. 11: 291-302.

Haraguchi N, Utsunomiya T, Inoue H, Tanaka F, Mimori K, Barnard GF, Mori M. Characterization of a side population of cancer cells from human gastrointestinal system. *Stem Cells*. 2006. 24:506-513.

He XC, Zhang J, Tong WG, Tawfik O, Ross J, Scoville DH, Tian Q, Zeng X, He X, Wiedemann LM, Mishina Y, Li L. BMP signaling inhibits intestinal stem cell self-renewal through suppression of Wnt-beta-catenin signaling. *Nat Genet*. 2004. 36:1117-1121.

He XC, Yin T, Grindley JC, Tian Q, Sato T, Tao WA, Dirisina R, Porter-Westpfahl KS, Hembree M, Johnson T, Wiedemann LM, Barrett TA, Hood L, Wu H, Li L. PTEN-deficient intestinal stem cells initiate intestinal polyposis. *Nat Genet*. 2007. 39:189-198.

Hermann, PC, Huber SL, Herrler T, Aicher A, Ellwart JW, Guba M, Bruns CJ, and Heeschen C Distinct Populations of Cancer Stem Cells Determine Tumor Growth and Metastatic Activity in Human Pancreatic Cancer. *Cell Stem Cell*. 2007. 1:313-323

Houghton J, Stoicov C, Nomura S, Rogers AB, Carlson J, Li H, Cai X, Fox JG, Goldenring JR, Wang TC. Gastric cancer originating from bone marrow-derived cells. *Science*. 2004. 306:1568-1571.

Iwama A, Oguro H, Negishi M, Kato Y, Morita Y, Tsukui H, Ema H, Kamijo T, Katoh-Fukui Y, Koseki H, van Lohuizen M, Nakauchi H. Enhanced self-renewal of hematopoietic stem cells mediated by the polycomb gene product Bmi-1. *Immunity*. 2004. 21:843-851.

Karam SM, John R, Alpers DH, Ponery AS. Retinoic acid stimulates the dynamics of mouse gastric epithelial progenitors. *Stem Cells*. 2005. 23:433-441.

Kern S, Hruban R, Hollingsworth MA, Brand R, Adrian TE, Jaffee E, Tempero MA. A white paper: the product of a pancreas cancer think tank. *Cancer Res*. 2001. 61:4923-4932.

Kondo M, Wagers AJ, Manz MG, Prohaska SS, Scherer DC, Beilhack GF, Shizuru JA, Weissman IL. Biology of hematopoietic stem cells and progenitors: implications for clinical application. *Annu Rev Immunol*. 2003. 21:759-806.

Kuang C, Xiao Y, Liu X, Stringfield TM, Zhang S, Wang Z, Chen Y. In vivo disruption of TGF-beta signaling by Smad7 leads to premalignant ductal lesions in the pancreas. *Proc Natl Acad Sci U S A*. 2006. 103:1858-1863.

Lavazais E, Pogu S, Sai P, Martignat L. Cytokine mobilization of bone marrow cells and pancreatic lesion do not improve streptozotocin-induced diabetes in mice by transdifferentiation of bone marrow cells into insulin-producing cells. *Diabetes Metab.* 2007. 33:68-78.

Li C, Heidt DG, Dalerba P, Burant CF, Zhang L, Adsay V, Wicha M, Clarke MF, Simeone DM. Identification of pancreatic cancer stem cells. *Cancer Res*. 2007. 67:1030-1037.

Lynch MD, Cariati M, Purushotham AD. Breast cancer, stem cells and prospects for therapy. *Breast Cancer Res*. 2006. 8:211.

Ma S, Chan KW, Hu L, Lee TK, Wo JY, Ng IO, Zheng BJ, Guan XY. Identification and characterization of tumorigenic liver cancer stem/progenitor cells. *Gastroenterology*. 2007. 132:2542-2556.

McCulloch EA, Till JE. Perspectives on the properties of stem cells. *Nat Med.* 2005. 11:1026-1028. Review.

Mills JC, Andersson N, Hong CV, Stappenbeck TS, Gordon JI. Molecular characterization of mouse gastric epithelial progenitor cells. *Proc Natl Acad Sci U S A.* 2002 Nov 12;99(23):14819-24.

Miura M, Miura Y, Padilla-Nash HM, Molinolo AA, Fu B, Patel V, Seo BM, Sonoyama W, Zheng JJ, Baker CC, Chen W, Ried T, Shi S. Accumulated chromosomal instability in murine bone marrow mesenchymal stem cells leads to malignant transformation. *Stem Cells*. 2006. 24:1095-1103.

Miyashita T, Ohta T, Fujimura T, Ninomiya I, Fushida S, Hattori T, Miwa K. Duodenal juice stimulates oesophageal stem cells to induce Barrett's oesophagus and oesophageal adenocarcinoma in rats. *Oncol Rep*. 2006. 15:1469-1475.

Modlin IM, Kidd M, Lye KD, Wright NA. Gastric stem cells: an update. *Keio J Med.* 2003. 52:134-137.

Murtaugh LC, Leach SD. A case of mistaken identity? Nonductal origins of pancreatic "ductal" cancers. *Cancer Cell.* 2007. 11:211-213.

Murtaugh LC, Stanger BZ, Kwan KM, Melton DA. Notch signaling controls multiple steps of pancreatic differentiation. *Proc Natl Acad Sci U S A*. 2003. 100:14920-14925.

Noguchi H, Xu G, Matsumoto S, Kaneto H, Kobayashi N, Bonner-Weir S, Hayashi S. Induction of pancreatic stem/progenitor cells into insulin-producing cells by adenoviral-mediated gene transfer technology. *Cell Transplant*. 2006. 15:929-938.

O'Brien CA, Pollett A, Gallinger S, Dick JE. A human colon cancer cell capable of initiating tumour growth in immunodeficient mice. *Nature*. 2007. 445:106-110.

Ohlstein B, Spradling A. Multipotent Drosophila intestinal stem cells specify daughter cell fates by differential notch signaling. *Science*. 2007. 315:988-992.

Olempska M, Eisenach PA, Ammerpohl O, Ungefroren H, Fandrich F, Kalthoff H. Detection of tumor stem cell markers in pancreatic carcinoma cell lines. *Hepatobiliary Pancreat Dis Int.* 2007. 6:92-97.

Oshima Y, Suzuki A, Kawashimo K, Ishikawa M, Ohkohchi N, Taniguchi H. Isolation of mouse pancreatic ductal progenitor cells expressing CD133 and c-Met by flow cytometric cell sorting. *Gastroenterology*. 2007. 132:720-732.

Park IK, Qian D, Kiel M, Becker MW, Pihalja M, Weissman IL, Morrison SJ, Clarke MF. Bmi-1 is required for maintenance of adult self-renewing haematopoietic stem cells. *Nature*. 2003. 423:302-305.

Ricci-Vitiani L, Lombardi DG, Pilozzi E, Biffoni M, Todaro M, Peschle C, De Maria R. Identification and expansion of human colon-cancer-initiating cells. *Nature*. 2007. 445:111-1115.

Schmelzer E, Zhang L, Bruce A, Wauthier E, Ludlow J, Yao HL, Moss N, Melhem A, McClelland R, Turner W, Kulik M, Sherwood S, Tallheden T, Cheng N, Furth ME, Reid LM. Human hepatic stem cells from fetal and postnatal donors. *J Exp Med*. 2007. 204:1973-87.

Seeberger KL, Dufour JM, Shapiro AM, Lakey JR, Rajotte RV, Korbutt GS. Expansion of mesenchymal stem cells from human pancreatic ductal epithelium. *Lab Invest*. 2006 86:141-153.

Shiras A, Chettiar ST, Shepal V, Rajendran G, Prasad GR, Shastry P. Spontaneous transformation of human adult nontumorigenic stem cells to cancer stem cells is driven by genomic instability in a human model of glioblastoma. *Stem Cells*. 2007. 25:1478-1489.

Singh SK, Hawkins C, Clarke ID, Squire JA, Bayani J, Hide T, Henkelman RM, Cusimano MD, Dirks PB. Identification of human brain tumour initiating cells. *Nature*. 2004. 432:396-401.

Stanger BZ, Stiles B, Lauwers GY, Bardeesy N, Mendoza M, Wang Y, Greenwood A, Cheng KH, McLaughlin M, Brown D, Depinho RA, Wu H, Melton DA, Dor Y. Pten constrains centroacinar cell expansion and malignant transformation in the pancreas. *Cancer Cell*. 2005. 8:185-195.

Suetsugu A, Nagaki M, Aoki H, Motohashi T, Kunisada T, Moriwaki H. Characterization of CD133+ hepatocellular carcinoma cells as cancer stem/progenitor cells. *Biochem Biophys Res Commun*. 2006. 351:820-824.

Suzuki A, Zheng Y, Kondo R, Kusakabe M, Takada Y, Fukao K, Nakauchi H, Taniguchi H. Flow-cytometric separation and enrichment of hepatic progenitor cells in the developing mouse liver. *Hepatology*. 2000. 32:1230-1239.

Suzuki A, Zheng Yw YW, Kaneko S, Onodera M, Fukao K, Nakauchi H, Taniguchi H. Clonal identification and characterization of self-renewing pluripotent stem cells in the developing liver. *J Cell Biol*. 2002. 156:173-184.

Todaro M, Alea MP, Di Stefano AB, Cammareri P, Vermeulen L, Iovino F, Tripodo C, Russo A, Gulotta G, Medema JP, Stassi G. Colon Cancer Stem Cells Dictate Tumor Growth and Resist Cell Death by Production of Interleukin-4. *Cell Stem Cells*. 2007. 1:389-402.

Wang X, Ge S, Gonzalez I, McNamara G, Rountree CB, Xi KK, Huang G, Bhushan A, Crooks GM. Formation of pancreatic duct epithelium from bone marrow during neonatal development. *Stem Cells*. 2006. 24:307-314.

Zou GM. Cancer Stem cells in leukemia, recent advances. *J Cell Physiol*. 2007. 213:440-444.

In: Stem Cell Applications in Disease and Health
Editors: W. Burnsides and R. Ellsley

ISBN: 978-1-60456-255-2

Chapter VI

In Vitro Approach of Germline Stem Cells in the Fly and the Mouse

Yuzo Niki*
Department of Sciences, Faculty of Science, Ibaraki University, Mito 310-8512, Japan

Abstract

The ability of self-renewal division of germline stem cells (GSCs) is responsible for continuous production of functional gametes in a gonad throughout an animal's lifetime. The developmental dynamics of germline cells are well studied using several model animals. Germline cells are specified by maternal inherited determinants (preformation) in the fly or by inductive signals (epigenesis) in the mouse during early embryonic development, respectively. After migrating into the presumptive gonadal regions, primordial germ cells (PGCs) acquire stem cell status. GSCs are located at the apical tip of the gonads by associating with neighboring somatic cells to form 'niche' in the fly. Growth factors produced by niche cells or surrounding tissues are believed to function for the maintenance and division of GSCs. In addition to the biological interest in understanding the cellular and molecular mechanisms of the stem cell system, understanding of the mechanisms of stem cell behavior is crucial to therapeutic usage of these cells, since malfunction of these processes would result in infertility in humans. Recent advances in the establishment of stable cell lines of GSCs would provide new opportunities to study GSCs and their niche cells in detail. Furthermore, these *in vitro* systems have enabled genetic modification and cryoreservation.

* E-mail address; yuzoniki@mx.ibaraki.ac.jp, Tele phone; +81-29-228-8380, Fax; +81-29-228-8380

Background

Tissue Stem Cells and Their Niche

Adult stem cells are undifferentiated cells that can renew themselves and can differentiate to yield the specialized cell types of the tissue or organ. The primary role of adult stem cells is to maintain the homeostasis in a living organism. Understanding the cellular and molecular mechanisms controlling stem cell function is crucial to the future use of stem cells in regenerative medicine, as well as in understanding aging, tumor formation and gametogenesis. The *Drosophila* germline stem cell (GSC) is the most conspicuous example that can be easily recognized and is now the most suitable model for stem cell biology. I will describe here properties of the germline development and formation of the germinal niche, signal transduction pathways and characters of the cell lines of GSCs in the fly and the mouse in this review.

1. Germline Development

It is well known that germline is segregated from the remainder of soma during early embryonic development. Germline segregation differs among species and is classified into two types of mechanism: one is by the maternal component (preformation) and the other is by instructive signals (epigenesis). The two mechanisms are not necessarily mutually exclusive (reviewed by Extavour and Akam, 2003; Extavour, 2007). Maternal components can be observed as a conspicuous cytoplasm, called the germ plasm in some of the former group. A number of histological studies and experimental studies of germ plasm have been reported since the 19th century in these animals (reviewed by Eddy, 1975). Precursors of germ cells (primordial germ cells, PGCs) incorporate germline-specific organelles, components of the germ plasm, into their cytoplasm and thus are distinguishable from soma. On the other hand, it is not easy to identify PGCs during embryonic stages in the latter group. Irrespective of the differences of the germ cell specification, germline cells have common mechanisms of expression of germline-specific genes, transcription repression, chromatin remodeling to preserve totipotency during development (reviewed by Seydoux and Braun, 2006).

1.1. The Fly

The germline of *Drosophila* is established from pole cells, or PGCs, which form at the posterior tip of the pre-blastoderm embryo (Figure 1). PGCs are determined by maternally derived germ plasm (or pole plasm in insect terminology) that contains essential germ cell-specific components, polar granules (reviewed by Mahowald, 2001; Thomson and Lasko, 2005) (Figure 1b). Polar granules are ribonucleoprotein complexes containing Oskar (Ephrussi *et al.*, 1991), Tudor (Bardsley *et al.*, 1993), Germ cell-less (Johngen, *et al.*, 1994), Vasa (Hay *et al.*, 1998), Aubergine (Harris and Macdonald, 2001), a non-coding RNA, *polar granule component* (*pgc*) (Nakamura *et al.*, 1996) and mitochondrial large and small ribosomal RNAs (Iida and Kobayashi, 1998; Kashikawa, *et al.*, 1999).

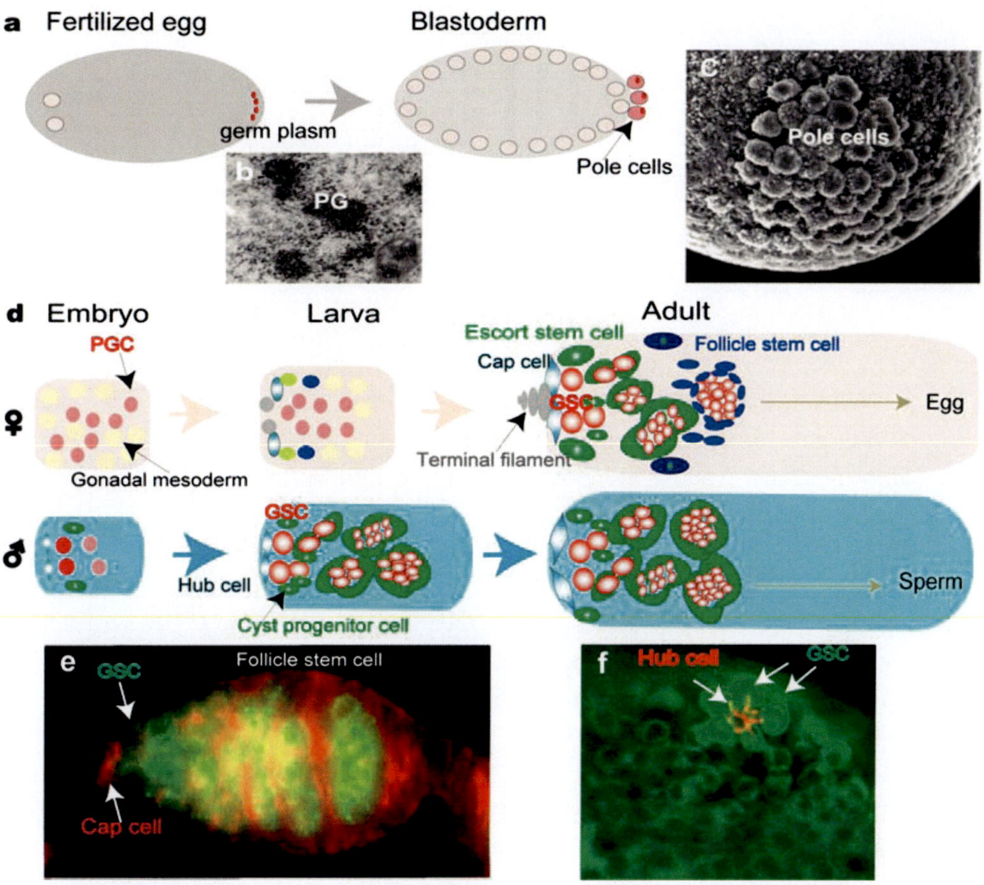

Figure 1. Schematic view of germline development of *Drosophila*. (a). Germ plasm and pole cell formation. (b). Transmission electron micrograph of polar granules (pg). (c). Scanning electron micrograph of pole cells at the blastoderm. (d). Development of gonads and the formation of germinal niche from embryonic to adult stages. (e). A germarium of adult ovary stained with anti-Vasa antibody and Rhodamine conjugated phalloidin. (f). Germinal niche of larval testes stained with anti-Fasciclin III and Vasa antibodies. The germ plasm is localized at the posterior pole of an embryo and includes electron-dense particles called polar granules. Pole cells bud off the embryo. PGCs associate with gonadal mesoderm to form gonads at late embryonic stage. In the male, the germinal niche is formed at the late embryonic stage. Hub cells (niche cells) locate at the proximal end of the testes. Spermatogonia are enclosed with a pair of the cyst progenitor cells. In the female, there is no sign of the formation of germinal niche during embryonic and early larval periods. Precursors of niche cells locate at the anterior region of gonads. PGCs that associate with the precursors of niche cells become GSCs during late larval to pupa stages. In an adult ovariole, cap cells locate at the anterior tip of an ovariole and the terminal filament stretches from the basal region of the cap cells. Escort stem cells locate just posteriorly to the cap cells and associate with GSCs and cap cells. Escort cells enclose the cystoblast and cystocytes and move them to the posterior of the germarium. Another set of somatic stem cells, follicle stem cells (FSCs) locate at the posterior region of a germarium and associate with 16 cystocytes to form an egg chamber.

pgc RNA functions to repress the activity of RNA polymerase II (Martinho, *et al.*, 2004). Aubergine is related to eukaryotic translation initiation factor 2C and functions for RNAi machinery to maintain stem cell status (reviewed by Carmell *et al.*, 2002). The *vasa* gene encodes ATP-dependent RNA hericase of the DEAD-box family, which contains the conserved motif Asp-Glu-Ala-Asp, and is typified by the eukaryotic translation initiation factor eIF4A (Lasko and Ashburner, 1998) and is required for promoting translation of at least two known mRNAs, *nanos* and *gurken* (Styhler *et al.*, 1998). Vasa homologs are found in many animals and have revealed a highly conserved role for Vasa protein among different organisms, as well as some important differences in its regulation (e.g., bird, Tsunekawa *et al.*, 2000; fish, Kobayashi *et al.*, 2002; ascidian, Sunanaga *et al.*, 2006; rat and mouse, Fujiwara *et al.*, 1994; Komiya and Tanigawa, 1995; Tanaka *et al.*, 2000). Nanos functions to migrate PGCs (Kobayashi *et al.*, 1996), to suppress somatic cell fate in pole cells (Hayashi *et al.*, 2004) and to suppress apoptosis (Sato *et al.*, 2007). Three homologs of Nanos are found in mice. Nanos2 and Nanos3 are essential for germ cell development and Nanos2 is involved in the male germ cell maintenance (Tsuda *et al.*, 2003).

PGCs divide several times during the blastoderm stage to form about 40 cells (Figure 1c), which subsequently are internalized in the midgut invagination. PGCs seem to be transcriptionally quiescent (Zalokar, 1976) and are arrested in the G2 phase of the cell cycle due to repression of Cyclin B production by Nanos and Pumilio (Asaoka-Taguchi *et al.*, 1999). At the onset of the gastrulation, PGCs begin to migrate through the midgut epithelium into the dorsal mesoderm where they aggregate with somatic gonadal precursors to form the embryonic gonad (Boyle *et al.*, 1997). Early events on the formation and migration of PGCs have been extensively studied (reviewed by Starz-Gaiano and Lehmann, 2001). PGCs reenter mitosis 16 hours after egg laying in a sexually dimorphic manner (Sonnenblick, 1941). The reentry of the division of PGCs is regulated by Wnt signal pathway (Sato *et al.*, 2008). Then, Decapentaplegic (Dpp), which belongs to the transforming growth factor-β (TGF-β) superfamily, a homolog of bone morphogenetic protein 4 (BMP4), acts as a main growth factor to regulate the mitosis of PGCs (Sato *et al.*, 2008) and represses the differentiation (Gilboa and Lehmann, 2004). Sex dimorphism becomes apparent before the first instar larva (Sonnenblick, 1941). In females, about twelve PGCs present at the first instar larva and continue to divide during larval periods. About one hundred PGCs populate in the pair of ovaries at the late third instar larval stage (King, 1970; Zhu and Xie, 2003; Gilboa and Lehmann, 2006). Among them, about 30 PGCs that associate with precursors of the cap cells are considered to become GSCs at the prepupal stage (King, 1970; Bhat and Schedl, 1997; Zhu and Xie, 2003; Asaoka and Lin, 2004) (Figure 1d). The remainder of PGCs differentiate into oocytes directly. In the male, PGCs resume division during early stages of gonad development (Wawersik *et al.*, 2005). The niche of male GSCs is formed at the late embryonic stage (Le Bras and Van Doren, 2006) (Figure 1d). Sevenless, a receptor tyrosine kinase, is required to ensure that the niche develops in the anterior region of male embryonic gonads (Kitadate *et al.*, 2007).

1.2. The Mouse

In mouse, the germ-cell lineage is induced by instructive signals. After the first 5 days of fertilization, the zygote becomes a blastocyst, and then begins the process of implantation (Figure 2a). At this time, the embryo consists of an inner cell mass (ICM), primitive endoderm, and trophectoderm. The ICM cells are the precursors for epiblast cells that give rise to the three germ layers (ectoderm, endoderm and mesoderm). Prior to the first identification of PGCs, their precursor cells originate in the proximal part of the epiblast near the extraembryonic tissues (Tam and Zhou, 1996) (Figure 2b). PGCs become recognizable in the posterior region of the extraembryonic mesoderm by their surface alkaline phosphatase activity (Chiquoine, 1954; Ginsburg *et al.*, 1990) and by predominant expression of a member of the interferon-induced transmembrane protein gene family, *Iftm3* (Tanaka and Matsui, 2002) at around Embryonic Day (E) 7.25. Cell labeling studies have shown that a population of posterior and posterolateral epiblast cells that enters the primitive streak proximally will give rise to this extra-embryonic mesoderm and to PGCs (Lawson and Hage, 1994; Parameswaran and Tam, 1995; Kinder *et al.*, 1999). PGCs are induced in response to bone morphogenetic proteins (BMPs) produced in the extra-embryonic ectoderm, but depend also on the presence of visceral endoderm (de Sousa Lopes *et al.*, 2004; Saiti and Lacham-Kaplan, 2007; Matsui and Okamura, 2005). The repertoire of BMP signaling proteins involved in PGC induction comprise BMPs (BMP2, -4 and -8b), a type I BMP receptor (Alk2) and two BMP-Smads (Smad1/5) (Lawson *et al.*, 1999; Ying *et al.*, 2000; Ying *et al.*, 2001; Ying and Zhao, 2001; Chang and Matzuk, 2001; Tremblay *et al.*, 2001; Hayashi *et al.*, 2002; de Sousa Lopes *et al.*, 2004; Okamura *et al.*, 2005). Blimp1, a potent transcriptional repressor of a histone methyltransferase subfamily, is thought to be induced by Bmp4 and is a key regulator of germ cell specification (Vincent *et al.*, 2005; Ohinata *et al.*, 2005). By analyzing single-cell cDNA libraries from founder PGCs at E7.5, two new genes, *fragilis* and *stella* are found to be specifically expressed whereas Hox genes show repressed expression in PGCs (Saitou *et al.*, 2002). Saitou *et al* propose that *stella*-positive and Hox-negative gene expression is fundamental for these cells to escape from the somatic cells. Interestingly, Blimp1 begins to express in the single layer of the most proximal epiblast cells in the E6.25 embryos at the onset of gastrulation but do not express in ICM or ES cells. Thus, the PGC specification might occur much earlier than previously thought in a restricted population of Blimp1-positive cells (Ohinata *et al.*, 2006). The cells expressing Blimp1 are the lineage-restricted germ cell precursors that subsequently go on to express *stella*. PGCs undergo extensive orderly reprogramming of epigenetic chromatin modifications soon after their fate is restricted (Seki *et al.*, 2005). In addition to BMP signaling, constant expression of E-cadherin in the precursor cells is necessary for the specification of PGCs (Okamura *et al.*, 2003). A subset of the precursors that loses E-cadherin expression migrates toward a more distal region and differentiates into allantois (Figure 2c). Cells maintaining E-cadherin expression are organized into clusters and are destined to form PGCs (reviewed by Matsui and Okamura, 2005). By blocking with a monoclonal antibody for E-cadherin, differentiation of PGCs is inhibited from the precursors of E6.75 embryos in culture. Subsequent PGC localization and survival also depend on the presence of Bmp4 in the extra-embryonic

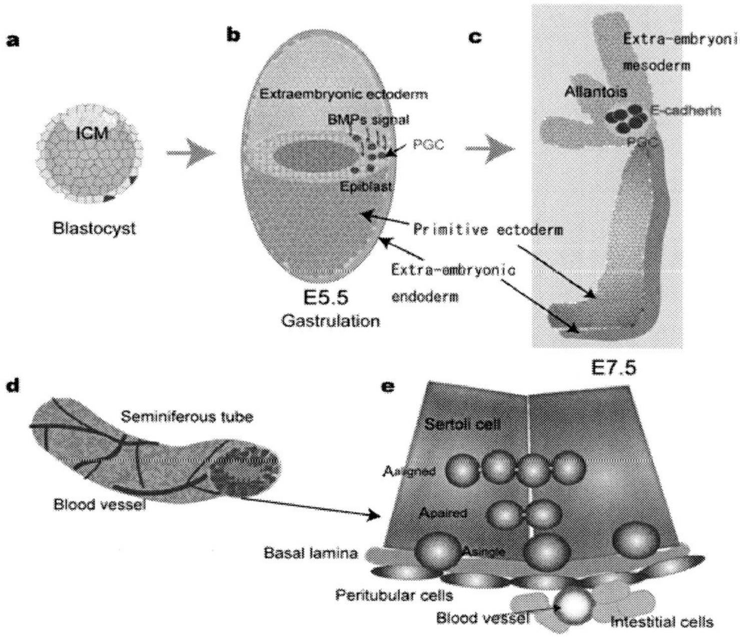

Figure 2. Schematic view of germline development in the mouse. (a) Blastocyst forming inner cell mass (ICM). (b). Gastrula embryo at Embryonic Day (E) 5.5. (c). PGCs appear in the extra-embryonic mesoderm at E7.5. (d). A seminiferous tube of adult testes. (e). The germinal niche of SSCs. In blastocyst, an embryo consists of an inner cell mass (ICM), primitive endoderm, and trophectoderm. The ICM is the precursor for epiblast cells that give rise to three germ layers and PGCs. Embryonic ectoderm cells release BMPs signals and some cells that receive them become precursors of PGCs at the posterior region of the extraembryonic mesoderm. Precursors of PGCs express several markers, alkaline phosphatase, *Iftm3* and Blimp1. Cells maintaining E-cadherin expression are organized into clusters and are destined to form PGCs. A subset of the precursors that loses E-cadherin expression migrates toward a more distal region and differentiates into allantois. In the adult mouse, spermatogenesis occurs within the seminiferous tubules, which are composed of germ cells, Sertoli cells and peritubular cells, and are surrounded by interstitial cells and blood vessels develop to make branches. All types of spermatogonia are localized on the peripheral basement membrane, and the subsequent cell types are arranged in a sequential order towards the lumen. The germinal niche is shown to be close to the blood vessel branch points.

mesoderm (Fujiwara *et al.*, 2001). From around E8.5, PGCs migrate into the embryonic mesoderm, through the hindgut endoderm and along the dorsal mesentery, and finally into the genital ridge at E10.5-11.5. PGCs continue to proliferate mitotically during the migration and after arrival at the genital ridge until E12.5-13.5. The processes of migration and differentiation of germ cells, as well as gonadal tissue formation in embryos, are largely regulated by growth factors and hormones, such as Steel factor, stromal cell-derived factor 1, anti-Müllerian hormone, or testosterone. At E12.5, developing gonads show the first sign of sexual dimorphism with testicular cord formation in the males. PGCs at this stage take different fates according to the sex of embryos. The sex-differentiation of the germ cells is directed by the sex of the gonadal somatic cells and not by that of the germ cells themselves (reviewed by McLaren, 2001). In males, PGCs are called gonocytes when entering in the fetal testis cord, cease mitosis and resume mitotic activity after birth. In the fetus, gonocytes are located in the center of the testis cord, but they either undergo apoptosis or migrate to the basement membrane and become GSCs during the first few days after birth (Spermatogonial Stem Cells, SSCs). In female, after arriving at the ovary, PGCs develop in clusters of cells (germ line cysts or oocyte nests). Subsequently, the oogonia enter meiosis and become oocytes. The oocyte nests break apart into individual cells and become packaged into primordial follicles. During this time, only a subset of oocytes ultimately survives and the remaining immature eggs undergo apoptosis. The adult ovary does not contain GSCs. Thus, the final number of oocytes available for reproduction of the next generation is defined at birth.

2. Germinal Niche and Signal Pathways of GSCs in the Fly

GSCs are seen in adult gonads of both sexes. GSCs locate at the anterior tip of gonads and are associated with niche cells via adherence junctions (Xie and Spradling, 2000; Spradling *et al*., 2001) (Figure 1d-f). The undifferentiated status and self-renewal mitosis of the GSCs are maintained by growth factors that are secreted from the niche and possibly other surrounding somatic cells. One daughter cell of GSC division that moves away from the niche is released from the repression by the growth factors, begins to express the differentiation genes and undergoes a series of differentiation steps. There are presented many common mechanisms of signal transduction pathways, niche formation and differentiation processes among them (reviewed by Xi *et al*., 2005; Kirilly and Xie, 2007; Fuller and Spradling, 2007).

2.1. Female GSCs and Their Niche

In the adult female, each ovary consists of approximately 15 ovarioles or egg tubes. At the anterior tip of each ovariole in the region called the germarium, a small number of somatic cells (terminal filament, cap and escort cells) form a germinal niche where 2 to 3 female GSCs are located (Wieschaus and Szabad, 1979) (Fig. 1e). In addition to fGSCs, two sets of somatic stem cells, i.e., escort stem cells (ESCs) (Decotto and Spradling, 2005) and somatic (follicle) stem cells (FSCs)* are located in the germarium of each ovariole (Margolis and Spradling, 1985). (*The term of somatic stem cells abbreviated as SSC would be mistakable for the mammals' Spermatogonial Stem Cell and I will use the follicle stem cell (FSC) in the present review.) Maintenance of fGSCs requires contact with the cap cells by means of adherens (Song *et al*., 2002) and gap (Tazuke *et al*., 2002; Gilboa *et al*., 2003) junctions, and the continuous secretion of Dpp by terminal filament and cap cells (Xie and Spradling, 1998). Dpp functions as a short-range growth factor that represses the expression of the *bag-of-marble* (*bam*) gene that initiates the differentiation process. One of the daughter cells of fGSCs moves away from the niche to become a cystoblast. The Dpp signal is released and *bam* is expressed in the cystoblast (Chen and McKearin, 2003). Cystoblast continues four additional divisions with incomplete cytokinesis to produce 16 cystocytes. It is noted that germline cells at each stage are identifiable by the presence of germline specific membranous structures called spectrosome (fGSCs and cystoblast) or fusome (cystocytes), which are enriched for membrane skeletal proteins such as α- and β-spectrin, the adducin-like Hts protein and ankyrin (Lin *et al.*, 1994; Lin and Spradling, 1995; de Cuevas and Spradling, 1998) (Fig. 3). Spectrosome and fusome anchor mitotic spindles during asymmetric germ cell divisions (Deng and Lin, 1997). When the cystoblast undergoes four synchronous nuclear divisions with incomplete cytokinesis, the spectrosome enlarges into the fusome that stretches through each of the cytoplasmic bridges connecting the 16 daughter cells (reviewed by de Cuevas *et al.*, 1997). Escort cells encase developing cysts, move them posteriorly at the germarium and then undergo apoptosis at the time of follicle (egg chamber) formation. Then the prefollicular cells, descendant of FSCs (Margolis and Spradling, 1995; Zhang and Kalderon, 2000), enclose the 16 cysts to form an egg chamber. One of the 16 cystocytes is determined as an oocyte. JAK/STAT signaling plays for the maintenance of escort cells (Decotto and Spradling, 2005). Proliferation of FSCs depends upon Hedgehog (Hh) secreted by the terminal filament and cap cells (Forbes *et al.*, 1996; Zhang and Kalderon, 2001) and maintenance of FSCs requires adherens junctions (Song and Xie, 2002). Dpp and Gbb (homolog of BMP5-8) also play a role in maintaining FSCs (Kirilly *et al.*, 2005). fGSCs are lost when signal transduction pathways are inhibited, whereas the niche cells can sustain after fGSCs are removed. By removing fGSCs, an 'empty' niche was induced experimentally, where cap cell activity persisted. In this empty niche, inner gonial sheath cells, including FSCs, enter the empty niche, and prefollicular cells proliferate as long as cap cells remain (Kai and Spradling, 2003). It is noteworthy that the determination process is reversible. Four- and eight-cell cystocytes can be converted into single stem-like cells when Dpp is overproduced (Kai *et al.*, 2004). The same reversion into GSCs is found in males (Brawley and Matunis, 2004).

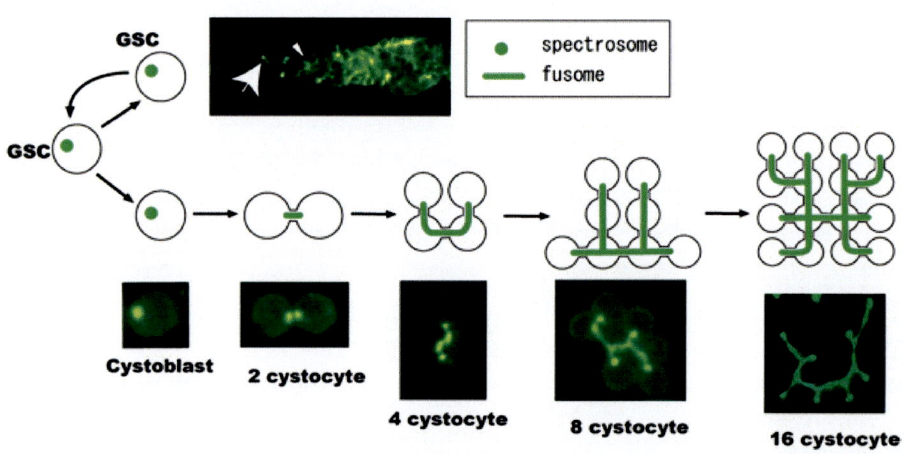

Figure 3. Morphological changes of germline-specific membranous structures, spectrosome and fusome. Spectrosome is dot structure in GSCs and cystoblast. As following incomplete cytokinesis of cystocytes, spectrosome becomes branched and is called fusome.

In addition, maintenance of stem cells requires a number of other germline functions, such as *nanos* and *pumilio* (Forbes and Lehmann1998), and *piwi* (Cox, *et al.*, 2000) although *piwi* plays a more important role in the soma, and a somatic niche, characterized by the expression of *fs(1)Yb* (King *et al.*, 2001). In each instance, mutations in these genes lead to a loss of GSCs.

2.2. Male GSCs and Their Niche

One of the conspicuous differences between sexes is that the male germinal niche is formed at the late embryonic stage (Fig. 1d). Thus, the number of male GSCs is determined until the larval stage. The adult testis is a long, coiled tube filled with cells at all stages of spermatogenesis (reviewed by Fuller, 1993). Approximately nine mGSCs present at the apical tip of the testis, forming a ring that closely surrounds a cluster of post-mitotic somatic cells called the hub (niche cells) (Fig. 1f). One of the daughter cells of mGSCs that moves away from the niche becomes gonialblast. The gonialblast undergoes four rounds of mitotic divisions with incomplete cytokinesis, creating a cluster of 16 interconnected spermatogonia. In addition, each spermatogonium is surrounded by a pair of the cyst progenitor cells (somatic stem cells). JAK/STAT signaling maintains mGSCs self-renewal (Kiger *et al.*, 2001; Rawlings *et al.*, 2004; Tulina and Matunis, 2001). The ligand that activates the *Drosophila* JAK/STAT pathway, Upd, is normally expressed in the hub cells. Ectopic expression of Upd in early germ cells leads to a dramatic increase in the number of mGSCs (Kiger *et al.*, 2001; Tulina and Matunis, 2001). The Upd ligand is glycosylated and is tightly associated with the extracellular matrix after secretion, which potentially limits its diffusion (Harrison *et al.*, 1998). The BMP signal produced by the hub cells also functions to control mGSC self-renewal by repressing *bam* expression (Shivdasani and Ingham, 2003; Schulz *et al.*, 2004; Kawase *et al.*, 2004). *bam* functions to direct the differentiation of mGSCs into spermatogonia as seen in fGSCs (Gonczy *et al.*, 1997; Schulz *et al.*, 2002). However, *bam* expression starts around 4-cell spermatogonia, and *bam* mutation (or dpp over expression) causes overproliferation of spermatogonia, which are interconnected each other (sometimes more than 16 cell spermatogonia, such as 32, 64 etc.) but not mGSCs. Thus, *dpp-bam* control point appears to be a bit different in the male germ line. A rhomboid homolog, *stet* organizes encapsulation by cyst progenitor cells. In *stet* testes, numerous mGSCs accumulate (Schulz *et al.*, 2002). Spindle orientation plays a crucial role in regulating stem cell behavior (Yamashita *et al.*, 2003). mGSCs divide perpendicular to the hub cells. Asymmetric mGSC divisions are controlled by a combination of extrinsic factor(s) that specify stem cell identity and intrinsic cellular machinery that acts at the centrosome and cell cortex to orient the cell division plane with respect to the signaling microenvironment (reviewed by Yamashita *et al.*, 2005). Developmentally programmed asymmetric behavior and inheritance of mother and daughter centrosomes underlie the stereotyped spindle orientation and asymmetric outcome of stem cell divisions. The mother centrosome remains anchored near the niche while the daughter centrosome migrates to the opposite side of the cell before spindle formation (Yamashita *et al.*, 2007).

2.3. Mouse SSCs and Their Niche

Spermatogenesis occurs within the seminiferous tubules, which are composed of germ cells, Sertoli cells, and peritubular cells (Figure 2d, e). During adult spermatogenesis in the mouse, the stem cell activity resides in a small, primitive set of spermatogonia referred as the undifferentiated spermatogonia, which correspond to A_{single}, A_{paired}, and $A_{aligned}$ spermatogonia (de Rooij and Russell, 2000). Besides self-renewing as a population, undifferentiated spermatogonia generate differentiating spermatogonia, meiotic spermatocytes, haploid spermatids and spermatozoa. In the seminiferous tubules, all types of spermatogonia are localized on the peripheral basement membrane, and the subsequent cell types are arranged in a sequential order towards the lumen (Figure 2e). The technique of spermatogonial transplantation makes it possible to study the functional aspect of spermatogonial stem cells (Brinster and Zimmermann, 1994; Ogawa, *et al.*, 2000). In addition, many new developments, such as cryopreservation, xenotranplantation, purification, and culturing of spermatogonial stem cells, have been achieved (reviewed by Ogawa, 2001 and Brinster, 2002). Xenotransplantation experiments suggest that external factors to support self-renewal of SSCs would be conserved among various mammalian species including human (Ogawa, *et al.*, 1999; Nagano *et al.*, 2002). Sertoli cells are the only somatic cells that interact directly with germ cells, therefore are supposed to constitute the primary cellular component of germinal niche. The lack of expression of membrane-bound Steel factor (SF) on Sertoli cells in infertile Steel (Sl) mutant mice prevents the differentiation of spermatogonia that express c-kit receptor, resulting in azoospermia (Flanagan *et al.*, 1991), whereas the germ cells in infertile white spotting (W) mice have mutations in the c-kit receptor, causing a similar condition (Geissler *et al.*, 1988). When SSCs from Sl mice are transplanted into the testes of W mice that have normal Sertoli cells, complete spermatogenesis occurs and offspring are produced from the transplanted Sl stem cells (Ogawa *et al.*, 2000). Furthermore, when Sertoli cells of W mice are transplanted into infertile Sl mice, the fertility is recovered (Kanatsu-Shinohara *et al.*, 2005a). Thus, the interaction between Sertoli cells and germ cells is critical for spermatogenesis. Sertoli cells produce glial cell line-derived neurotrophic factor (GDNF) and disruption of GDNF expression results in spermatogenic defects and infertility. GDNF stimulates SSC self-renewal (Meng *et al.*, 2000). Murine SSCs can be expanded *in vitro* by adding GDNF in the culture media (Kanatsu-Shinohara *et al.*, 2003, Kubota *et al.*, 2004). Clearly, GDNF-induced cell signaling plays a central role in SSC self-renewal. Using time-lapse imaging of Ngn3/EGFP–labeled undifferentiated spermatogonia and three-dimensional reconstitution, undifferentiated spermatogonia are found to preferentially located in restricted portions of the basal compartment within the seminiferous tubule, which are adjacent to the blood vessels and interstitium and are particularly close to the blood vessel branch points. Thus, the germinal niche seems to form as an indirect consequence of vasculature formation, which is much different from those determined by the developmental program in the fly (Yoshida *et al.*, 2007).

3. EG and GS Cell Lines in the Mouse

Many attempts to culture mouse PGCs have been reported. All available culture methods for mouse PGCs rely upon the presence of cell feeder layers (reviewed by De Felici, 1998). By adding basic fibroblast growth factor (bFGF) in culture media, mouse PGCs have been successfully cultured and named as EG cells (Matsui *et al.*, 1992). The EG cells continue to proliferate and then produce pluripotent ES cells with characteristics resembling the cells derived from the inner cell mass (Matsui *et al.*, 1992; Resnick *et al.*, 1992). Besides soluble growth factors and compounds, adhesion to cell monolayer *in vitro* and neighboring cells *in vivo* is critical for optimal PGC growth (reviewed by De Felici *et al.*, 2004). A cocktail of soluble growth factors, SF, leukemia inhibitory factor (LIF), BMP-4, stroma derived factor-1, bFGF and compounds (N-acetyl-cysteine, forskolin, retinoic acid) are needed to sustain the survival and self-renewal of mouse PGCs in the absence of somatic cell support (Farini *et al.*, 2005). A stable cell line of SSCs (GS cell) has been also established by culturing enzymatically dispersed testes in a medium containing GDNF, bFGF, epidermal growth factor, and LIF(Kanatsu-Shinohara *et al.*, 2003). The GS cells can be maintained with self-renewing divisions even without feeder cells for a long-term culture. The GS cells are positive for SSC markers and negative for PGC marker. During long time culture, the GS cells retain the functional ability but their telomeres gradually shorten, suggesting that they are not immortal (Kanatsu-Shinohara *et al.*, 2005b). Interestingly, there appear ES-like cells, termed multi-potent GS cells (mGSCs) showing multi-potent differentiating ability, in the culture of GS cells (Kanatsu-Shinohara, *et al.*, 2004). The mGSCs are able to spontaneously differentiate into derivatives of the three embryonic germ layers *in vitro* and generate teratomas in immuno-deficient mice. When injected into an early blastocyst, mGSCs contribute to the development of various organs and show germline transmission. Thus, the capacity to form multipotent cells persists in the adult mouse testis. Similar multipotent adult germline stem cells have been also established from the adult mouse testis by two independent groups (Guan *et al.*, 2006; Seandel *et al.*, 2007).

4. Origin of ES Cells and PGC-Like Cells in Adult Tissue

ES cells can be induced to differentiate into germ cells (Hübner *et al.*, 2003; Toyooka *et al.*, 2003). At least some of the germ cell-specific genes expressed by ES cells, and not by primitive ectoderm cells, are essential for the long-term maintenance of the pluripotent state. A hypothesis is proposed that ES, EC and EG cells are originated from PGCs by their common properties (Zwaka and Thomson, 2005).

In the female, GSCs have been believed not to present in the adult ovary. However, there is a report indicating the presence of GSCs in adult females (Johnson *et al.*, 2005). When cells of bone marrow were transplanted into sterilized mice, fertility was restored. They emphasize that GSCs present in bone marrow and differentiate into functional germ cells. There arise many critical commentaries for and against their work (reviewed by McLean, 2006; Tilly and Johnson, 2007). Recently, bone marrow stem cells are shown to differentiate into male germ cells (Nayernia *et al.*, 2006). Ratajczak and co-workers reported the presence of PGC-like cells in bone marrow and named them very small embryonic-like (VSEL) stem cells (reviewed by Kunica *et al.*, 2007 and Ratajczak *et al.*, 2007). VSEL cells express several markers of PGCs (fetal-type alkaline phosphatase, Oct-4, SSEA-1, CXCR4, Mvh, Stella, Fragilis, Nobox and Hdac6) and display several features typical for ES cells such as small size (2-4μm in diameter), large nuclei surrounded by a narrow rim of cytoplasm and euchromatin. VSEL cells are released from bone marrow and circulate in peripheral blood during tissue/organ injuries (Kucia *et al.*, 2004, 2006; Wojakowski *et al.*, 2004). Similar cells that express embryonic PGC marker have been found in various anatomical locations (reviewed by Kucia *et al.*, 2007 and Ratajczak *et al.*, 2007). These lines of evidence would suggest that PGCs have multiple functions to become tissue or cancer stem cells in adult tissues and not only produce functional germ cells.

5. Establishment of fGSCs in the Fly

As described above, the *in vitro* systems are very useful for the study of germ cells in the mouse but were not fully developed to be available in the research of PGCs and GSCs in the fly. One of the impediments to the *in vitro* study of fly germ cells is the difficulty of collecting sufficient number of germline cells for cell cultures, though many attempts have been reported to culture embryonic and imaginal somatic cells (Echalier, 1997). PGCs at the blastoderm stage have been collected with a density gradient and checked for morphological criteria (Allis *et al.*, 1977). After culturing, PGCs showed normal metabolic properties but did not sustain to be alive for a long-term culture (Allis *et al.*, 1979). Another approach for collecting purified germline cells is to use tumorous gonads by mutation. As described above, fGSCs continue to grow and the resulting daughter cells retain the stem cell status in *bam* mutant ovaries. Interestingly, ESC and FSC cells also continue to divide and thus numerous fGSCs, escort and prefollicular cells accumulate in expanding *bam* tumorous ovaries (Figure 4b). More than 10,000 cells of fGSCs and almost the same number of somatic cells accumulate in *bam* ovarioles of 30-day-old females (Niki unpublished). *bam* fGSCs can associate with precursors of gonadal mesoderm and populate at embryonic gonads when transplanted into recipient embryos. Furthermore, donor *bam* fGSCs can differentiate into functional germ cells when wild-type allele of *bam* gene is expressed (Niki and Mahowald, 2003). Thus, *bam* fGSCs can be 'rejuvenated' as PGCs in embryonic environment to become again functional fGSCs to produce offspring in adult ovaries. As *in vivo* experiments, Dpp is most effective for the growth of *bam* fGSCs and sonic Hedgehog (sHh) for the soma in culture. The mixture of Dpp and sHh enhances the growth of both of soma and *bam* GSCs. Although *bam* fGSCs and soma stably sustain for many months, cell death takes place frequently in culture media containing Dpp and sHh. By adding fly extract instead of growth factors, stable cell lines from *bam* ovarioles is established and named fGS/OSS (female Germline Stem cells/ Ovarian Somatic Sheet) (Niki *et al.*, 2006) (Figure 4c). OSS cells are descendants of FSCs and associate with fGSCs via DE-cadherin and β-catenin, indicative of the presence of adherens junctions. Both *bam* fGSCs and the accompanying somatic cells can be cultured stably more than three years and require splitting every 5-7 days. The germ cells in fGS/OSS cultures continue to express Vasa and Nanos and show the dot spectrosome characteristics of fGSCs *in vivo*. OSS cells produce Dpp and pMAD appear in the nuclei of the germ cells indicating that Dpp signal pathway occurs between them. That is, germinal niche signal transduction pathways seen *in vivo* occur in fGS/OSS, though the niche cells are replaced with prefollicular cells instead of cap cells (Figure 4d). OSS cells would be very useful for culturing fGSCs of various genotypes. In addition, OSS can sustain the maintenance and division of mGSCs for long-term culture (Niki, unpublished). This would provide opportunities to study the commonality and difference of GSCs and the germinal niche between sex at the cellular and molecular level. Furthermore, it would be possible to give insight into evolutional aspects of conserved genes that function in both the mouse and fly.

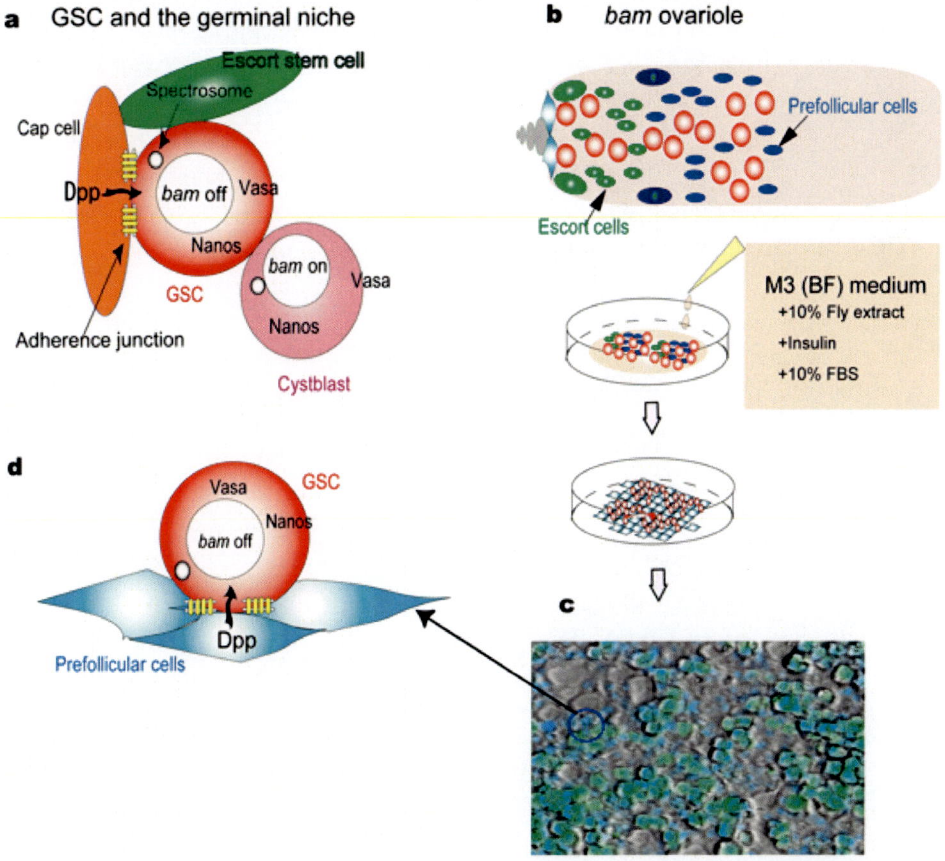

Figure 4. Signal transduction pathway between fGSCs and cap cells, and establishment of a stable cell line of fGS/OSS. (a). Signal transduction and adherence junctions between fGSCs and cap cells. (b). Schematic view of *bam* ovariole. (c). Living fGS/OSS cells marked with *vasa*-GFP. (d). Signal transduction and association between prefollicular cells and fGSCs. Cap cells produce Dpp (homology of BMP4) and *bam* is repressed to express in fGSCs. There adherence junctions present between cap cells and fGSCs. In *bam* ovarioles, escort and prefollicular cells are overproduced in addition to GSCs. By culturing cell masses of *bam* ovarioles in media with fly extract, insulin and fetal bovine serum, a stable cell line, fGS/OSS, has been established. In fGS/OSS, prefollicular cells and fGSCs associated via DE-cadherin and Dpp signal pathway occurs as seen at the niche *in vivo*.

Finally, extensive studies of molecular profiling of fGSCs (Kai *et al.*, 2005), mGSC (Terry *et al.*, 2006) and embryonic PGCs (Shigenobu *et al.*, 2006) have been carried out. Molecular profiling of fGSCs and mGSCs in our cultures is being conducted (Shigenobu *et al*).

6. Conclusion

At present, GSCs and their niche become accessible to analyze *in vitro* at the molecular level in both the fly and mouse. The stable cell lines of GSCs would provide opportunities for genetic modifications such as gene targeting and RNAi technique and cryoreservation of germline cells of various genotypes. In the mouse, *in vitro* retroviral-mediated gene delivery into SSCs of both adult and immature mice results in stable integration and expression of a transgene in SSCs (Nagano *et al.*, 2001). Techniques of transplantation and *in vitro* culture are requisite for an application of human SSCs in medicine for restoration of fertility in patients who have defects in producing functional germ cells due to diseases or drug or radiated treatments. Another potential clinical application using human SSCs is germline gene therapy for many genetic disorders (reviewed by Kubota and Brinster, 2006; Brinster, 2007). PGC-like cells would be also useful for germline gene therapy.

Acknowledgments

I thank Drs. Toshie Kai at Temasek Life Sciences Laboratory at Singapore, Satoru Kobayashi at Okazaki Institute for Integrative Bioscience, and Takehiko Ogawa at Dept. of Urology, Yokohama City Univ.; Yukiko Yamashita at Life Science Inst., Michigan Univ. and Shosei Yoshida at Dept. of Pathology and Tumor Biol., Kyoto Univ. for their careful reading of the manuscript and their suggestions. I am especially grateful to Dr. Anthony P. Mahowald at Dept. Mol. Genet. & Cell Biol., Univ. of Chicago, for his continuing encouragement and helpful suggestions. The research reported here was supported by the Naito Foundation and Japan Society for the Promotion of Science and Technology in Japan.

References

Allis, C. D., Waring, G. L. and Mahowald, A. P. (1977) Mass isolation of pole cells from *Drosophila melanogaster*. *Dev. Biol.* 56, 372–381.

Allis, C. D., Underwood, E. M., Caulton, J. H. and Mahowald, A. P. (1979) Pole cells of *Drosophila melanogaster* in culture. Normal metabolism, ultrastructure, and functional capabilities. *Dev. Biol.* 69, 451–465.

Asaoka-Taguchi, M., Yamada, M., Nakamura, A., Hanyu, K. and Kobayashi, S. (1999) Maternal Pumilio acts together with Nanos in germline development in *Drosophila* embryos. *Nature Cell Biol.* 1, 431–437.

Asaoka, M. and Lin, H. (2004) Germline stem cells in the *Drosophila* ovary descend from pole cells in the anterior region of the embryonic gonad. *Development* 131, 5079–5089.

Bardsley, A., McDonald, K. and Boswell, R. E. (1993) Distribution of *tudor* protein in the *Drosophila* embryo suggests separation of functions based on site of localization. *Development* 119, 207–219.

Bellve, A. R., Cavicchia, J. C., Millette, C. F., O'Brien, D. A, Bhatnagar, Y. M. and Dym, M. (1977) Spermatogenic cells of the prepuberal mouse. Isolation and morphological characterization. *J. Cell Biol.* 74, 68–85.

Bhat, K. M. and Schedl, P. (1997) Establishment of stem cell identity in the *Drosophila* germline. *Dev. Dyn.* 210, 371–382.

Boyle, M., Bonini, N. and DiNardo, S. (1997) Expression and function of *clift* in the development of somatic gonadal precursors within the *Drosophila* mesoderm. *Development* 124, 971–982.

Brawley, C. and Matunis, E. (2004) Regeneration of male germline stem cells by spermatogonial dedifferentiation *in vivo*. *Science* 304, 1331–1334.

Brinster, R. L. and Zimmermann, J. W. (1994) Spermatogenesis following male germ cell transplantation. *Proc. Natl. Acad. Sci. USA* 91, 11298–11302.

Brinster, R. L. (2002) Germline stem cell transplantation and transgenesis. *Science* 296, 2174–2176.

Brinster, R. L. (2007) Male Germline Stem Cells: From mice to men. *Science* 316, 404–405.

Carmell, M. A., Xuan, Z., Zhang, M. Q. and Hannon, G. J. (2002) The Argonaute family: tentacles that reach into RNAi, developmental control, stem cell maintenance, and tumorigenesis. *Genes Dev.* 16, 2733–2742.

Chang, H. and Matzuk, M. M. (2001) Smad5 is required for mouse primordial germ cell development. *Mech. Dev.* 104, 61–67.

Chen, D. and McKearin, D. M. (2003) Dpp signaling silences *bam* transcription directly to establish asymmetric divisions of germline stem cells. *Curr. Biol.* 13, 1786–1791.

Chiquoine, A. D. (1954) The identification, origin, and migration of the primordial germ cells of the mouse embryo. *Anat. Rec.* 118, 135–146.

Cox, D. N., Chao, A., Baker, J., Chang, L., Qiao, D. and Lin, H. (1998) A novel class of evolutionarily conserved genes defined by *piwi* are essential for stem cell self-renewal. *Genes Dev.* 12, 3715–3727.

Cox, D. N., Chao, A. and Lin, H. (2000) *piwi* encodes a nucleoplasmic factor whose activity modulates the number and division rate of germline stem cells. *Development* 127, 503–514.

Decotto, E. and Spradling, A. C. (2005) The *Drosophila* ovarian and testis stem cell niches: similar somatic stem cells and signals. *Devel. Cell* 9, 501–510.

de Cuevas, M., Lilly, M. A. and Spradling, A. C. (1997) Germline cyst formation in *Drosophila*. *Annu. Rev. Genet.* 31, 405–428.

de Cuevas, M. and Spradling, A. C. (1998) Morphogenesis of the *Drosophila* fusome and its implications for oocyte specification. *Development* 125, 2781–2789.

De Felici, M. (1998) Isolation and culture of germ cells from the mouse embryo. In *Cell Biology: A Laboratory Handbook*. Vol. 1. pp. 73–85. Academic Press, San Diego.

De Felici, M., Scaldaferri, M. L., Lobascio, M., Iona, S., Nazzicone, V., Klinger, F. G. and Farini, D. (2004) Experimental approaches to the study of primordial germ cell lineage and proliferation. *Hum. Reprod. Updat.* 10, 197–206.

Deng, W. and Lin, H. (1997) Spectrosomes and fusomes anchor mitotic spindles during asymmetric germ cell divisions and facilitate the formation of a polarized microtubule array for oocyte specification in *Drosophila. Dev. Biol.* 189, 79–94.

de Rooij, D. G. and Russell, L. D. (2000) All you wanted to know about spermatogonia but were afraid to ask. *J. Andro.* 21, 788–798.

de Sousa Lopes, S. M., Roelen, B. A., Monteiro, R. M., Emmens, R, Lin, H. Y., Li, E., Lawson, K. A. and Mummery, C. L. (2004) BMP signaling mediated by ALK2 in the visceral endoderm is necessary for the generation of primordial germ cells in the mouse embryo. *Genes Devel.* 18, 1838–1849.

Echalier, G. (1997) *Drosophila* cells in culture. Academic Press, San Diego.

Eddy, E. M. (1975) Germ plasm and the differentiation of the germ cell line. *Int. Rev. Cytol.* 43, 229–280.

Ephrussi, A., Dickinson, L. K. and Lehmann, R. (1991) Oskar organizes the germ plasm and directs localization of the posterior determinant *nanos*. *Cell* 66, 37–50.

Extavour, C. G. M. (2007) Evolution of the bilaterian germ line: lineage origin and modulation of specification mechanisms. *Integr. Comp. Biol.* 47, 770–785.

Extavour, C. G. M. and Akam, M. (2003) Mechanisms of germ cell specification across the metazoans: epigenesis and preformation. *Development* 130, 5869–5884.

Farini, D., Scaldaferri, M. L., Iona, S., Sala, G. L. and De Felici, M. (2005) Growth factors sustain primordial germ cell survival, proliferation and entering into meiosis in the absence of somatic cells. *Dev. Biol.* 285, 49–56.

Flanagan, J. G., Chan, D. C. and Leder, P. (1991) Transmembrane form of the *Kit* ligand growth factor is determined by alternative splicing and is missing in the *Sld* mutant. *Cell* 65, 1025–1035.

Forbes, A. J., Lin, H., Ingham, P. W. and Spradling, A. C. (1996) Hedgehog is required for the proliferation and specification of ovarian somatic cells prior to egg chamber formation in *Drosophila. Development* 122, 1125–1135.

Forbes, A. and Lehmann, R. (1998) Nanos and Pumilio have critical roles in the development and function of *Drosophila* germline stem cells. *Development* 125, 679–690.

Fujiwara, Y., Komiya, T., Kawabata, H., Sato, M., Fujimoto, H., Furusawa, M. and Noce, T. (1994) Isolation of a DEAD-family protein gene that encodes a murine homolog of *Drosophila vasa* and its specific expression in germ cell lineage. *Proc. Natl. Acad. Sci. USA* 91, 12258–12262.

Fujiwara, T., Dunn, N. R. and Hogan, B. L. M. (2001) Bone morphogenetic protein 4 in the extraembryonic mesoderm is required for allantois development and the localization and survival of primordial germ cells in the mouse. *Proc. Natl. Acad. Sci. USA* 98, 13739–13744.

Fuller, M. T. (1993) Spermatogenesis. In *The Development of* Drosophila melanogaster. (ed. M. Bate and A. Martinez-Arias). Vol. 1, pp.71–147. Cold Spring Harbor Press, New York.

Fuller, M. T. and Spradling, A. C. (2007) Male and female *Drosophila* germline stem cells: Two versions of immortality. *Science* 316, 402–404.

Geissler, E. N., Ryan, M. A. and Housman, D. E. (1988) The dominant-white spotting (W) locus of the mouse encodes the c-kit proto-oncogene. *Cell* 55, 185–192.

Gilboa, L., Forbes, A., Tazuke, S. I., Fuller, M. T. and Lehmann, R. (2003) Germ line stem cell differentiation in *Drosophila* requires gap junctions and proceeds via an intermediate state. *Development* 130, 6625–6634.

Gilboa, L. and Lehmann, R. (2004) Repression of primordial germ cell differentiation parallels germ line stem cell maintenance. *Curr. Biol.* 14, 981–986.

Gilboa, L. and Lehmann, R. (2006) Soma-germline interactions coordinate homeostasis and growth in the *Drosophila* gonad. *Nature* 443, 97–100.

Ginsburg, M., Snow, M. H. and McLaren, A. (1990) Primordial germ cells in the mouse embryo during gastrulation. *Development* 110, 521–528.

Gonczy, P., Matunis, E. and DiNardo, S. (1997) *bag-of-marbles* and *benign gonial cell neoplasm* act in the germline to restrict proliferation during *Drosophila* spermatogenesis. *Development* 124, 4361–4371.

Guan, K., Nayernia, K., Maier, L. S., Wagner, S., Dressel, R., Lee, J. H. and Nolte, J. (2006) Pluripotency of spermatogonial stem cells from adult mouse testis. *Nature* 440, 1199–1203.

Harris, A. N. and Macdonald, P. M. (2001) *aubergine* encodes a *Drosophila* polar granule component required for pole cell formation and related to eIF2C. *Development* 128, 2823–2832.

Harrison, D. A., McCoon, P. E., Binari, R., Gilman, M. and Perrimon, N. (1998) *Drosophila* unpaired encodes a secreted protein that activates the JAK signaling pathway. *Genes Dev.* 12, 3252–326.

Hay, B., Jan, L. and Jan, Y. (1988) A protein component of *Drosophila* polar granules is encoded by *vasa* and has extensive sequence similarity to ATP-dependent helicases. *Cell* 55, 577–587.

Hayashi, K., Kobayashi, T., Umino, T., Goitsuka, R., Matsui, Y. and Kitamura, D. (2002) SMAD1 signaling is critical for initial commitment of germ cell lineage from mouse epiblast. *Mech. Dev.* 118, 99–109.

Hayashi, Y., Hayashi, M. and Kobayashi, S. (2004) Nanos suppresses somatic cell fate in *Drosophila* germ line. *Proc. Natl. Acad. Sci. USA* 101, 10338–10342.

Hübner, K., Fuhrmann, G., Christenson, L. K., Kehler, J., Reinbold, R., De LaFuente, R., Wood, J., Strauss, J. F. 3rd., Boiani, M. and Scholer, H. R. (2003) Derivation of oocytes from mouse embryonic stem cells. *Science* 300, 1251–1256.

Iida, T. and Kobayashi, S. (1998) Essential role of mitochondrially encoded large rRNA for germ-line formation in *Drosophila* embryos. *Proc. Natl. Acad. Sci. USA* 95, 11274–11278.

Jongens, T. A., Ackerman, L. D., Swedlow, J. R., Jan, L. Y. and Jan, Y. N. (1994) *Germ cell-less* encodes a cell type specific nuclear pore-associated protein and functions early in the germ-cell specification pathway of *Drosophila. Genes Dev.* 8, 2123–2136.

Johnson, J., Bagley, J., Skaznik-Wikiel, M., Lee, H.J., Adams, G. B., Niikura, Y., Tschudy, K. S., Tilly, J.C., Cortes, M. L., Forkert, R., Spitzer, T., Iacomini, J., Scadden, D. T. and

Tilly, J. L. (2005) Oocyte generation in adult mammalian ovaries by putative germ cells in bone marrow and peripheral blood. *Cell* 122, 303–315.

Kai, T. and Spradling, A. (2003) An empty *Drosophila* stem cell niche reactivates the proliferation of ectopic cells. *Proc. Natl. Acad. Sci. USA* 100, 4633–4638.

Kai, T. and Spradling, A. (2004) Differentiating germ cells can revert into functional stem cells in *Drosophila melanogaster* ovaries. *Nature* 428, 564–569.

Kai, T., Williams, D. and Spradling, A. C. (2005) The expression profile of purified *Drosophila* germline stem cells. *Dev. Biol.* 283, 486–502.

Kanatsu-Shinohara, M., Ogonuki, N., Inoue, K., Miki, H., Ogura, A., Toyokuni, S. and Shinohara, T. (2003) Long-term proliferation in culture and germline transmission of mouse male germline stem cells. *Biol. Reprod.* 69, 612–616.

Kanatsu-Shinohara, M., Inoue, K., Lee, J., Yoshimoto, M., Ogonuki, N., Miki, H., Baba, S., Kato, T. Kazuki, Y., Toyokuni, S., Toyoshima, M., Niwa, O., Oshimura, M., Heike, T., Nakahata, T., Ishino, F., Ogura, A. and Shinohara, T. (2004) Generation of pluripotent stem cells from neonatal mouse testis. *Cell* 119, 1001–1012.

Kanatsu-Shinohara, M., Miki, H., Inoue, K., Ogonuki, N., Toyokuni, S., Ogura, A. and Shinohara, T. (2005a) Germline niche transplantation restores fertility in infertile mice. *Hum. Repro.* 20, 2376–2382.

Kanatsu-Shinohara, M., Ogonuki, N., T. Iwano, T., Lee, J., Kazuki, Y., Inoue, K., Miki, H., Takehashi, M., Toyokuni, S., Shinkai, Y., Oshimura, M., Ishino, F., Ogura, A. and Shinohara, T. (2005b) Genetic and epigenetic properties of mouse male germline stem cells during long-term culture. *Development* 132, 4155–4163.

Kashikawa, M., Amikura, R., Nakamura, A. and Kobayashi, S. (1999) Mitochondrial small ribosomal RNA is present on polar granules in early cleavage embryos of *Drosophila melanogaster*. *Dev. Grow Diff.* 41, 495–502.

Kawase, E., Wong, M. D., Ding, B. C. and Xie, T. (2004) Gbb/Bmp signaling is essential for maintaining germline stem cells and for repressing *bam* transcription in the *Drosophila* testis. *Development* 131, 1365–1375.

Kiger, A. A., Jones, D. L., Schulz, C., Rogers, M. B. and Fuller, M. T. (2001) Stem cell self-renewal specified by JAK-STAT activation in response to a support cell cue. *Science* 294, 2542–2545.

Kinder, S. J., Tsang, T. E., Quinlan, G. A., Hadjantonakis, A.-K., Nagy, A. and Tam, P. P. L. (1999) The orderly allocation of mesodermal cells to the extraembryonic structures and the anteroposterior axis during gastrulation of the mouse embryo. *Development* 126, 4691–4701.

King, R. C. (1970) *Ovarian Development in* Drosophila melanogaster. Academic Press, New York.

King, F. J., Szakmary, A., Cox, D. and Lin, H. (2001) Yb modulates the divisions of both germline and somatic stem cells through *piwi*-and *hh*-mediated mechanisms in the *Drosophila* ovary. *Mol. Cell* 7, 497–508.

Kirilly, D., Spana, E. P., Perrimon, N., Padgett, R. W. and Xie, T. (2005) BMP signaling is required for controlling somatic stem cell self-renewal in the *Drosophila* ovary. *Dev. Cell* 9, 651–662.

Kirilly, D. and Xie, T. (2007) The *Drosophila* ovary: an active stem cell community. *Cell Res.* 17, 15–25.

Kitadate, Y., Shigenobu, S., Arita, K. and Kobayashi, S. (2007) Boss/Sev signaling from germline to soma restricts germline-stem-cell-niche formation in the anterior region of *Drosophila* male gonads. *Dev. Cell* 1, 151–159.

Kobayashi, S., Yamada, M., Asaoka, M. and Kitamura, T. (1996) Essential role of the posterior morphogen *nanos* for germline development in *Drosophila. Nature* 380, 708–711.

Kobayashi, T., Kajiura-Kobayashi, H. and Nagahama, Y. (2002) Two isoforms of *vasa* homologs in a teleost fish: their differential expression during germ cell differentiation. *Mech. Dev.* 111, 167–171.

Komiya, T. and Tanigawa, Y. (1995) Cloning of a gene of the DEAD box protein family which is specifically expressed in germ cells in rats. *Biochem. Biophys. Res. Commun.* 207, 405–410.

Kubota, H., Avarbock, M. R. and Brinster, R. L. (2004) Growth factors essential for self-renewal and expansion of mouse spermatogonial stem cells. *Proc. Natl. Acad. Sci. USA.* 101, 16489–16494.

Kubota, H. and Brinster, R. L. (2006) Technology Insight: *in vitro* culture of spermatogonial stem cells and their potential therapeutic uses. *Nat. Clin. Pract. Endocrinol. Metab.* 2, 99–108.

Kucia, M., Dawn, B., Hunt, G., Guo, Y., Wysoczynski, M., Majka, M., Ratajczak, J., Rezzoug, F., Ildstad, S. T., Bolli, R. and Ratajczak, M. Z. (2004) Cells expressing early cardiac markers reside in the bone marrow and are mobilized into the peripheral blood after myocardial infarction. *Circ. Res.* 95, 1191–1199.

Kucia, M., Machalinski, B. and Ratajczak, M. Z. (2006) The developmental deposition of epiblast/germ cell-line derived cells in various organs as a hypothetical explanation of stem cell plasticity. *Act. Neurobiol. Exp.* 66, 331–341.

Kucia, M., Wysoczynski, B. and Ratajczak, M. Z. (2008) Identification of very small embryonic like (VSEL) stem cells in bone marrow. *Cell Tissue Res.* 331, 125–134.

Lasko, P. and Ashburner, M. (1988) The product of the *Drosophila* gene *vasa* is very similar to eukaryotic initiation factor-4A. *Nature* 335, 611–617.

Lawson, K. A. and Hage, W. J. (1994) Clonal analysis of the origin of primordial germ cells in the mouse. *CIBA Found. Symp.* 182, 68–91.

Lawson, K. A., Dunn, N. R., Roelen, B. A., Zeinstra, L. M., Davis, A. M., Wright, C. V. E., Korving, J. P. W. F. M. and Hogan, B. L. M. (1999) Bmp4 is required for the generation of primordial germ cells in the mouse embryo. *Genes Dev.* 13, 424–436.

Le Bras, S. and Van Doren, M. (2006) Development of the male germline stem cell niche in *Drosophila. Dev. Biol.* 294, 92–103.

Lin, H., Yue, L. and Spradling, A. C. (1994) The *Drosophila* fusome, a germline-specific organelle, contains membrane skeletal proteins and functions in cyst formation. *Development* 120, 947–956.

Lin, H. and Spradling, A. C. (1995) Fusome asymmetry and oocyte determination in *Drosophila. Devel. Genet.* 16, 6–12.

Mahowald, A. P. (2001) Assembly of the *Drosophila* germ plasm. *Int. Rev. Cytol.* 203, 187–213.

Margolis, J. and Spradling, A. C. (1995) Identification and behaviour of epithelial stem cells in *Drosophila* ovaries. *Development* 121, 3797–3807.

Martinho, R. G., Kunwar, P. S., Casanova, J. and Lehmann, R. (2004) A noncoding RNA is required for the repression of RNA polII-dependent transcription in primordial germ cells. *Curr. Biol.* 14, 159–165.

Matsui, Y., Zsebo, K. and Hogan, B. L. M. (1992) Derivation of pluripotential embryonic stem cells from murine primordial germ cells in culture. *Cell* 70, 841–847.

Matsui, Y. and Okamura, D. (2005) Mechanisms of germ-cell specification in mouse embryos. *BioEssays* 27, 136–143.

McLaren, A. (2001) Mammalian germ cells: birth, sex, and immortality. *Cell Str. Funct.* 26, 119–122.

McLean, D. J. (2006) Vertebrate reproductive stem cells: Recent insights and technological advances. *Seminars in Cell & Dev. Biol.* 17, 534–539.

Meng, X., Lindahl, M., Hyvo¨nen, M. E., Parvinen, M., de Rooij, D. G., Hess, M. W., Raatikainen-Ahokas, A., Sainio, K., Rauvala, H., Lakso, M., Pichel, J. G., Westphal, H., Saarma, M. and Sariola, H. (2000) Regulation of cell fate decision of undifferentiated spermatogonia by GDNF. *Science* 287, 1489–1493.

Nakamura, A., Amikura, R., Mukai, M., Kobayashi, S. and Lasko, P. F. (1996) Requirement for a noncoding RNA in *Drosophila* polar granules for germ cell establishment. *Science* 274, 2075–2079.

Nagano, M., Brinster, C. J., Orwig, K. E., Ryu, B.-Y., Avarbock, M. R. and Brinster, R. L. (2001) Transgenic mice produced by retroviral transduction of male germ-line stem cells. *Proc. Natl. Acad. Sci. USA* 98, 13090–13095.

Nagano, M., Patrizio, P. and Brinster, R. L. (2002) Long-term survival of human spermatogonial stem cells in mouse testes. *Fertility and Sterility* 78, 1225–1233.

Nayernia, K., Lee, J. H., Drusenheimer, N., Nolte, J., Wulf, G., Dressel, R. Gromoll, J. and Engel, W. (2006) Derivation of male germ cells from bone marrow stem cells. *Lab. Invest.* 86, 654–663.

Niki, Y. and Mahowald, A. P. (2003) Ovarian cystocytes can repopulate the embryonic germ line and produce functional gametes. *Proc. Natl. Acad. Sci. USA*. 100, 14042–14045.

Niki, Y., Yamaguchi, T. and Mahowald, A. P. (2006) Establishment of stable cell lines of *Drosophila* germ-line stem cells. *Proc. Natl. Acad. Sci. USA* 103, 16325–16330.

Ogawa, T., Dobrinski, I., Avarbock, M. R. and Brinster, R. L. (1999) Xenogeneic spermatogenesis following transplantation of hamster germ cells to mouse testes. *Biol. Reprod.* 60, 515–521.

Ogawa, T., Dobrinski, I., Avarbock, M. R. and Brinster, R. L. (2000) Transplantation of male germ line stem cells restores fertility in infertile mice. *Nat. Med.* 6, 29–34.

Ogawa, T. (2001) Spermatogonial transplantation: the principle and possible applications. *J. Mol. Med.* 79, 368–374.

Okamura, D., Kimura, T., Nakano, T. and Matsui, Y. (2003) Cadherin-mediated cell interaction regulates germ cell determination in mice. *Development* 130, 6423–6430.

Okamura, D., Hayashi, K. and Matsui, Y. (2005) Mouse epiblasts change responsiveness to BMP4 signal required for PGC formation through functions of extraembryonic ectoderm. *Mol. Repro. Dev.* 70, 20–29.

Ohinata, Y., Payer, B., O'Carroll, D., Ancelin, K., Ono, Y., Sano, M., Barton, S. C., Obukhanych, T., Nussenzweig, M., Tarakhovsky, A., Saitou, M. and Surani, M. A. (2005) Blimp1 is a critical determinant of the germ cell lineage in mice. *Nature* 436, 207–213.

Ohinata, Y., Seki, Y., Payer, B., O'Carroll, D., Surani, M.A. and Saitou, M. (2006) Germline recruitment in mice: a genetic program for epigenetic reprogramming. *Ernst Schering Res. Found. Workshop* 60, 143–174.

Parameswaran, M. and Tam, P. P. (1995) Regionalisation of cell fate and morphogenetic movement of the mesoderm during mouse gastrulation. *Dev. Genet.* 17, 16–28.

Ratajczak, M. Z., Machalinski, B., Wojakowski, W. and Kucia, M. (2007) A hypothesis for an embryonic origin of pluripotent Oct-4^{+} stem cells in adult bone marrow and other tissues. *Leukemia* 21, 860–867.

Rawlings, J. S., Rosler, K. M. and Harrison, D. A. (2004) The JAK/STAT signaling pathway. *J. Cell Sci.* 117, 1281–1283.

Resnick, J. L., Bixler, L. S., Cheng, L. and Donovan, P. J. (1992) Long-term proliferation of mouse primordial germ cells in culture. *Nature* 359, 550–551.

Saiti, D. and Lacham-Kaplan, O. (2007) Mouse germ cell development *in-vivo* and *in-vitro. Biomarker Insights* 2, 241–252.

Saitou, M., Barton, S. C. and Surani, M. A. (2002) A molecular programme for the specification of germ cell fate in mice. *Nature* 418, 293–300.

Sato, K., Hayashi, Y., Ninomiya, Y., Shigenobu, S., Arita, K., Mukai, M. and Kobayashi, S. (2007) Maternal Nanos represses *hid/skl*-dependent apoptosis to maintain the germ line in *Drosophila* embryos. *Proc. Natl. Acad. Sci. USA* 104, 7455–7460.

Sato, T., Ueda, S. and Niki, Y. (2008) Wingless signaling initiates mitosis of primordial germ cells during development in *Drosophila. Mech. Dev.* 125,

Schulz, C., Wood, C. G., Jones, D. L., Tazuke, S. I. and Fuller, M. T. (2002) Signaling from germ cells mediated by the rhomboid homolog *stet* organizes encapsulation by somatic support cells. *Development* 129, 4523–4534.

Schulz, C., Kiger, A. A., Tazuke, S. I., Yamashita, Y. M., Pantalena-Filho, L. C., Jones, D. L., Wood, C. G. and Fuller, M. T.(2004) A misexpression screen reveals effects of *bag-of-marbles* and TGF beta class signaling on the *Drosophila* male germ-line stem cell lineage. *Genetics* 167, 707–723.

Seandel, M., James, D., Shmelkov, S. V., Falciatori, I., Kim, J., Chavala, S., Scherr, D. S., Zhang, F., Torres, R., Gale, N. W., Yancopoulos, G. D., Murphy, A., Valenzuela, D. M., Hobbs, R. M., Pandolfi, P. P. and Rafii, S. (2007) Generation of functional multipotent adult stem cells from GPR125^{+} germline progenitors. *Nature* 449, 346–350.

Seki, Y., Hayashi, K., Itoh, K., Mizugaki, M., Saitou, M. and Matsui, Y. (2005) Extensive and orderly reprogramming of genome-wide chromatin modifications associated with specification and early development of germ cells in mice. *Dev. Biol.* 278, 440–458.

Seydoux, G. and E. Braun, R. (2006) Pathway to totipotency: Lessons from germ cells. *Cell* 127, 892–904.

Shigenobu, Y., Kitadate, K., Noda, C. and S. Kobayashi (2006) Molecular characterization of embryonic gonads by gene expression profiling in *Drosophila melanogaster*. *Proc. Natl. Acad. Sci., USA* 103, 13728–13733.

Shivdasani, A. A. and Ingham, P. W. (2003) Regulation of stem cell maintenance and transit amplifying cell proliferation by tgf-beta signaling in *Drosophila* spermatogenesis. *Curr. Biol.* 13, 2065–2072.

Song, X., Zhu, C. H., Doan, C. and Xie, T. (2002) Germline stem cells anchored by adherens junctions in the *Drosophila* ovary niches. *Science* 296, 1855–1857.

Song, X. and Xie, T. (2002) DE-cadherin-mediated cell adhesion is essential for maintaining somatic stem cells in the *Drosophila ovary*. *Proc. Natl. Acad. Sci. USA* 99, 14813–14818.

Sonnenblick, B. P. (1941) Germ cell movements and sex differentiation of the gonads in the *Drosophila* embryo. *Proc. Natl. Acad. Sci. USA* 27, 484–488.

Spradling, A. C., Drummond-Barbosa, D. and Kai, T. (2001) Stem cells find their niche. *Nature* 414, 98–104.

Starz-Gaiano, M. and Lehmann, R. (2001) Moving towards the next generation. *Mech. Dev.* 105, 5–18.

Styhler, S., Nakamura, A., Swan, A., Suter, B. and Lasko, P. (1998) *vasa* is required for GURKEN accumulation in the oocyte, and is involved in oocyte differentiation and germline cyst development. *Development* 125, 1569–1578.

Sunanaga, T., Watanabe, A. and Kawamura, K. (2006) Involvement of *vasa* homolog in germline recruitment from coelomic stem cells in budding tunicates. *Dev. Genes Evo*.217, 1–11.

Tam, P. P. L. and Zhou, S. X. (1996) The allocation of epiblast cells to ectodermal and germ-line lineages is influenced by the position of the cells in the gastrulating mouse embryo. *Dev. Biol.* 178, 124–132.

Tanaka, S. S., Toyooka, Y., Akasu, R., Katoh-Fukui, Y., Nakahara, Y., Suzuki, R., Yokoyama, M. and Noce, T. (2000) The mouse homolog of *Drosophila* Vasa is required for the development of male germ cells. *Gene Dev.* 14, 841–853.

Tanaka, S. S. and Matsui, Y. (2002) Developmentally regulated expression of *mil-1* and *mil-2*, mouse interferon-induced transmembrane protein like genes, during formation and differentiation of primordial germ cells. *Mech. Dev.* 119S, S261–S267.

Tazuke, S. I., Schulz, C., Gilboa, L., Fogarty, M., Mahowald, A. P., Guichet, A., Ephrussi, A., Wood, C. G., Lehmann, R. and Fuller, M. T. (2002) A germline-specific gap junction protein required for survival of differentiating early germ cells. *Development* 129, 2529–2539.

Terry, N. A., Tulina, N., Matunis, E. and DiNardo, S. (2006) Novel regulators revealed by profiling *Drosophila* testis stem cells within their niche. *Dev. Biol.* 294, 246–257.

Thomson, T. and Lasko, P. (2005) Tudor and its domains: germ cell formation from a Tudor perspective. *Cell Res.* 15, 281–291.

Tilly, J. L. and Johnson, J. (2007) Recent arguments against germ cell renewal in the adult human ovary: Is an absence of marker gene expression really acceptable evidence of an absence of oogenesis? *Cell Cycle* 6, 879–883.

Toyooka, Y., Tsunekawa, N., Akasu, R. and Noce, T. (2003) Embryonic stem cells can form germ cells *in vitro*. *Proc. Natl. Acad. Sci. USA* 100, 11457–11462.

Tsunekawa, N., Naito, M., Sakai, Y., Nishida, T. and Noce, T. (2000) Isolation of chicken *vasa* homolog gene and tracing the origin of primordial germ cells. *Development* 127, 2741–2750.

Tsuda, M., Sasaoka, Y., Kiso, M., Abe, K., Haraguchi, S., Kobayashi, S. and Saga, Y. (2003) Conserved role of *nanos* proteins in germ cell development. *Science* 301, 1239–1241.

Tremblay, K. D., Dunn, N. R. and Robertson, E. J. (2001) Mouse embryos lacking Smad1 signals display defects in extra-embryonic tissues and germ cell formation. *Development* 128, 3609–3621.

Tulina, N. and Matunis, E. (2001) Control of stem cell self-renewal in *Drosophila* spermatogenesis by JAK-STAT signaling. *Science* 294, 2546–2549.

Vincent, S. D., Dunn, N. R., Sciammas, R., Shapiro-Shalef, M., Davis, M. M., Calame, K., Bikoff, E. K. and Robertson, E. J. (2005) The zinc finger transcriptional repressor Blimp1/Prdm1 is dispensable for early axis formation but is required for specification of primordial germ cells in the mouse. *Development* 132, 1315–1325.

Wawersik, M., Milutinovich, A., Casper, A. L., Matunis, E., Williams, B. and Van Doren, M. (2005) Somatic control of germline sexual development is mediated by the JAK/STAT pathway. *Nature* 436, 563–567.

Wieschaus, E. and Szabad, J. (1979) The development and function of the female germ line in *Drosophila melanogaster*: a cell lineage study. *Devel. Biol.* 68, 29–46.

Wojakowski, W., Tendera, M., Michalowska, A., Majka, M., Kucia, M., Maslankiewicz, K., Wyderka, R., Ochala, A., and Ratajczak, M. Z. (2004) Mobilization of CD34/CXCR4$^+$, CD34/CD117$^+$, c-met$^+$ stem cells, and mononuclear cells expressing early cardiac, muscle, and endothelial markers into peripheral blood in patients with acute myocardial infarction. *Circulation* 110, 3213–3220.

Xie, T. and Spradling, A. C. (1998) Decapentaplegic is essential for the maintenance and division of the germline stem cells in the *Drosophila* ovary. *Cell* 94, 251–260.

Xie, T. and Spradling, A. C. (2000) A niche maintaining germ line stem cells in the *Drosophila* ovary. *Science* 290, 328–330.

Xi, R., Kirilly, D. and Xie, T. (2005) Molecular mechanisms controlling germline and somatic stem cells: Similarities and differences. *Curr. Opin. Genet. Dev.* 15, 381–387.

Yamashita, Y., Jones, D. L. and Fuller, M. T. (2003) Orientation of asymmetric stem cell division by the APC tumor suppressor and centrosome. *Science* 301, 1547–1550.

Yamashita, Y., Fuller, M. T. and Jones, D. L. (2005) Signaling in stem cell niches: lessons from the *Drosophila* germline. *J. Cell Sci.* 118, 665–672.

Yamashita, Y. M., Mahowald, A. P., Perlin, J. R. and Fuller, M. T. (2007) Asymmetric inheritance of mother versus daughter centrosome in stem cell division. *Science* 315, 518–521.

Ying, Y., Liu, X. M., Marble, A., Lawson, K. A. and Zhao, G. Q. (2000) Requirement of Bmp8b for the generation of primordial germ cells in the mouse. *Mol. Endocrinol.* 14, 1053–1063.

Ying, Y. and Zhao G-Q. (2001) Cooperation of endoderm-derived BMP2 and extraembyonic ectoderm-derived BMP4 in primordial germ cell generation in mouse. *Dev. Biol.* 232, 484–492.

Ying, Y., Qi, X. and Zhao, G. Q. (2001) Induction of primordial germ cells from murine epiblasts by synergistic action of BMP4 and BMP8B signaling pathways. *Proc. Natl. Acad. Sci. USA* 98, 7858–7862.

Yoshida, S., Sukeno, M. and Nabeshima, Y. (2007) A vasculature-associated niche for undifferentiated spermatogonia in the mouse testis. *Science* 317, 1722–1726.

Zalokar, M. (1976) Autoradiographic study of protein and RNA formation during early development of *Drosophila* eggs. *Dev. Biol.* 49, 425–437.

Zhang, Y. and Kalderon, D. (2000) Regulation of cell proliferation and patterning in *Drosophila* oogenesis by Hedgehog signaling. *Development* 127, 2165–2176.

Zhang, Y. and Kalderon, D. (2001) Hedgehog acts as a somatic stem cell factor in *Drosophila* ovary. *Nature* 410, 599–604.

Zhu, C. H. and Xie, T. (2003) Clonal expansion of ovarian germline stem cells during niche formation in *Drosophila. Development* 130, 2579–2588.

Zwaka, T. P. and Thomson, J. A. (2005) A germ cell origin of embryonic stem cells? *Development* 132, 227–233.

In: Stem Cell Applications in Disease and Health
Editors: W. Burnsides and R. Ellsley
ISBN: 978-1-60456-255-2

Chapter VII

Limbal Stem Cell Deficiency and Its Management: From Lab to Limbus

***Sajjad Ahmad*[1, 2, 3,*]*, Majlinda Lako*[1, 2]*, Francisco Figueiredo*[1, 3] *and Sai Kolli*[1, 2, 3]**

[1]North East England Stem Cell Institute and
[2]Institute of Human Genetics, Newcastle University, Central Parkway, Newcastle upon Tyne, NE1 3BZ, United Kingdom, Tel: +44 191 214 8618, Fax: +44 191 214 8666
[3]Department of Ophthalmology, Royal Victoria Infirmary, Queen Victoria Road, Newcastle upon Tyne, NE1 4LP, United Kingdom, Tel: +44 191 282 5582
Fax: +44 191 282 0733
*Email: sajjad.ahmad@ncl.ac.uk

Abstract

The cornea forms the clear front of the eye and its clarity is critical for normal visual perception. The corneal surface is covered by an epithelium which is continuous with that of the surrounding conjunctiva. The corneal epithelium is renewed by stem cells located in the transition zone between the cornea and conjunctiva known as the limbus. A loss of these so-called limbal stem cells results in the disease of limbal stem cell deficiency, most commonly caused by chemical burns to the eye. This disease is characterised by failure of corneal epithelial maintenance and encroachment of the corneal surface by the phenotypically different conjunctival epithelium and its associated blood vessels. Both these characteristics result in a painful and blinding disease. In this chapter, the theory of limbal stem cell maintenance of the corneal epithelium, limbal stem cell deficiency and its causes, the diagnosis of limbal stem cell deficiency and its various management options (including the use of ex vivo expanded limbal epithelium) will be discussed. In particular, the role played by the basic science of limbal stem cell biology in the management of limbal stem cell deficiency will be highlighted.

Introduction

The cornea forms the clear front of the eye and enables the transmission and focussing of light to the retina at the back of the eye. Its clarity therefore plays an important role in its normal functioning. The surface of the cornea is made up of a 5-7 cell layer thick epithelium. This is continuous with the epithelium of the surrounding conjunctiva (Figure 1). The conjunctival epithelium differs from the corneal epithelium in two main ways. Firstly, the conjunctival epithelium has an approximately 10% population of goblet cells which are normally absent in corneal epithelium. Secondly, the stroma underlying the conjunctival epithelium is highly vascularised, whereas that of the cornea is normally avascular. There is a 360° narrow band of epithelium, approximately 1.5 mm wide, which encircles the corneal epithelium and acts as a barrier between the corneal and conjunctival epithelia. It is now known that the putative stem cells for the corneal epithelium reside in the limbal epithelium, the so-called limbal stem cells [1]. The corneal epithelium itself is devoid of its own stem cells.

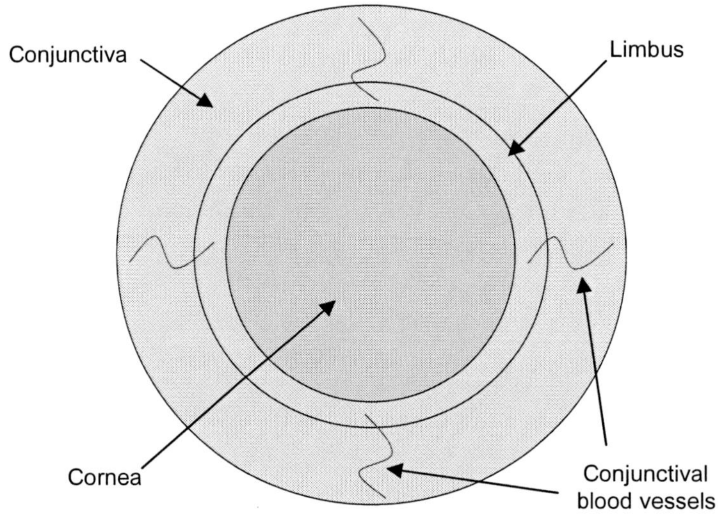

Figure 1. Schematic illustration of the front of the eye. The cornea is the clear front of the eye and its surface is continuous with the conjunctiva peripherally. The limbus is a narrow band of tissue which acts as a barrier between the cornea and conjunctiva. The conjunctiva is a vascular tissue and the cornea is avascular.

In this chapter, aspects of limbal stem cell biology relevant to normal corneal epithelial maintenance will be highlighted. Failure of such maintenance leads to the clinical condition of limbal stem cell deficiency which is a painful and blinding disease. It is a disease which is often difficult to treat. In this chapter, mechanisms by which normal corneal epithelial maintenance fails in limbal stem cell deficiency will be discussed. In addition, the symptoms, signs and diagnostic laboratory tests for this disease will be presented. The remainder of the chapter will then discuss the various management options currently available for limbal stem cell deficiency, highlighting the importance of limbal stem cell biology in identifying these treatments.

Maintenance of the Corneal Epithelium

There are various pieces of in vitro, animal and clinical evidence that point to the basal layer of the limbal epithelium as harbouring stem cells for the corneal epithelium. In 1971, Davanger and Evensen were the first to propose the importance of the limbus in corneal epithelial maintenance [2]. They studied the healing of corneal epithelial wounds in guinea pigs. They noted that in the guinea pig eye, where the limbal epithelium is pigmented, the corneal epithelial wound healed with pigmented epithelium migrating from the limbus. They therefore concluded that the limbus was a generative organ for the corneal epithelium.

In 1988, Ebato and co-workers showed that human limbal epithelium had a higher proliferative capacity in culture than corneal epithelium [3]. Stem cells have a very high proliferative potential in comparison to their more differentiated progeny. The findings of Ebato and co-workers therefore provided further evidence for the limbal location of stem cells for the corneal epithelium. In 1989, Cotsarelis and co-workers studied corneal epithelial turnover in mouse eyes by radioactive labelling of cells in S phase of the cell cycle [4]. They showed that epithelial cells in the basal layer of the limbal epithelium retained the radioactive label for longer and were therefore going through the cell cycle much more slowly than other cells of the limbal and corneal epithelia. However, upon corneal epithelial wounding, these label retaining cells could be made to cycle and divide much more rapidly. Slow cell cycling during homeostasis and rapid cycling during wounding are properties associated with stem cells. Cotsarelis and co-workers therefore concluded that epithelial cells in the basal layer of the limbal epithelium were the stem cells for the corneal epithelium.

In 1989, Kenyon and Tseng were the first to provide clinical evidence for the limbal location of corneal epithelial stem cells by highlighting the success of limbal transplantation [5]. In limbal stem cell deficiency the corneal epithelium cannot be maintained and the corneal surface becomes covered with the surrounding conjunctival epithelium and its blood vessels [1]. Kenyon and Tseng showed that the corneal epithelium in these patients could be successfully reconstituted by transplanting healthy limbal epithelium. In 1991, Tseng's group studied corneal epithelial wound healing in rabbits with varying degrees of limbal epithelial damage [6]. They showed that removal of the basal layers of the limbal epithelium prevented normal corneal epithelial regeneration and resulted in limbal stem cell deficiency. This rabbit model of limbal stem cell deficiency again indicated the importance of the basal layer of the limbal epithelium in corneal epithelial maintenance.

The above in vitro, animal and clinical studies reinforce the X, Y, Z hypothesis of corneal epithelial maintenance originally proposed by Thoft and Friend in 1983 (Figure 2) [7]. The corneal epithelium is exposed to the external environment and therefore sheds aged differentiated epithelial cells from its central surface. This is the Z component in the hypothesis. In order to maintain constant corneal epithelial mass, these lost cells must be replaced. The hypothesis states that these shed cells are replaced by movement of cells from the lower layers of the corneal epithelium. This is the X component. The cells of the basal layer of the corneal epithelium which have migrated superficially are then replaced by cells originating from the periphery of the cornea and limbus. This is the Y component. Corneal epithelial maintenance can therefore be defined by the equation $X+Y=Z$. This formula states that if the corneal epithelium is to be maintained, then cell loss must be balanced by cell replacement. We now know that ultimately corneal epithelial cell replacement can be traced back to limbal stem cell division in the basal layer of the limbal epithelium [1].

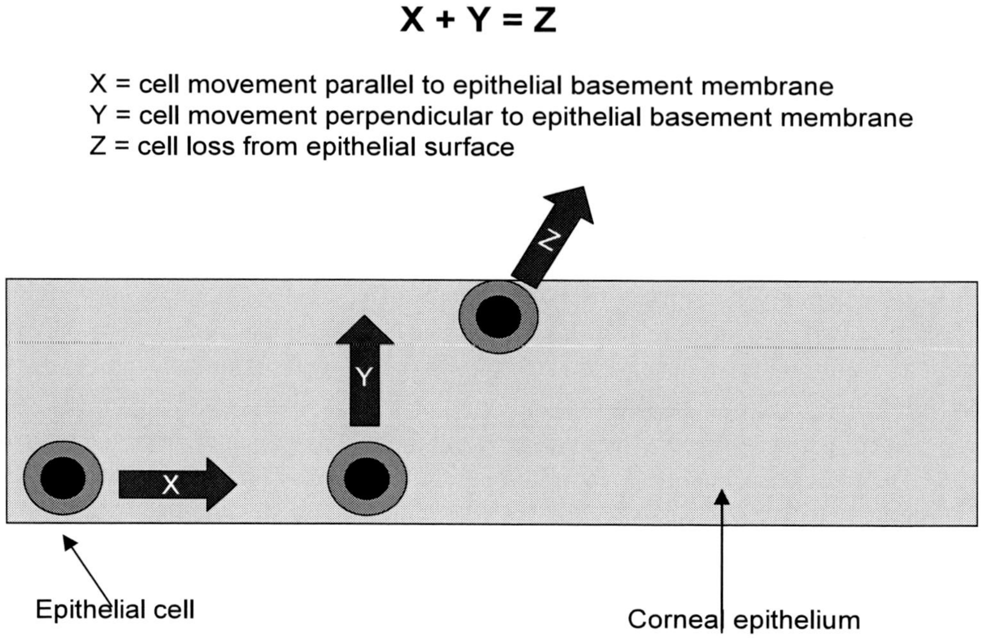

Figure 2. A schematic illustration of the X, Y, Z hypothesis of corneal epithelial maintenance. X is the movement of epithelial cells across the corneal surface parallel to the corneal epithelial basement membrane. Y is the movement of epithelial cells towards the corneal epithelial surface in a direction perpendicular to the basement membrane. The sum of X and Y is equal to Z which represents the loss of corneal epithelial cells from the corneal surface.

In humans, the palisades of Vogt at the limbus are thought to harbour the limbal stem cells [2, 8]. The palisades of Vogt consist of a series of radially orientated fibrovascular ridges especially concentrated along the superior and inferior limbus. The region between these ridges is occupied by epithelium where the limbal stem cells reside.

The Pathophysiology and Causes of Limbal Stem Cell Deficiency

Limbal stem cell deficiency results from the loss or dysfunction of limbal stem cells [9]. This loss or dysfunction could be primary or secondary. Primary limbal stem cell deficiency occurs when the limbal stem cells themselves are directly affected. Secondary limbal stem cell deficiency results when the environment in which the limbal stem cells reside is affected, and this then results in secondary loss or abnormal functioning of the limbal stem cells. Either way however, the main functions of the limbal epithelium break down. The corneal epithelium cannot be maintained due to the lack of limbal stem cells, and the surrounding conjunctival epithelium and its blood vessels then encroach on to the corneal surface.

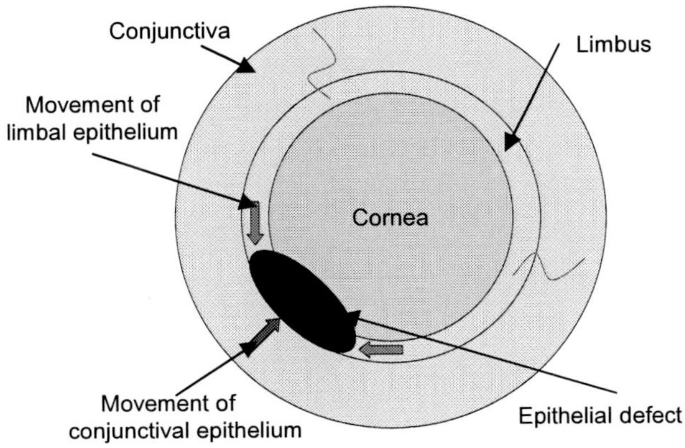

Figure 3. A diagram illustrating the barrier formation function of the limbal epithelium. If the limbal epithelium is wounded, the ensuing space can only be covered by the surrounding healthy limbal epithelium or the conjunctival epithelium. If the limbal epithelium cannot form a barrier quickly enough, the cornea is at considerable risk of being covered by the conjunctival epithelium and its blood vessels.

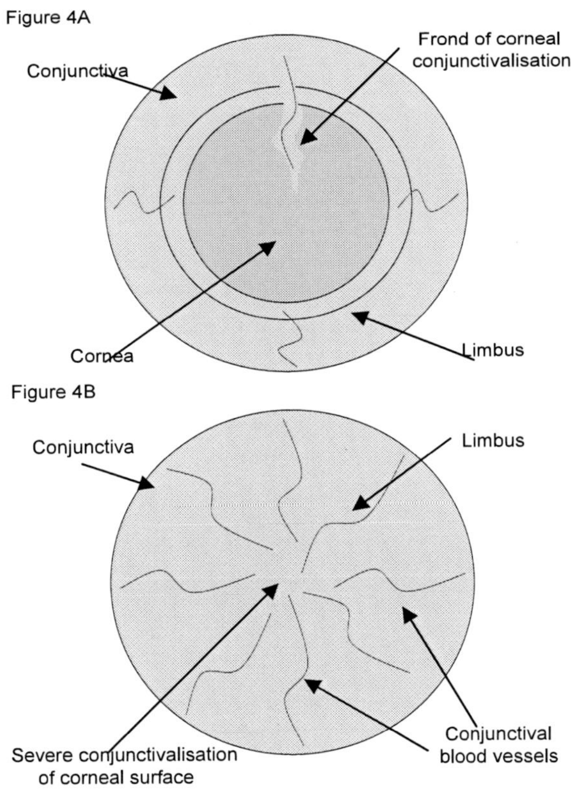

Figure 4. Different extents of limbal stem cell deficiency. A. In partial or focal limbal stem cell deficiency, only a segment or a few discrete segments of the corneal surface are covered by conjunctival epithelium with or without its blood vessels. The conjunctivalisation may or may not affect the visual axis. B. In total or diffuse limbal stem cell deficiency, the majority of the corneal surface is conjunctivalised.

The limbus can be affected by limbal stem cell deficiency to varying degrees. Limbal stem cell deficiency can then be described as partial or total depending on the extent of limbus affected [1]. In 1990, Dua and Forrester studied the healing of corneal epithelial wounds in patients where the limbal epithelium was also affected (Figure 3) [10]. They showed that in these cases the damaged limbal epithelium grows circumferentially to reform a barrier between the conjunctival epithelium and healing corneal epithelium. If this circumferential healing is delayed, for example due to the extent of limbal epithelium affected, then the conjunctiva grows into this defect and on to the corneal epithelium. If only part of the corneal surface becomes conjunctivalised then limbal stem cell deficiency can be described as partial or focal (Figure 4A). If the majority of the corneal surface is affected then limbal stem cell deficiency can be described as total or diffuse (Figure 4B). Although it is most likely for total limbal stem cell deficiency to be symptomatic, partial limbal stem cell deficiency can also result in symptoms depending on the extent of the cornea affected, especially if the central corneal surface is involved [11].

Table 1.

LIMBAL STEM CELL DEFICIENCY		
Hereditary	**Acquired**	**Idiopathic**
Aniridia	Chemical burn	No cause identified
Ectodermal dysplasia	Thermal burn	
Multiple endocrine deficiency related keratitis	Extensive limbal trauma	
	Extensive limbal surgery	
	Extensive limbal cryotherapy	
	Limbal irradiation	
	Cytotoxic agents	
	Contact lens related keratitis	
	Stevens-Johnson syndrome	
	Ocular cicatricial pemphigoid	
	Atopic keratitis	
	Vernal keratitis	

Shows the various causes of limbal stem cell deficiency. There are many known hereditary and acquired causes of limbal stem cell deficiency as well as the entity of idiopathic limbal stem cell deficiency.

There are various causes of limbal stem cell deficiency (Table 1). Hereditary causes include aniridia, ectodermal dysplasias and keratitis associated with multiple endocrine deficiency [1, 9]. The cause of limbal stem cell deficiency in these conditions is poorly understood. However in aniridia, *PAX6* mutations are thought to play a major role probably by affecting the environment of the limbal stem cells [12]. Acquired causes of limbal stem cell deficiency include chemical or thermal burns or mechanical trauma to the eye. Acquired limbal stem cell deficiency can also result from iatrogenic causes, such as extensive limbal surgery (for example for tumours) [13], extensive limbal cryotherapy, therapeutic irradiation at or near the limbus [9], and the use of cytotoxic agents on the surface of the eye (for example mitomycin C and 5-fluorouracil) [14]. Contact lens use can also result in acquired limbal stem cell deficiency. There are many reasons for this. The prolonged hypoxic and mechanical insults to the limbus, as well as the chronic exposure to the limbus of contact lens solutions result in limbal stem cell loss [5, 15]. Chronic ocular surface inflammation, for example in Stevens-Johnson syndrome, ocular cicatricial pemphigoid, and chronic atopic or vernal limbitis, can also result in acquired limbal stem cell deficiency [16]. In addition to hereditary and acquired causes of limbal stem cell deficiency, idiopathic cases have also been described in whom no specific cause of the disease can be identified [17].

Diagnosing Limbal Stem Cell Deficiency: Symptoms, Signs and Investigation

The diagnosis of limbal stem cell deficiency is made mainly on clinical grounds (Table 2). The symptoms of limbal stem cell deficiency can be understood by considering limbal stem cell function in the eye. Firstly, the corneal epithelium cannot be maintained due to a lack of limbal stem cells, resulting in recurrent corneal epithelial breakdown and inflammation. This causes the symptoms of ocular pain and discomfort, redness, photophobia, tearing, blepharospasm and reduced vision. Secondly in limbal stem cell deficiency, the corneal surface is replaced by conjunctival epithelium and its blood vessels, a process known as conjunctivalisation. This impairs corneal clarity significantly and results in impaired vision. Limbal stem cell deficiency is therefore both a painful and blinding disease [18, 19].

Table 2.

LIMBAL STEM CELL DEFICIENCY		
Symptoms	**Signs**	**Investigations**
Ocular discomfort and pain	Dilated conjunctival vessels	Corneal impression cytology
Red eye	Corneal vascularisation	In vivo confocal microscopy
Photophobia	Corneal haze	
Excess tearing	Corneal epithelial defects	
Blepharospasm	Delayed fluorescein staining	
Reduced vision	Loss of palisades of Vogt	

Shows the main symptoms, signs and investigations in limbal stem cell deficiency. The diagnosis of limbal stem cell deficiency is very much made on clinical grounds and this is confirmed usually by central corneal impression cytology.

The signs of limbal stem cell deficiency play an important role in diagnosing the disease (Table 2). The eye often appears chronically red with engorged and dilated conjunctival blood vessels. Conjunctivalisation of the corneal surface often results in superficial corneal vascularisation. Fluoroscein staining of the corneal surface can also provide useful clues. Corneal epithelial cells are held together by tight junctions making the normal corneal surface impermeable to hydrophilic molecules such as fluoroscein. However, defects in the corneal epithelium result in early and often intense fluoroscein staining. There are two main causes of the chronic and often persistent epithelial defects seen in limbal stem cell deficiency. Firstly, due to a lack of limbal stem cells, the corneal epithelium cannot be renewed normally. Secondly, unlike corneal epithelial cells, conjunctival epithelial cells are held much more loosely both to each other and to the underlying corneal stroma in limbal stem cell deficiency [20]. This therefore also makes the conjunctivalised corneal epithelial surface prone to epithelial breakdown. In addition, because conjunctival epithelial cells are much more loosely attached to each other and more widely spaced than corneal epithelial cells, this makes the conjunctival epithelium more permeable to fluoroscein. In limbal stem cell deficiency, in addition to the early intense fluoroscein staining due to epithelial defects, delayed or late faint fluoroscein staining of the corneal surface can also be seen due to the presence of conjunctival epithelium [21]. Another important but often subtle sign in limbal stem cell deficiency is the loss of normal limbal architecture noted by absence of the palisades of Vogt [19]. These are deep structures often difficult to visualise and so their absence is not a confirmation of limbal stem cell deficiency but a helpful additional sign.

Confirmation of a diagnosis of limbal stem cell deficiency can be made by corneal impression cytology (Table 2). Impression cytology is a relatively non-invasive technique which involves gently pressing cellulose acetate paper on the ocular surface [22]. This removes the superficial layers of the surface epithelium, allowing histological examination. In this way, for limbal stem cell deficiency, corneal impressions can then be analysed for conjunctivalisation. This histological analysis can be done by identifying goblet cell presence on the corneal surface, indicative of conjunctival epithelium. This is done by periodic acid Schiff staining the impression [9]. Cytokeratin 19 is present in conjunctival epithelial cells but not in normal superficial corneal epithelial cells [23]. Immuno-histochemistry for cytokeratin 19 on corneal impressions has also therefore been used to confirm conjunctivalisation. Another recently identified investigative tool in limbal stem cell deficiency is the use of confocal microscopy to examine the limbus [24]. Because the palisades of Vogt are deep structures, imaging them previously has been difficult, mainly due to image degradation. Using *in vivo* white light confocal microscopy, it has become possible to study these structures in the living eye (Table 2). Athough not routinely used at present, this may provide an additional routine investigation for limbal stem cell deficiency.

The Management of Limbal Stem Cell Deficiency

The remainder of this chapter will concentrate on the treatment options available for limbal stem cell deficiency. The particular management route taken depends on the severity of the disease symptoms and the extent of the signs of limbal stem cell deficiency. It also depends on whether the patient has a healthy eye from which to obtain healthy limbal epithelium and stem cells. In particular, the role of newer techniques in limbal stem cell deficiency management will be discussed, including the *ex vivo* expansion of limbal epithelium. It must be noted that most of the management options originate from our understanding of limbal stem cell biology. One should also take into account the presence of other ocular surface problems, e.g. lid abnormalities, and the correction of those should always be performed prior to any LSC graft.

Conservative Management Options

Limbal stem cell deficiency is associated with an irregular corneal epithelial surface prone to epithelial defects and inflammation. Due to this, the ocular surface is also prone to infection. The conservative or medical management of limbal stem cell deficiency is therefore centred on alleviating these problems (Figure 5) [19]. Conservative management is in the form of preservative free topical treatment to the eye. Ocular lubricants (drops, gels or ointments) are used on a frequent basis to alleviate ocular irritation. Topical steroids are used to suppress the sometimes severe and chronic ocular inflammation associated with limbal stem cell deficiency. In cases where the corneal epithelium is prone to persistent breakdown, broad spectrum topical antibiotics are helpful in preventing secondary ocular surface infections. Autologous serum drops, produced from the patient's own blood, have been used successfully to treat chronic ocular surface problems and have an important role in the management of limbal stem cell deficiency [25]. Autologous serum drops have been used in several ocular surface conditions including severe dry eye [26], persistent epithelial defects [27] and superior limbic keratoconjunctivitis [28]. Autologous serum has been shown to contain epidermal growth factor, transforming growth factor β1 and vitamin A in particular, all of which have been shown to promote epithelial wound healing [27, 29]. In addition to the use of topical drops, bandage contact lenses can also be useful in alleviating the symptoms of limbal stem cell deficiency and promoting epithelial healing and integrity. In more severe cases of limbal stem cell deficiency, conservative management alone may not be the best option and so surgical measures have to be employed.

Sequential Conjunctival Epitheliectomy

Sequential conjunctival epitheliectomy was first described by Dua and co-workers in 1994 [11]. It involves the removal of conjunctival epithelium from the surface of the cornea, enabling this area to be covered by the surrounding normal corneal epithelium. This may have to be repeated several times (sequentially) in order to replace the encroaching conjunctival epithelium with corneal epithelium. This technique is particularly useful in treating mild partial limbal stem cell deficiency where a frond of conjunctival epithelium is encroaching on to the central corneal surface, which would impair the patient's vision (Figure 6).

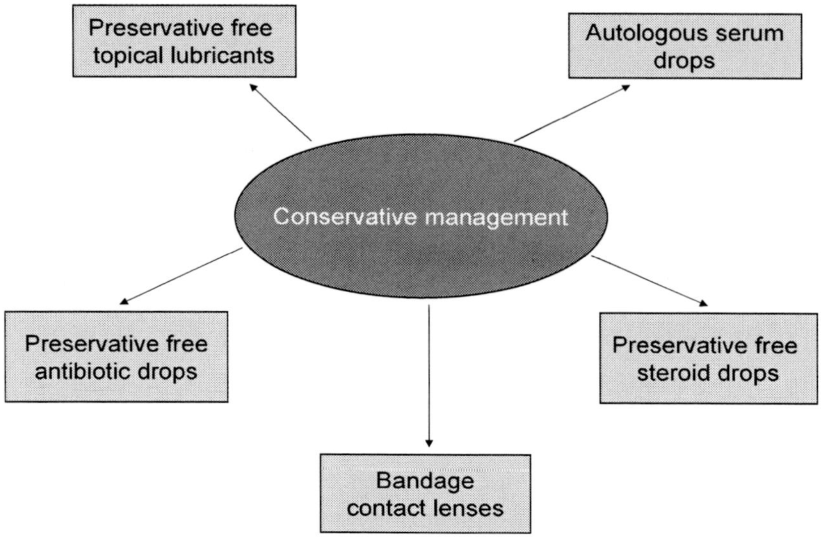

Figure 5. A schematic illustration indicating the various conservative management options for limbal stem cell deficiency. Conservative or medical treatment options can be used for all forms of limbal stem cell deficiency and also in conjunction with the surgical options.

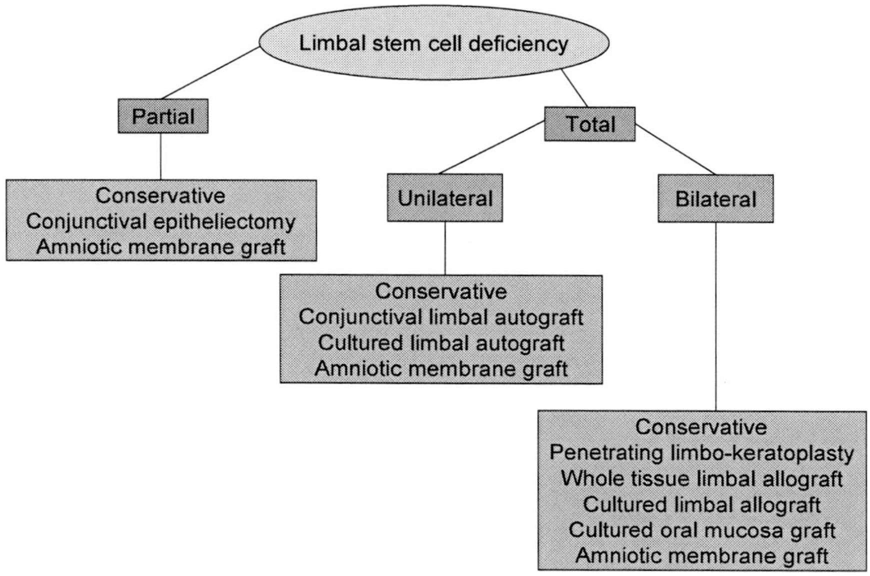

Figure 6. A flow diagram showing the various management options for limbal stem cell deficiency. As indicated in the diagram, conservative management options can be pursued for all forms of limbal stem cell deficiency. The surgical treatment of limbal stem cell deficiency depends on whether the disease is partial or total and whether the total limbal stem cell deficiency is unilateral or bilateral. Amniotic membrane transplantation can be useful in all forms of limbal stem cell deficiency often in conjunction with the other surgical treatment modalities.

Amniotic Membrane Transplantation

Human amniotic membrane has been widely used for many years now for the management of severe ocular surface disorders including limbal stem cell deficiency. Amniotic membrane is the innermost layer of the placenta and consists of three main layers: an avascular stroma, a thick basement membrane and an epithelial monolayer.

Although the precise mechanisms by which amniotic membrane encourages ocular surface healing are not known, studies have reported several favourable attributes. Amniotic membrane epithelium and stroma contain several growth factors, including epidermal growth factor, hepatocyte growth factor and keratinocyte growth factor, which are known to promote normal corneal epithelialisation [30]. The thick basement membrane of amniotic membrane, composed of collagen IV, laminin and fibronectin, also promotes epithelial healing by providing a suitable substrate for corneal epithelial cells to adhere to and expand [31]. In addition, amniotic membrane is known to inhibit corneal scarring by suppressing transforming growth factor β1 signalling which induces proliferation and myofibroblast differentiation of corneal stromal fibroblasts (and hence scar formation) [32]. The amniotic membrane stroma reduces ocular surface inflammation by stimulating inflammatory cells into rapid apoptosis and by producing protease inhibitors [33]. In addition, amniotic membrane is known to hinder corneal vascularisation. Most importantly, amniotic membrane lacks most of the major histocompatibility complexes, thereby inciting little immunological reaction and making it relatively immune privileged [34]. Its transplantation on to the ocular surface therefore requires no recipient immune suppression.

With regard to limbal stem cell deficiency, Tseng and co-workers have shown that amniotic membrane transplantation alone can be used successfully in cases of partial limbal stem cell deficiency with persistent epithelial defects (Figure 6) [35]. For more severe cases of limbal stem cell deficiency, the use of amniotic membrane transplantation in conjunction with other surgical treatment modalities can also be used (Figure 6). The adjunct transplantation of human amniotic membrane therefore plays an important role in the management of limbal stem cell deficiency.

Penetrating Keratoplasty and Penetrating Limbo-Keratoplasty

In severe limbal stem cell deficiency, conjunctivalisation of the corneal surface and any associated underlying stromal scarring results in marked visual loss. Penetrating keratoplasty (full thickness corneal transplantation) is the standard surgery for replacing opaque corneas. The success of this procedure however relies on the host producing a healthy corneal epithelium to replace the donor epithelium which is shed from the corneal graft over a period of time. Since replenishing of corneal epithelium is not possible in limbal stem cell deficiency, the corneal transplant inevitably fails due to conjunctivalisation of the cornea and subsequent associated graft rejection. This failure of standard penetrating keratoplasty in limbal stem cell deficiency has been shown by several studies [36-38].

There are only two main situations where standard penetrating keratoplasty has a role in the management of limbal stem cell deficiency. The first is after the treatment and subsequent remission of limbal stem cell deficiency in order to replace any underlying stromal opacification. The second is where there is progressive corneal thinning leading to perforation, where penetrating keratoplasty is used solely to maintain the integrity of the eye, i.e. for tectonic reasons.

There is a role for allogeneic penetrating limbo-keratoplasty in the primary management of bilateral severe limbal stem cell deficiency (Figure 6) [39, 40]. In this procedure, large donor corneal grafts composed of the entire limbus or eccentric donor corneal grafts composed of approximately 40% of the limbus are transplanted. In this way, although the recipient themselves cannot replace the corneal epithelium, as the donor graft is composed of donor limbus as well, this is no longer an issue. The major drawback of this procedure is the allogeneic nature of the tissue. Standard corneal transplantation does not require recipient immune suppression mainly due to the avascularity of the cornea. However, due to the vascularity of the limbus, amongst other important factors such as the presence of Langerhans cells at the limbus, penetrating limbo-keratoplasty is prone to immune rejection. Potent immune suppression of the recipient is therefore advocated for the success of this procedure as well as recipient-donor tissue matching. In one study, where 48 patients underwent allogeneic penetrating keratoplasties, the central corneas were clear in only 30% of the grafts after 5 years [40]. The majority of these clear grafts were tissue matched.

Whole Tissue Limbal Transplantation

The management of limbal stem cell deficiency was revolutionised following the early advances in limbal stem cell biology described earlier. In 1989, Kenyon and Tseng were the first to propose the definitive treatment of total limbal stem cell deficiency by the transplantation of healthy limbal tissue to the diseased limbus [5]. There are various techniques available for limbal transplantation and the exact technique used depends firstly on whether the limbal stem cell deficiency is unilateral or bilateral, and secondly on the source of the donor limbal tissue.

In cases of unilateral limbal stem cell deficiency, the other healthy eye can be used as a source of limbal tissue [5, 41]. This type of graft is known as a conjunctival limbal autograft and the patient does not require immune suppression due to the autologous nature of the tissue (Figure 6). In this transplant procedure, conjunctival tissue on the cornea and limbus is firstly removed from the limbal stem cell deficienct eye by performing a superficial keratectomy and then two limbal autografts from the other healthy eye, measuring approximately 2-3 clock hours each, are immediately transplanted to the limbus of the recipient eye (Figure 7). In the original case series presented by Kenyon and Tseng in 1989, an improvement in visual acuity was noted in 17 of the 21 patients and a stable ocular surface was noted in all but one patient [5]. However, the main drawback of this procedure is the potential risk of inducing limbal stem cell deficiency in the healthy other eye of the patient. For obvious reasons, patients are therefore often reluctant to undergo this procedure.

In cases of bilateral limbal stem cell deficiency, there are two potential sources of limbal tissue. Limbal tissue can either be obtained from a living related donor, a living related conjunctival limbal allograft, or from a cadaveric donor, a keratolimbal allograft (Figure 6) [41]. The technique of living related conjunctival limbal allograft transplantation is identical to the autograft procedure described above, and the amount of donor tissue removed is minimised in order to prevent iatrogenic limbal stem cell deficiency in the donor eye. In the Daya and Ilari case series, improvement in visual acuity occurred in 7 out of the 9 patients treated with living related conjunctival limbal allografts [42]. In the series presented by Rao and co-workers, improvement in visual acuity occurred in only 2 of the 7 patients treated [43]. In both these studies, tissue matching was indicated as an important prognostic factor for the success of the procedure. In keratolimbal allograft transplantation the amount of donor tissue used is not limited due to the non-living source of the tissue [44]. In this procedure, following superficial keratectomy, the whole of the recipient limbus is transplanted with cadaveric peripheral corneal and limbal (keratolimbal) tissue. The short term results of this procedure are favourable with an improvement in visual acuity of 82-100% [45, 46]. However, longer term follow up of approximately 3 to 5 years shows a drop in this improvement to 48-82% [47, 48]. The major drawback of both living related conjunctival limbal and keratolimbal allografts is the requirement of potent recipient immune suppression to maintain success of the procedure [49].

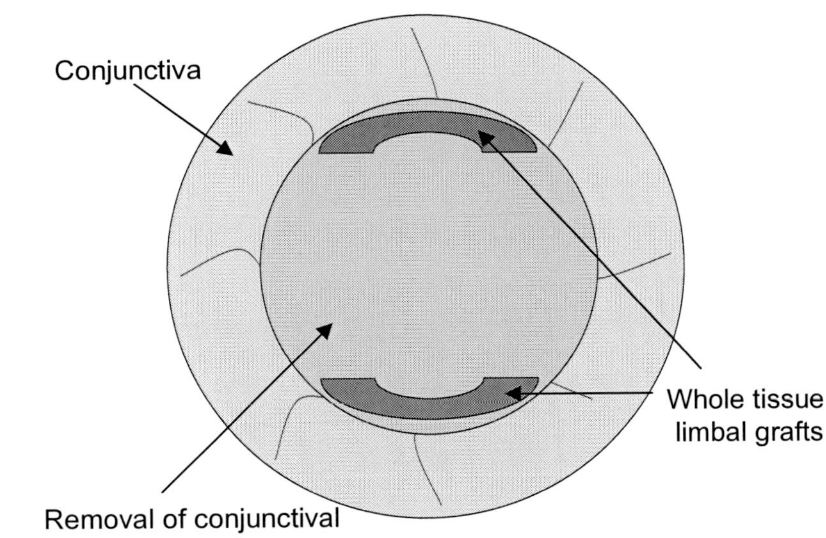

Figure 7. A schematic illustration of whole tissue limbal transplantation. The conjunctival tissue is firstly removed from the cornea and limbus by performing an extended superficial keratectomy. Usually two relatively large pieces of donor limbal tissue are harvested from the donor eye and transplanted to the limbus of the recipient eye. This is often combined with amniotic membrane transplantation as well.

Cultured Limbal Transplantation

There are two major problems associated with whole tissue limbal transplants which have been alluded to earlier [18, 50, 51]. Firstly, for conjunctival limbal grafts (both autografts and living related allografts), because the donor eyes are live eyes, there is a limit to the amount of tissue that can be removed for the procedure due to the risk of iatrogenic limbal stem cell deficiency. Secondly, using both living related conjunctival limbal and keratolimbal allografts, due to the allogeneic nature of the tissue, potent immune suppression is required in the recipient, even in cases of tissue matching. These are real issues for a subgroup of patients with severe limbal stem cell deficiency. In 1997, Pellegrini and co-workers showed that small biopsies of limbal epithelial tissue, measuring approximately 1 mm by 1 mm, could be grown in culture and expanded (ex vivo expansion) to produce corneal epithelial sheets [50]. The sheets could then be transplanted to the limbal stem cell deficienct eye following superficial keratectomy (Figure 8). Pellegrini and co-workers presented two successful cases where the patients' healthy eyes were used as the donor source and their other limbal stem cell deficient eyes were the treated eyes. In these cases the two problems associated with whole tissue grafts were therefore avoided. There was a negligible risk of iatrogenic limbal stem cell deficiency in the donor eye, and recipient immune suppression was not required due to the autologous nature of the tissue.

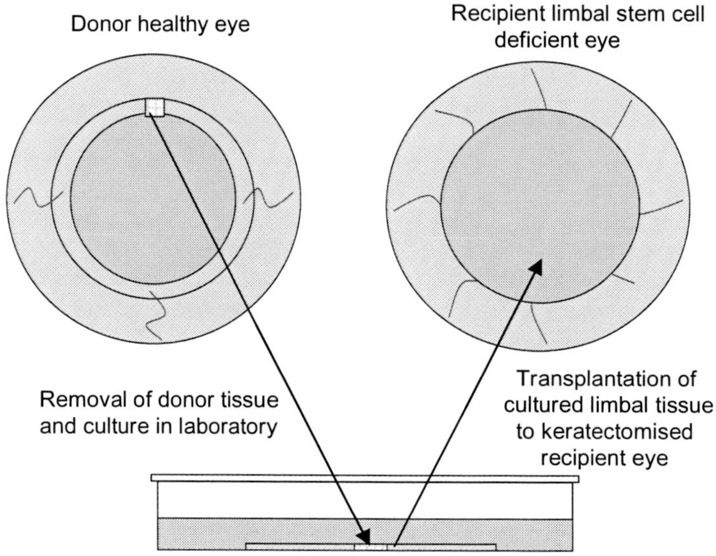

Figure 8. A schematic illustration of cultured limbal transplantation. A small sample of limbal tissue is taking from the healthy donor eye. This is taken to the laboratory and expanded in culture (ex vivo expansion). The conjunctival tissue is removed from the cornea and limbus of the limbal stem cell deficient eye. The expanded limbal epithelium is then transplanted to this prepared recipient eye.

Since the initial cases presented by Pellegrini and co-workers, different donor sources have been used and different culture techniques for limbal epithelium have been used to treat limbal stem cell deficiency with success. As with whole tissue grafts, the donor source of the limbal tissue can be autologous, i.e. the patient's other healthy eye, or allogeneic, i.e. the healthy eye of a living related or cadaveric donor (Figure 6). Obviously the use of allogeneic tissue will still require recipient immune suppression. Variations in the culture of limbal epithelium for clinical application have also been described. The original culture technique described by Pellegrini and co-workers involves the culture of human limbal epithelium with 3T3 mouse fibroblasts (3T3 co-culture) [50]. Other techniques avoid the use of these animal cells by culturing the limbal epithelium on human amniotic membrane [51]. A combination of these two culture methods, i.e. growing the limbal epithelium on amniotic membrane in the presence of 3T3 fibroblasts, is also used [52]. The main drawback with cultured limbal transplants is the use of animal cells (3T3 mouse fibroblasts) and animal products (foetal calf serum) in the culture process [53]. These issues are currently under investigation.

The results of cultured limbal transplantation are promising. In the original cases presented by Pellegrini and co-workers, both patients had an improvement in visual acuity following the culture of autologous limbal tissue [50]. In 2000, Tsai and co-workers, using the amniotic membrane culture technique, showed that 5 out of the 6 patients studied had an improvement in visual acuity, again using autologous tissue [51]. In 2003, Sangwan and co-workers presented a large case series of 125 patients showing a 100% success rate for cultured autologous limbal transplants [54]. The success rates for cultured allogeneic limbal transplants are also promising. In 2001, Koizumi and co-workers presented 13 cases of cultured cadaveric limbal transplants with an improvement in visual acuity in 10 cases [52]. In 2005, Daya and co-workers presented 10 cases of both cadaveric and living related donor cultured limbal transplants with an improvement in visual acuity in 40% of the cases [55]. From these case series it can certainly be seen that the results for cultured limbal transplants are very promising. In addition, with a 10 year history, their safety record is also good. The main benefit of cultured limbal transplantation is clearly in patients with unilateral severe limbal stem cell deficiency where the other eye is healthy and can be used as the donor source. In this case, the risk of iatrogenic limbal stem cell deficiency is made negligible and the requirement of immune suppression for graft survival is eliminated.

Cultured Oral Mucosa Transplantation

In cases of bilateral severe limbal stem cell deficiency, the use of autologous limbal tissue for either whole tissue or cultured limbal transplantation is not possible. The use of allogeneic sources of tissue for these patients therefore requires potent immune suppression. It has therefore been shown by several groups that the patient's own oral mucosa can be cultured using the amniotic membrane technique and this transplanted to the limbal stem cell deficient eyes in the same way as cultured limbal transplants (Figure 6) [56-58]. Due to the autologous nature of the tissue, the requirement of immune suppression for graft survival is eliminated. The results from the studies are very promising for the treatment of bilateral severe limbal stem cell deficiency. In 2004, Nakamura and co-workers presented 6 eyes which had undergone cultured oral mucosa transplantation and showed that visual acuity improved in all the eyes [56]. In 2006, Inatomi and co-workers presented the results of 15 eyes showing an improvement in visual acuity in 10 of the eyes [58]. In both these studies however the high incidence of peripheral corneal vascularisation is of concern. Longer term results will therefore be useful in assessing whether this becomes a long term sight reducing complication.

Conclusion

Limbal stem cell deficiency is a painful and blinding disease which is often very difficult to manage. Limbal stem cell biology has advanced considerably over the last 20 years in particular and our current management of limbal stem cell deficiency draws upon these advances. The result is one of being able to now manage these difficult to treat patients with much greater success. In particular, the use of whole tissue limbal, amniotic membrane and cultured limbal transplants has enabled the regeneration of a stable corneal surface in patients with severe limbal stem cell deficiency. This chapter highlights the importance of understanding a particular tissue's adult stem cell biology if we want to successfully manage patients where normal tissue regeneration fails. Limbal stem cell biology is no longer a pure laboratory based science and the management of limbal stem cell deficiency is not limited to the clinic. Indeed there is a huge overlap between the two and the next few years will certainly be interesting for both our understanding and management of corneal epithelial maintenance.

References

[1] Ahmad, S., F. Figueiredo, and M. Lako, Corneal epithelial stem cells: characterization, culture and transplantation. *Regen Med*, 2006. 1(1): p. 29-44.

[2] Davanger, M. and A. Evensen, Role of the pericorneal papillary structure in renewal of corneal epithelium. *Nature*, 1971. 229(5286): p. 560-1.

[3] Ebato, B., J. Friend, and R.A. Thoft, Comparison of limbal and peripheral human corneal epithelium in tissue culture. *Invest Ophthalmol Vis Sci*, 1988. 29(10): p. 1533-7.

[4] Cotsarelis, G., et al., Existence of slow-cycling limbal epithelial basal cells that can be preferentially stimulated to proliferate: implications on epithelial stem cells. *Cell*, 1989. 57(2): p. 201-9.

[5] Kenyon, K.R. and S.C. Tseng, Limbal autograft transplantation for ocular surface disorders. *Ophthalmology*, 1989. 96(5): p. 709-22; discussion 722-3.

[6] Huang, A.J. and S.C. Tseng, Corneal epithelial wound healing in the absence of limbal epithelium. In*vest Ophthalmol Vis Sci*, 1991. 32(1): p. 96-105.

[7] Thoft, R.A. and J. Friend, The X, Y, Z hypothesis of corneal epithelial maintenance. *Invest Ophthalmol Vis Sci,* 1983. 24(10): p. 1442-3.

[8] Townsend, W.M., The limbal palisades of Vogt. *Trans Am Ophthalmol Soc*, 1991. 89: p. 721-56.

[9] Puangsricharern, V. and S.C. Tseng, Cytologic evidence of corneal diseases with limbal stem cell deficiency. *Ophthalmology*, 1995. 102(10): p. 1476-85.

[10] Dua, H.S. and J.V. Forrester, The corneoscleral limbus in human corneal epithelial wound healing. *Am J Ophthalmol*, 1990. 110(6): p. 646-56.

[11] Dua, H.S., J.A. Gomes, and A. Singh, Corneal epithelial wound healing. *Br J Ophthalmol*, 1994. 78(5): p. 401-8.

[12] Ramaesh, K., et al., Evolving concepts on the pathogenic mechanisms of aniridia related keratopathy. *Int J Biochem Cell Biol*, 2005. 37(3): p. 547-57.

[13] Copeland, R.A., Jr. and D.H. Char, Limbal autograft reconstruction after conjunctival squamous cell carcinoma. *Am J Ophthalmol*, 1990. 110(4): p. 412-5.

[14] Knapp, A., et al., Serious corneal complications of glaucoma filtering surgery with postoperative 5-fluorouracil. *Am J Ophthalmol*, 1987. 103(2): p. 183-7.

[15] Clinch, T.E., K.M. Goins, and L.M. Cobo, Treatment of contact lens-related ocular surface disorders with autologous conjunctival transplantation. *Ophthalmology*, 1992. 99(4): p. 634-8.

[16] Samson, C.M., et al., Limbal stem cell transplantation in chronic inflammatory eye disease. *Ophthalmology*, 2002. 109(5): p. 862-8.

[17] Espana, E.M., et al., Idiopathic limbal stem cell deficiency. *Ophthalmology*, 2002. 109(11): p. 2004-10.

[18] Ahmad, S., F. Figueiredo, and M. Lako, Corneal epithelial stem cells: characterization, culture and transplantation. *Regen Med*, 2006. 1(1): p. 29-44.

[19] Dua, H.S. and A. Azuara-Blanco, Limbal stem cells of the corneal epithelium. *Surv Ophthalmol*, 2000. 44(5): p. 415-25.

[20] Buck, R.C., Ultrastructure of conjunctival epithelium replacing corneal epithelium. *Curr Eye Res,* 1986. 5(2): p. 149-59.

[21] Dua, H.S., The conjunctiva in corneal epithelial wound healing. *Br J Ophthalmol*, 1998. 82(12): p. 1407-11.

[22] Egbert, P.R., S. Lauber, and D.M. Maurice, A simple conjunctival biopsy. *Am J Ophthalmol*, 1977. 84(6): p. 798-801.

[23] Donisi, P.M., et al., Analysis of limbal stem cell deficiency by corneal impression cytology. *Cornea*, 2003. 22(6): p. 533-8.

[24] Patel, D.V., T. Sherwin, and C.N. McGhee, Laser scanning in vivo confocal microscopy of the normal human corneoscleral limbus. *Invest Ophthalmol Vis Sci*, 2006. 47(7): p. 2823-7.

[25] Geerling, G., S. Maclennan, and D. Hartwig, Autologous serum eye drops for ocular surface disorders. *Br J Ophthalmol*, 2004. 88(11): p. 1467-74.

[26] Tsubota, K., et al., Treatment of dry eye by autologous serum application in Sjogren's syndrome. *Br J Ophthalmol*, 1999. 83(4): p. 390-5.

[27] Tsubota, K., et al., Treatment of persistent corneal epithelial defect by autologous serum application. *Ophthalmology*, 1999. 106(10): p. 1984-9.

[28] Goto, E., et al., Treatment of superior limbic keratoconjunctivitis by application of autologous serum. *Cornea*, 2001. 20(8): p. 807-10.

[29] Poon, A.C., et al., Autologous serum eyedrops for dry eyes and epithelial defects: clinical and in vitro toxicity studies. *Br J Ophthalmol*, 2001. 85(10): p. 1188-97.

[30] Koizumi, N.J., et al., Growth factor mRNA and protein in preserved human amniotic membrane. *Curr Eye Res*, 2000. 20(3): p. 173-7.

[31] Cooper, L.J., et al., An investigation into the composition of amniotic membrane used for ocular surface reconstruction. *Cornea*, 2005. 24(6): p. 722-9.

[32] Choi, T.H. and S.C. Tseng, In vivo and in vitro demonstration of epithelial cell-induced myofibroblast differentiation of keratocytes and an inhibitory effect by amniotic membrane. *Cornea*, 2001. 20(2): p. 197-204.
[33] Kim, J.S., et al., Amniotic membrane patching promotes healing and inhibits proteinase activity on wound healing following acute corneal alkali burn. *Exp Eye Res*, 2000. 70(3): p. 329-37.
[34] Shimmura, S., et al., Antiinflammatory effects of amniotic membrane transplantation in ocular surface disorders. *Cornea*, 2001. 20(4): p. 408-13.
[35] Tseng, S.C., et al., Amniotic membrane transplantation with or without limbal allografts for corneal surface reconstruction in patients with limbal stem cell deficiency. *Arch Ophthalmol*, 1998. 116(4): p. 431-41.
[36] Brown, S.I., S.E. Bloomfield, and D.B. Pearce, A follow-up report on transplantation of the alkali-burned cornea. *Am J Ophthalmol*, 1974. 77(4): p. 538-42.
[37] Kremer, I., et al., Results of penetrating keratoplasty in aniridia. *Am J Ophthalmol*, 1993. 115(3): p. 317-20.
[38] Tugal-Tutkun, I., Y.A. Akova, and C.S. Foster, Penetrating keratoplasty in cicatrizing conjunctival diseases. *Ophthalmology*, 1995. 102(4): p. 576-85.
[39] Reinhard, T., et al., Homologous penetrating central limbo-keratoplasty (HPCLK) in bilateral limbal stem cell insufficiency. *Acta Ophthalmol Scand*, 1999. 77(6): p. 663-7.
[40] Reinhard, T., et al., Long-term results of allogeneic penetrating limbo-keratoplasty in total limbal stem cell deficiency. *Ophthalmology*, 2004. 111(4): p. 775-82.
[41] Holland, E.J. and G.S. Schwartz, The evolution of epithelial transplantation for severe ocular surface disease and a proposed classification system. *Cornea*, 1996. 15(6): p. 549-56.
[42] Daya, S.M. and F.A. Ilari, Living related conjunctival limbal allograft for the treatment of stem cell deficiency. Ophthalmology, 2001. 108(1): p. 126-33; discussion 133-4.
[43] Rao, S.K., et al., Limbal allografting from related live donors for corneal surface reconstruction. *Ophthalmology*, 1999. 106(4): p. 822-8.
[44] Espana, E.M., et al., Keratolimbal allograft in corneal reconstruction. *Eye*, 2004. 18(4): p. 406-17.
[45] Tsai, R.J. and S.C. Tseng, Human allograft limbal transplantation for corneal surface reconstruction. *Cornea*, 1994. 13(5): p. 389-400.
[46] Tsubota, K., et al., Reconstruction of the corneal epithelium by limbal allograft transplantation for severe ocular surface disorders. *Ophthalmology*, 1995. 102(10): p. 1486-96.
[47] Solomon, A., et al., Long-term outcome of keratolimbal allograft with or without penetrating keratoplasty for total limbal stem cell deficiency. *Ophthalmology*, 2002. 109(6): p. 1159-66.
[48] Ilari, L. and S.M. Daya, Long-term outcomes of keratolimbal allograft for the treatment of severe ocular surface disorders. *Ophthalmology*, 2002. 109(7): p. 1278-84.
[49] Holland, E.J., A.R. Djalilian, and G.S. Schwartz, Management of aniridic keratopathy with keratolimbal allograft: a limbal stem cell transplantation technique. *Ophthalmology*, 2003. 110(1): p. 125-30.

[50] Pellegrini, G., et al., Long-term restoration of damaged corneal surfaces with autologous cultivated corneal epithelium. *Lancet*, 1997. 349(9057): p. 990-3.

[51] Tsai, R.J., L.M. Li, and J.K. Chen, Reconstruction of damaged corneas by transplantation of autologous limbal epithelial cells. *N Engl J Med,* 2000. 343(2): p. 86-93.

[52] Koizumi, N., et al., Cultivated corneal epithelial stem cell transplantation in ocular surface disorders. *Ophthalmology*, 2001. 108(9): p. 1569-74.

[53] Ramaesh, K. and B. Dhillon, Ex vivo expansion of corneal limbal epithelial/stem cells for corneal surface reconstruction. *Eur J Ophthalmol*, 2003. 13(6): p. 515-24.

[54] Sangwan, V.S., et al., Successful reconstruction of damaged ocular outer surface in humans using limbal and conjuctival stem cell culture methods. *Biosci Rep*, 2003. 23(4): p. 169-74.

[55] Daya, S.M., et al., Outcomes and DNA analysis of ex vivo expanded stem cell allograft for ocular surface reconstruction. *Ophthalmology*, 2005. 112(3): p. 470-7.

[56] Nakamura, T., et al., Transplantation of cultivated autologous oral mucosal epithelial cells in patients with severe ocular surface disorders. *Br J Ophthalmol*, 2004. 88(10): p. 1280-4.

[57] Nishida, K., et al., Corneal reconstruction with tissue-engineered cell sheets composed of autologous oral mucosal epithelium. *N Engl J Med,* 2004. 351(12): p. 1187-96.

[58] Inatomi, T., et al., Midterm results on ocular surface reconstruction using cultivated autologous oral mucosal epithelial transplantation. *Am J Ophthalmol*, 2006. 141(2): p. 267-275.

In: Stem Cell Applications in Disease and Health
Editors: W. Burnsides and R. Ellsley
ISBN: 978-1-60456-255-2

Chapter VIII

Stem Cell Applications in Diseases: Stem Cells for the Treatment of Heart Failure

Shunsuke Ohnishi*[*] *and Noritoshi Nagaya
Department of Regenerative Medicine & Tissue Engineering,
National Cardiovascular Center Research Institute, Osaka, Japan

Abstract

Heart failure is one of the most important cardiovascular health problems throughout the world, with high mortality, and there is a need to develop more effective therapeutic strategies to replace specialized treatment such as mechanical circulatory support or cardiac transplantation.

Cell therapy is currently attracting growing interest as a potential new means of improving the prognosis of patients with heart failure. Autologous skeletal myoblasts have been the first to be tested in clinical trials; however, cardiovascular researchers have recently explored many other cell types, including bone marrow cells, endothelial progenitor cells, mesenchymal stem cells, embryonic stem cells, and resident cardiac stem cells. While recent experimental studies and early-phase clinical trials seem to support the concept that cell therapy may enhance cardiac repair, many challenges remain before achieving this goal. Further studies should focus on finding the optimal donor cells and cell numbers for transplantation, the mechanism by which engrafted cells improve cardiac function, controlling the survival, proliferation and differentiation of transplanted cells, and the development of more efficient cell delivery techniques.

[*] Correspondence to: Shunsuke Ohnishi M.D., Ph.D.
Fujishirodai 5-7-1, Suita, Osaka 565-8565, Japan
Email: sonishi@ri.ncvc.go.jp
Phone: +81-6-6833-5012
Fax: +81-6-6833-9865

Introduction

Heart failure is a major cardiovascular health problem worldwide. About 5.2 million patients in the US have heart failure, and the lifetime risk of developing chronic heart failure is 1 in 5 [1]. Coronary artery disease is the most common cause of heart failure, followed by idiopathic dilated cardiomyopathy and valvular heart disease [2, 3]. Myocardial infarction causes necrosis of the myocardium, followed by infiltration of inflammatory cells. Then a scar forms, leading to the loss of cardiac function, ventricular remodeling and progressive dysfunction, and finally, congestive heart failure [4-6]. Drugs commonly used for the treatment of chronic heart failure include loop diuretics, angiotensin-converting enzyme inhibitors, β-adrenergic receptor blockers, aldosterone antagonists, angiotensin II receptor blockers and digitoxin; however, patients with severe heart failure require invasive treatment such as ventricular assist devices, continuous inotropic infusion, and heart transplantation [7-9]. Therefore, there is a need to develop more effective, less invasive therapeutic strategies for heart failure.

Cell therapy for heart failure has the potential to restore cardiac function by inducing neovascularization, and regenerating and protecting cardiomyocytes [4, 10]. Skeletal myoblasts, bone marrow-derived mononuclear cells and endothelial progenitor cells have been applied for ischemic cardiovascular disease in human studies [11-14]. Recently, bone marrow-derived mesenchymal stem cells, a subpopulation of mononuclear cells, have emerged as a new therapeutic cell source because they possess multipotency and can be easily expanded in culture [15]. In addition, other types of stem cells such as embryonic stem cells and cardiac stem cells have received attention for future applications [16, 17]. This article reviews from basic and preclinical studies to clinical trials on stem cell therapy for heart failure.

Cell Sources for Cardiac Repair: Basic Research to Preclinical Study

Several kinds of stem cells have the potential to be applied for cardiac repair: pluripotent embryonic stem cells and somatic stem/progenitor cells.

Embryonic Stem Cells

Embryonic stem cells are pluripotent cells derived from the inner cell mass of blastocysts, and can differentiate into various cell types from all three germ layers including cardiomyocytes. Early studies in animal models have demonstrated that transplanted mouse and human embryonic stem cells or embryonic stem cell-derived cardiomyocytes can engraft and survive in the myocardium, and contributed to a beneficial outcome in the infracted heart [18-20]. However, the major obstacle to the clinical use of embryonic stem cells is the possibility of rejection reaction and tumor formation such as teratoma and teratocarcinoma in the recipients, as wells as ethical considerations [16, 21, 22].

Skeletal Myoblasts

Skeletal myoblasts are muscle progenitor cells which normally lie in a quiescent state under the basal lamina of muscular fibers, and are mobilized and proliferate to repair muscle after injury [23, 24]. They can be easily expanded in culture in an undifferentiated state, and are highly resistant to tissue ischemia. *In vivo* studies have demonstrated that myoblasts differentiate into myotubes, retain skeletal muscle properties, sustain engraftment in post-infarcted scars, and improve left ventricular function after transplantation into an infracted myocardium, although myotubes do not couple with resident cardiomyocytes electromechanically [25-27].

Bone Marrow Cells

Bone marrow cells include several types of stem/progenitor cells such as hematopoietc stem cells, endothelial progenitor cells and mesenchymal stem cells, although the vast majority of bone marrow cells are hematopoietic cells [28, 29]. Because of the extensive clinical experience in bone marrow transplantation for hematological diseases over the past decades and the fact that large numbers of bone marrow cells can be easily obtained, the study of bone marrow cell transplantation for treating heart failure has moved quickly from small animals to human studies. Transplantation of bone marrow mononuclear cells or hematopoietic stem cells induced therapeutic angiogenesis and myogenesis, and improved ventricular function mostly through paracrine effects [30-35]. However, it is yet to be clarified whether the optimal bone marrow cell population is hematopoietic stem cells, endothelial progenitor cells, mesenchymal stem cells, or abundant committed cells.

Endothelial Progenitor Cells

Endothelial progenitor cells, a rare subpopulation of bone marrow cells with a similar phenotype and function to those of fetal angioblasts, have originally been defined by their cell surface antigen expression of the hematopoietic markers CD34 and CD133 and the endothelial marker VEGF receptor-2 [36]. Endothelial progenitor cells have been involved in neovascularization after myocardial infarction, reduced collagen deposition and apoptosis of cardiomyocytes, and improved cardiac function after experimental myocardial infarction [37]. In addition, it has been demonstrated that granulocyte colony-stimulating factor (G-CSF) and stem cell factor can mobilize endothelial progenitor cells [38]. Clinical trials studying the ability of G-CSF to mobilize stem/progenitor cells in patients with coronary artery disease did not reach a conclusion on efficacy, while concerns have been raised on safety in relation to arterial restenosis and plaque destabilization [39, 40]. HMG-CoA reductase inhibitors [41], estrogens [42], exercise [43], and non-smoking [44] are more practical than growth factor or chemokine administration to enhance the number of circulating endothelial progenitor cells in patients.

Cardiac Stem Cells

In the past few years, the search for stem/progenitor cells in the heart has intensified. Evidence is accumulating that cardiac stem cells reside at specific locations in the adult heart. Anversa's group described the isolation of lineage-negative, c-Kit-positive cells from the heart of adult rats and humans, and showed that these cells were able to differentiate into cardiomyocytes, smooth muscle cells, and endothelial cells [17, 45]. In addition, Sca-1-positive cells were isolated from the adult mouse heart [46]. This Sca-1-positive subpopulation differentiated *in vitro* into cardiomyocytes in response to 5-azacytidine, and expressed several cardiac specific markers. In addition, Sca-1-positive cells differentiated into cardiomyocytes after intravenous injection. Messina et al. identified a subpopulation of cardiosphere-forming cells after culturing human atrial or ventricular biopsies and embryonic, fetal or postnatal mouse hearts [47]. The mouse cardiospheres started to beat spontaneously, whereas human cardiospheres only started to beat after co-culture with rat neonatal cardiomyocytes. Furthermore, cardiospheres expressed endothelial markers and were positive for Sca-1 and c-Kit. Recently, another subpopulation of cardiac stem cells has been described [48]. These cardiac stem cells were Isl1-positive and were identified in rat, mouse and human adult hearts. Isl1-positive cardiac stem cells were positive for cardiac transcriptional factors Nkx2.5 and GATA4, while transcriptional factors associated with mature cardiomyocytes were absent. However, Sca-1 and c-Kit were not expressed on these cells. When co-cultured with neonatal rat cardiomyocytes, 30% of the Isl1-positive cardiac stem cells differentiated into cardiomyocytes. Preclinical studies demonstrated that injection of cardiac stem cells significantly improved the cardiac function in mice with acute myocardial infarction [49].

Mesenchymal Stem Cells

Mesenchymal stem cells reside not only in bone marrow [50] and adipose tissue [51], but also in other tissues such as synovium [52], periosteum [53], muscle [54], dental pulp [55], periodontal ligament [56], placenta [57] and umbilical cord blood [58]. A recent study suggests that mesenchymal stem cells reside in virtually all post-natal organs and tissues, and may be localized to vessel walls [59]. Mesenchymal stem cells can differentiate into various types of cells including cardiomyocytes and vascular endothelial cells [15, 60-63]. Bone marrow-derived mesenchymal stem cells, which are most investigated, are a rare subpopulation of bone marrow cells (approximately 0.001-0.01%), but can easily expand in culture [15]. In addition, the immune phenotype of MSC is widely described as positive for major histocompatibility complex (MHC class I), and negative for MHC class II and costimulatory factors (CD40, CD80 and CD86), and thus regarded as non-immunogenic and might be used in an allogeneicsetting [64]. *In vivo* studies have demonstrated that directly or intravenously injected mesenchymal stem cells differentiated into endothelial cells, and were involved in angiogenesis as well as myogenesis in animal models of myocardial infarction [65-67]. Mesenchymal stem cells exert their effect on cardiac regeneration not only by differentiation into specific cell types, but also through paracrine actions. *In vitro* studies have demonstrated that mesenchymal stem cells can secrete a variety of angiogenic, antiapoptotic and mitogenic factors such as vascular endothelial growth factor (VEGF), hepatocyte growth factor (HGF), adrenomedullin (AM) and insulin-like growth factor-1 (IGF-1) [68, 69]. Interestingly, administration of conditioned medium obtained from mesenchymal stem cells culture exerted cytoprotective effects on the myocardium in an animal model of myocardial infarction [70], and enhanced collateral flow recovery in mice with hindlimb ischemia [68].

Myocardial injection of mesenchymal stem cells improved cardiac function in a rat model of dilated cardiomyopathy, a myocardial disease characterized by a loss of cardiomyocytes and an increase in fibroblasts [71, 72], possibly through induction of angiogenesis and myogenesis, as well as by inhibition of myocardial fibrosis [69]. In addition, intravenous administration of mesenchymal stem cells improved inflammatory changes and cardiac function in rats with acute myocarditis, suggesting an anti-inflammatory effect of mesenchymal stem cells [73].

Clinical Applications of Stem Cell Therapy

Stem cell therapy has the potential to be applied for the treatment of acute or chronic, ischemic or non-ischemic, heart failure, although the required mechanism might be different depending on the diseases as well as their status. Here we summarize the results of clinical trials applied for the treatment of heart failure including acute myocardial infarction and cardiomyopathy.

Acute Myocardial Infarction

Acute myocardial infarction involves a series of inflammatory, apoptotic, remodeling and healing events that could have implications for cell homing and for differential effects according to timing of cell delivery [74]. Most of the clinical studies performed intracoronary infusion of bone marrow mononuclear cells, and observed absolute increases (> 6%) in ejection fraction (Table 1). However, study protocols such as cell preparation, cell number, follow-up period and main endpoint are different among study groups, and several studies did not observe an improvement in ejection fraction. Nevertheless, cell therapy for acute myocardial infarction may be more beneficial to patients with more severely depressed ejection fraction [75].

Chronic Heart Failure

On the contrary to acute myocardial infarction, large areas of myocardium may need to be treated. Angiogenic effects could be beneficial for ischemic cardiomyopathy; however, myocyte regeneration would be required for the treatment of end-stage dilated cardiomyopathy. Patients undergoing thoracic surgery such as coronary artery bypass graft or left ventricular assist device may provide good opportunities for targeted cell delivery

Table 1. Clinical trials of cell therapy in acute myocardial infarction

Study	n	Cell type	Dose ($\times 10^6$)	Time after AMI (days)	Delivery method	Follow-up (months)	Result of LVEF	Adverse events
Strauer et al. [77]	20	MNC	28	5-9	IC	3	Not improved	None reported
Chen et al. [78]	69	MSC	48000-60000	18	IC	6	Improved (18%)	None reported
BOOST [79, 80]	60	MNC	2460	5	IC	18	Improved (6%)	None reported
Fernandez-Aviles et al. [31]	20	MNC	78	14	IC	6	Improved (6%)	None reported
Janssens et al. [81]	67	MNC	304	1	IC	4	Not improved	None reported
TOPCARE-AMI [82, 83]	59	MNC vs EPC	213 (MNC) 16 (EPC)	3-7	IC	12	Improved (9%)	None reported
ASTAMI [84]	100	MNC	68	4-8	IC	6	Not improved	None reported
REPAIR-AMI [75]	204	MNC	236	3-7	IC	4	Improved (6%)	None reported

MNC, bone marrow-derived mononuclear cells; EPC, endothelial progenitor cells; MSC, mesenchymal stem cells, IC, intracoronary infusion; LVEF, left ventricular ejection fraction

Table 2. Clinical trials of cell therapy in cardiomyopathy

Study	n	Cell type	Dose ($\times 10^6$, mean or range)	Delivery method	Follow-up (months)	Result of LVEF	Adverse events
Menasche et al. [85]	10	SKM	871	CABG+EC	11	Improved (8%)	SVT
Hagege et al. [86]	9	SKM	62-1100	CABG+EC	52	Improved (4%)	SVT, NSVT
Herreros et al. [87]	12	SKM	221	CABG+EC	3	Improved (18%)	NSVT
Siminiak et al. [88]	10	SKM	0.4-50	CABG+EC	12	Improved (7%)	SVT
POZNAN [89]	9	SKM	17-106	TCV	6	Improved (3-8%) in 6/9	None reported
Chachques et al. [90]	20	SKM	>300	CABG+EC	14	Improved (24%)	None reported
Perin et al. [76, 91]	21	MNC	26	TE	12	Improved (9%)	None reported
Fuchs et al. [92]	10	MNC	33	TE	3	Not improved	None reported
Tse et al. [13]	8	MNC	?	TE	3	Not improved	None reported
Stamm et al. [93]	40	MNC (CD133+)	7.5	CABG+EC	6	Improved (10%)	None reported
Archundia et al. [94]	5	MNC	400	CABG+EC	6	Improved (≥10%)	None reported
Hendrikx et al. [95]	20	MNC	60	CABG+EC	4	Not improved	None reported
Brehm et al. [96]	18	MNC	?	IC	3	Improved (15%)	None reported
Patel et al. [97]	20	EPC	22	CABG+EC	6	Improved (17%)	None reported
Erbs et al. [98]	26	EPC	69	IC	3	Improved (14%)	None reported
TOPCARE-CHD [99]	75	MNC vs EPC	205 (MNC) 22 (EPC)	IC	3	Improved (3%) by MNC Not improved by EPC	None reported
Chen et al.	45	MSC	?	IC	12	Not improved	None reported

SKM, skeletal myoblasts; MNC, bone marrow-derived mononuclear cells; EPC, endothelial progenitor cells; MSC, mesenchymal stem cells, IC, intracoronary infusion; LVEF, left ventricular ejection fraction; CABG, coronary artery bypass graft; EC, epicardial injection; TCV, transcoronary-venous injection; TE, transendocardial injection; IC, intracoronary infusion; LVEF, left ventricular ejection fraction SVT, sustained VT; NSVT, nonsustained VT.

(epicardial injection); however, these combination therapies might complicate interpretation of a beneficial effect (Table 2). Recently developed left ventricular electromechanical mapping system enabled guiding targeted transendocardial injection of cells [13, 76].

Conclusion

Stem cell therapy has emerged as a new tool for the treatment of heart failure, based on basic research and a number of preclinical studies; however, which type of cells, how many of cells, which method of delivery have significant value in the treatment of heart failure, is currently being explored in clinical trials. Furthermore, the underlying mechanisms responsible for cardiac repair and cardioprotection still need to be elucidated.

References

[1] Rosamond, W., Flegal, K., Friday, G., Furie, K., Go, A., Greenlund, K., Haase, N., Ho, M., Howard, V., Kissela, B., Kittner, S., Lloyd-Jones, D., McDermott, M., Meigs, J., Moy, C., Nichol, G., O'Donnell, C.J., Roger, V., Rumsfeld, J., Sorlie, P., Steinberger, J., Thom, T., Wasserthiel-Smoller, S., and Hong, Y. (2007). Heart disease and stroke statistics--2007 update: a report from the American Heart Association Statistics Committee and Stroke Statistics Subcommittee. *Circulation,* 115(5):e69-171.

[2] Fox, K.F., Cowie, M.R., Wood, D.A., Coats, A.J., Gibbs, J.S., Underwood, S.R., Turner, R.M., Poole-Wilson, P.A., Davies, S.W., and Sutton, G.C. (2001). Coronary artery disease as the cause of incident heart failure in the population. *Eur Heart J.* 22(3):228-236.

[3] He, J., Ogden, L.G., Bazzano, L.A., Vupputuri, S., Loria, C., and Whelton, P.K. (2001). Risk factors for congestive heart failure in US men and women: NHANES I epidemiologic follow-up study. *Arch Intern Med.* 161(7):996-1002.

[4] Fraser, J.K., Schreiber, R.E., Zuk, P.A., and Hedrick, M.H. (2004). Adult stem cell therapy for the heart. *Int J Biochem Cell Biol.* 36(4):658-666.

[5] Tiyyagura, S.R. and Pinney, S.P. (2006). Left ventricular remodeling after myocardial infarction: past, present, and future. *Mt Sinai J Med.* 73(6):840-851.

[6] Sharpe, N. (2004). Cardiac remodeling in coronary artery disease. *Am J Cardiol.* 93(9A):17B-20B.

[7] Hunt, S.A., Abraham, W.T., Chin, M.H., Feldman, A.M., Francis, G.S., Ganiats, T.G., Jessup, M., Konstam, M.A., Mancini, D.M., Michl, K., Oates, J.A., Rahko, P.S., Silver, M.A., Stevenson, L.W., Yancy, C.W., Antman, E.M., Smith, S.C., Jr., Adams, C.D., Anderson, J.L., Faxon, D.P., Fuster, V., Halperin, J.L., Hiratzka, L.F., Jacobs, A.K., Nishimura, R., Ornato, J.P., Page, R.L., and Riegel, B. (2005). ACC/AHA 2005 Guideline Update for the Diagnosis and Management of Chronic Heart Failure in the Adult: a report of the American College of Cardiology/American Heart Association Task Force on Practice Guidelines (Writing Committee to Update the 2001 Guidelines

for the Evaluation and Management of Heart Failure): developed in collaboration with the American College of Chest Physicians and the International Society for Heart and Lung Transplantation: endorsed by the Heart Rhythm Society. *Circulation,* 112(12):e154-235.

[8] Bukharovich, I.F. and Kukin, M. (2006). Optimal medical therapy for heart failure. *Prog Cardiovasc Dis.* 48(5):372-385.

[9] Adorisio, R., De Luca, L., Rossi, J., and Gheorghiade, M. (2006). Pharmacological treatment of chronic heart failure. *Heart Fail Rev.* 11(2):109-123.

[10] Wollert, K.C. and Drexler, H. (2005). Clinical applications of stem cells for the heart. *Circ Res.* 96(2):151-163.

[11] Menasche, P., Hagege, A.A., Scorsin, M., Pouzet, B., Desnos, M., Duboc, D., Schwartz, K., Vilquin, J.T., and Marolleau, J.P. (2001). Myoblast transplantation for heart failure. *Lancet,* 357(9252):279-280.

[12] Tateishi-Yuyama, E., Matsubara, H., Murohara, T., Ikeda, U., Shintani, S., Masaki, H., Amano, K., Kishimoto, Y., Yoshimoto, K., Akashi, H., Shimada, K., Iwasaka, T., and Imaizumi, T. (2002). Therapeutic angiogenesis for patients with limb ischaemia by autologous transplantation of bone-marrow cells: a pilot study and a randomised controlled trial. *Lancet,* 360(9331):427-435.

[13] Tse, H.F., Kwong, Y.L., Chan, J.K., Lo, G., Ho, C.L., and Lau, C.P. (2003). Angiogenesis in ischaemic myocardium by intramyocardial autologous bone marrow mononuclear cell implantation. *Lancet,* 361(9351):47-49.

[14] Iwami, Y., Masuda, H. and Asahara, T. (2004). Endothelial progenitor cells: past, state of the art, and future. *J Cell Mol Med.* 8(4):488-497.

[15] Pittenger, M.F., Mackay, A.M., Beck, S.C., Jaiswal, R.K., Douglas, R., Mosca, J.D., Moorman, M.A., Simonetti, D.W., Craig, S., and Marshak, D.R. (1999). Multilineage potential of adult human mesenchymal stem cells. *Science,* 284(5411):143-147.

[16] Siu, C.W., Moore, J.C. and Li, R.A. (2007). Human embryonic stem cell-derived cardiomyocytes for heart therapies. *Cardiovasc Hematol Disord Drug Targets;* 7(2):145-152.

[17] Bearzi, C., Rota, M., Hosoda, T., Tillmanns, J., Nascimbene, A., De Angelis, A., Yasuzawa-Amano, S., Trofimova, I., Siggins, R.W., Lecapitaine, N., Cascapera, S., Beltrami, A.P., D'Alessandro, D.A., Zias, E., Quaini, F., Urbanek, K., Michler, R.E., Bolli, R., Kajstura, J., Leri, A., and Anversa P. (2007). Human cardiac stem cells. *Proc Natl Acad Sci U S A.;* 104(35):14068-14073.

[18] Hodgson, D.M., Behfar, A., Zingman, L.V., Kane, G.C., Perez-Terzic, C., Alekseev, A.E., Puceat, M., and Terzic, A. (2004). Stable benefit of embryonic stem cell therapy in myocardial infarction. *Am J Physiol Heart Circ Physiol.* 287(2):H471-479.

[19] Kehat, I., Khimovich, L., Caspi, O., Gepstein, A., Shofti, R., Arbel, G., Huber, I., Satin, J., Itskovitz-Eldor, J., and Gepstein, L. (2004). Electromechanical integration of cardiomyocytes derived from human embryonic stem cells. *Nat Biotechnol.* 22(10):1282-1289.

[20] Singla, D.K., Hacker, T.A., Ma, L., Douglas, P.S., Sullivan, R., Lyons, G.E., and Kamp, T.J. (2006). Transplantation of embryonic stem cells into the infarcted mouse heart: formation of multiple cell types. *J Mol Cell Cardiol.* 40(1):195-200.

[21] Lerou, P.H. and Daley, G.Q. (2005). Therapeutic potential of embryonic stem cells. *Blood Rev.* 19(6):321-331.

[22] Wu, D.C., Boyd, A.S. and Wood, K.J. (2007). Embryonic stem cell transplantation: potential applicability in cell replacement therapy and regenerative medicine. *Front Biosci.* 12:4525-4535.

[23] Menasche, P. (2007). Skeletal myoblasts as a therapeutic agent. *Prog Cardiovasc Dis.* 50(1):7-17.

[24] Siminiak, T., Kalmucki, P. and Kurpisz, M. (2004). Autologous skeletal myoblasts for myocardial regeneration. *J Interv Cardiol.* 17(6):357-365.

[25] Dowell, J.D., Rubart, M., Pasumarthi, K.B., Soonpaa, M.H., and Field, L.J. (2003). Myocyte and myogenic stem cell transplantation in the heart. *Cardiovasc Res.* 58(2):336-350.

[26] Murry, C.E., Wiseman, R.W., Schwartz, S.M., and Hauschka, S.D. (1996). Skeletal myoblast transplantation for repair of myocardial necrosis. *J Clin Invest.* 98(11):2512-2523.

[27] Taylor, D.A., Atkins, B.Z., Hungspreugs, P., Jones, T.R., Reedy, M.C., Hutcheson, K.A, Glower, D.D., and Kraus, W.E. (1998). Regenerating functional myocardium: improved performance after skeletal myoblast transplantation. *Nat Med.* 4(8):929-933.

[28] Abkowitz, J.L., Catlin, S.N., McCallie, M.T., and Guttorp, P. (2002). Evidence that the number of hematopoietic stem cells per animal is conserved in mammals. *Blood,* 100(7):2665-2667.

[29] Pittenger, M.F. and Martin, B.J. (2004). Mesenchymal stem cells and their potential as cardiac therapeutics. *Circ Res.* 95(1):9-20.

[30] Kinnaird, T., Stabile, E., Burnett, M.S., and Epstein, S.E. (2004). Bone-marrow-derived cells for enhancing collateral development: mechanisms, animal data, and initial clinical experiences. *Circ Res.* 95(4):354-363.

[31] Fernandez-Aviles, F., San Roman, J.A., Garcia-Frade, J., Fernandez, M.E., Penarrubia, M.J., de la Fuente, L., Gomez-Bueno, M., Cantalapiedra, A., Fernandez, J., Gutierrez, O., Sanchez, P.L., Hernandez, C., Sanz, R., Garcia-Sancho, J., and Sanchez, A. (2004). Experimental and clinical regenerative capability of human bone marrow cells after myocardial infarction. *Circ Res.* 95(7):742-748.

[32] Orlic, D., Kajstura, J., Chimenti, S., Jakoniuk, I., Anderson, S.M., Li, B., Pickel, J., McKay, R., Nadal-Ginard, B., Bodine, D.M., Leri, A., and Anversa, P. (2001). Bone marrow cells regenerate infarcted myocardium. *Nature,* 410(6829):701-705.

[33] Kajstura, J., Rota, M., Whang, B., Cascapera, S., Hosoda, T., Bearzi, C., Nurzynska, D., Kasahara, H., Zias, E., Bonafe, M., Nadal,-Ginard, B., Torella, D., Nascimbene, A., Quaini, F., Urbanek, K., Leri, A., and Anversa, P. (2005). Bone marrow cells differentiate in cardiac cell lineages after infarction independently of cell fusion. *Circ Res.* 96(1):127-137.

[34] Murry, C.E., Soonpaa, M.H., Reinecke, H., Nakajima, H., Nakajima, H.O., Rubart, M., Pasumarthi, K.B., Virag, J.I., Bartelmez, S.H., Poppa, V., Bradford, G., Dowell, J.D., Williams, D.A., and Field, L.J. (2004). Haematopoietic stem cells do not transdifferentiate into cardiac myocytes in myocardial infarcts. *Nature,* 428(6983):664-668.

[35] Balsam, L.B., Wagers, A.J., Christensen, J.L., Kofidis, T., Weissman, I.L., and Robbins, R.C. (2004). Haematopoietic stem cells adopt mature haematopoietic fates in ischaemic myocardium. *Nature,* 428(6983):668-673.

[36] Asahara, T. and Kawamoto, A. (2004). Endothelial progenitor cells for postnatal vasculogenesis. *Am J Physiol Cell Physiol.* 287(3):C572-579.

[37] Kocher, A.A., Schuster, M.D., Szabolcs, M.J., Takuma, S., Burkhoff, D., Wang, J., Homma, S., Edwards, N.M., and Itescu, S. (2001). Neovascularization of ischemic myocardium by human bone-marrow-derived angioblasts prevents cardiomyocyte apoptosis, reduces remodeling and improves cardiac function. *Nat Med.* 7(4):430-436.

[38] Aicher, A., Zeiher, A.M. and Dimmeler, S. (2005). Mobilizing endothelial progenitor cells. *Hypertension,* 45(3):321-325.

[39] Kang, H.J., Kim, H.S., Zhang, S.Y., Park, K.W., Cho, H.J., Koo, B.K., Kim, Y.J., Soo Lee, D., Sohn, D.W., Han, K.S., Oh, B.H., Lee, M.M., and Park, Y.B. (2004). Effects of intracoronary infusion of peripheral blood stem-cells mobilised with granulocyte-colony stimulating factor on left ventricular systolic function and restenosis after coronary stenting in myocardial infarction: the MAGIC cell randomised clinical trial. *Lancet,* 363(9411):751-756.

[40] Matsubara, H. (2004). Risk to the coronary arteries of intracoronary stem cell infusion and G-CSF cytokine therapy. *Lancet,* 363(9411):746-747.

[41] Dimmeler, S., Aicher, A., Vasa, M., Mildner-Rihm, C., Adler, K., Tiemann, M., Rutten, H., Fichtlscherer, S., Martin, H., and Zeiher, A.M. (2001). HMG-CoA reductase inhibitors (statins) increase endothelial progenitor cells via the PI 3-kinase/Akt pathway. *J Clin Invest.* 108(3):391-397.

[42] Iwakura, A., Luedemann, C., Shastry, S., Hanley, A., Kearney, M., Aikawa, R., Isner, J.M., Asahara, T., and Losordo, D.W. (2003). Estrogen-mediated, endothelial nitric oxide synthase-dependent mobilization of bone marrow-derived endothelial progenitor cells contributes to reendothelialization after arterial injury. *Circulation*; 108(25):3115-3121.

[43] Laufs, U., Werner, N., Link, A., Endres, M., Wassmann, S., Jurgens, K., Miche, E., Bohm, M., and Nickenig, G. (2004). Physical training increases endothelial progenitor cells, inhibits neointima formation, and enhances angiogenesis. *Circulation*; 109(2):220-226.

[44] Kondo, T., Hayashi, M., Takeshita, K., Numaguchi, Y., Kobayashi, K., Iino, S., Inden, Y., and Murohara, T. (2004). Smoking cessation rapidly increases circulating progenitor cells in peripheral blood in chronic smokers. *Arterioscler Thromb Vasc Biol.* 24(8):1442-1447.

[45] Beltrami, A.P., Barlucchi, L., Torella, D., Baker, M., Limana, F., Chimenti, S., Kasahara, H., Rota, M., Musso, E., Urbanek, K., Leri, A., Kajstura, J., Nadal-Ginard, B., and Anversa, P. (2003). Adult cardiac stem cells are multipotent and support myocardial regeneration. *Cell,* 114(6):763-776.

[46] Oh, H., Bradfute, S.B., Gallardo, T.D., Nakamura, T., Gaussin, V., Mishina, Y., Pocius, J., Michael, L.H., Behringer, R.R., Garry, D.J., Entman, M.L., and Schneider, M.D. (2003). Cardiac progenitor cells from adult myocardium: homing, differentiation, and fusion after infarction. *Proc Natl Acad Sci U S A.;* 100(21):12313-12318.

[47] Messina, E., De Angelis, L., Frati, G., Morrone, S., Chimenti, S., Fiordaliso, F., Salio, M., Battaglia, M., Latronico, M.V., Coletta, M., Vivarelli, E., Frati, L., Cossu, G., and Giacomello, A. (2004). Isolation and expansion of adult cardiac stem cells from human and murine heart. *Circ Res.* 95(9):911-921.

[48] Laugwitz, K.L., Moretti, A., Lam, J., Gruber, P., Chen, Y., Woodard, S., Lin, L.Z., Cai, C.L., Lu, M.M., Reth, M., Platoshyn, O., Yuan, J.X., Evans, S., and Chien, K.R. (2005). Postnatal isl1+ cardioblasts enter fully differentiated cardiomyocyte lineages. *Nature,* 433(7026):647-653.

[49] Smith, R.R., Barile, L., Cho, H.C., Leppo, M.K., Hare, J.M., Messina, E., Giacomello, A., Abraham, M.R., and Marban, E. (2007). Regenerative potential of cardiosphere-derived cells expanded from percutaneous endomyocardial biopsy specimens. *Circulation,* 115(7):896-908.

[50] Friedenstein, A.J., Gorskaja, J.F. and Kulagina, N.N. (1976). Fibroblast precursors in normal and irradiated mouse hematopoietic organs. *Exp Hematol.* 4(5):267-274.

[51] Gimble, J. and Guilak, F. (2003). Adipose-derived adult stem cells: isolation, characterization, and differentiation potential. *Cytotherapy,* 5(5):362-369.

[52] De Bari, C., Dell'Accio, F., Tylzanowski, P., and Luyten, F.P. (2001). Multipotent mesenchymal stem cells from adult human synovial membrane. *Arthritis Rheum.* 44(8):1928-1942.

[53] Nakahara, H., Bruder, S.P., Goldberg, V.M., and Caplan, A.I. (1990). In vivo osteochondrogenic potential of cultured cells derived from the periosteum. *Clin Orthop Relat Res.* (259):223-232.

[54] Asakura, A., Komaki, M. and Rudnicki, M. (2001). Muscle satellite cells are multipotential stem cells that exhibit myogenic, osteogenic, and adipogenic differentiation. *Differentiation,* 68(4-5):245-253.

[55] Gronthos, S., Mankani, M., Brahim, J., Robey, P.G., and Shi, S. (2000). Postnatal human dental pulp stem cells (DPSCs) in vitro and in vivo. *Proc Natl Acad Sci U S A.;* 97(25):13625-13630.

[56] Seo, B.M., Miura, M., Gronthos, S., Bartold, P.M., Batouli, S., Brahim, J., Young, M., Robey, P.G., Wang, C.Y., and Shi, S. (2004). Investigation of multipotent postnatal stem cells from human periodontal ligament. *Lancet,* 364(9429):149-155.

[57] Fukuchi, Y., Nakajima, H., Sugiyama, D., Hirose, I., Kitamura, T., and Tsuji, K. (2004).Human placenta-derived cells have mesenchymal stem/progenitor cell potential. *Stem Cells,* 22(5):649-658.

[58] Erices, A., Conget, P. and Minguell, J.J. (2000). Mesenchymal progenitor cells in human umbilical cord blood. *Br J Haematol.* 109(1):235-242.

[59] da Silva Meirelles, L., Chagastelles, P.C. and Nardi, N.B. (2006). Mesenchymal stem cells reside in virtually all post-natal organs and tissues. *J Cell Sci.* 119(Pt 11):2204-2213.

[60] Prockop, D.J. (1997). Marrow stromal cells as stem cells for nonhematopoietic tissues. *Science*; 276(5309):71-74.

[61] Makino, S., Fukuda, K., Miyoshi, S., Konishi, F., Kodama, H., Pan, J., Sano, M., Takahashi, T., Hori, S., Abe, H., Hata, J., Umezawa, A., and Ogawa, S. (1999).

Cardiomyocytes can be generated from marrow stromal cells in vitro. *J Clin Invest.* 103(5):697-705.

[62] Wakitani, S., Saito, T. and Caplan, A.I. (1995). Myogenic cells derived from rat bone marrow mesenchymal stem cells exposed to 5-azacytidine. *Muscle Nerve*; 18(12):1417-1426.

[63] Xu, W., Zhang, X., Qian, H., Zhu, W., Sun, X., Hu, J., Zhou, H., and Chen, Y. (2004). Mesenchymal stem cells from adult human bone marrow differentiate into a cardiomyocyte phenotype in vitro. *Exp Biol Med.* (Maywood) 229(7):623-631.

[64] Chamberlain, G., Fox, J., Ashton, B., and Middleton, J. (2007). Mesenchymal Stem Cells: their Phenotype, Differentiation Capacity, Immunological Features and Potential for Homing. *Stem Cell.*

[65] Nagaya, N., Fujii, T., Iwase, T., Ohgushi, H., Itoh, T., Uematsu, M., Yamagishi, M., Mori, H., Kangawa, K., and Kitamura, S. (2004). Intravenous administration of mesenchymal stem cells improves cardiac function in rats with acute myocardial infarction through angiogenesis and myogenesis. *Am J Physiol Heart Circ Physiol.* 287(6):H2670-2676.

[66] Liu, J., Hu, Q., Wang, Z., Xu, C., Wang, X., Gong, G., Mansoor, A., Lee, J., Hou, M., Zeng, L., Zhang, J.R., Jerosch-Herold, M., Guo, T., Bache, R.J., and Zhang, J. (2004). Autologous stem cell transplantation for myocardial repair. *Am J Physiol Heart Circ Physiol.* 287(2):H501-511.

[67] Davani, S., Marandin, A., Mersin, N., Royer, B., Kantelip, B., Herve, P., Etievent, J.P., and Kantelip, J.P. (2003). Mesenchymal progenitor cells differentiate into an endothelial phenotype, enhance vascular density, and improve heart function in a rat cellular cardiomyoplasty model. *Circulation,* 108 Suppl 1:II253-258.

[68] Kinnaird, T., Stabile, E., Burnett, M.S., Lee, C.W., Barr, S., Fuchs, S., and Epstein, S.E. (2004). Marrow-derived stromal cells express genes encoding a broad spectrum of arteriogenic cytokines and promote in vitro and in vivo arteriogenesis through paracrine mechanisms. *Circ Res.* 94(5):678-685.

[69] Nagaya, N., Kangawa, K., Itoh, T., Iwase, T., Murakami, S., Miyahara, Y., Fujii, T., Uematsu, M., Ohgushi, H., Yamagishi, M., Tokudome, T., Mori, H., Miyatake, K., and Kitamura, S. (2005). Transplantation of mesenchymal stem cells improves cardiac function in a rat model of dilated cardiomyopathy. *Circulation,* 112(8):1128-1135.

[70] Gnecchi, M., He, H., Liang, O.D., Melo, L.G., Morello, F., Mu, H., Noiseux, N., Zhang, L., Pratt, R.E., Ingwall, J.S., and Dzau, V.J. (2005). Paracrine action accounts for marked protection of ischemic heart by Akt-modified mesenchymal stem cells. *Nat Med.* 11(4):367-368.

[71] Cooper, L.T., Jr. and Gersh, B.J. (2002). Viral infection, inflammation, and the risk of idiopathic dilated cardiomyopathy: can the fire be extinguished? *Am J Cardiol.* 90(7):751-754.

[72] Hein, S. and Schaper, J. (1996). Pathogenesis of dilated cardiomyopathy and heart failure: insights from cell morphology and biology. *Curr Opin Cardiol.* 11(3):293-301.

[73] Ohnishi, S., Yanagawa, B., Tanaka, K., Miyahara, Y., Obata, H., Kataoka, M., Kodama, M., Ishibashi-Ueda, H., Kangawa, K., Kitamura, S., and Nagaya, N. (2007).

Transplantation of mesenchymal stem cells attenuates myocardial injury and dysfunction in a rat model of acute myocarditis. *J Mol Cell Cardiol.* 42(1):88-97.

[74] Burke, A.P. and Virmani, R. (2007). Pathophysiology of acute myocardial infarction. *Med Clin North Am.* 91(4):553-572; ix.

[75] Schachinger, V., Erbs, S., Elsasser, A., Haberbosch, W., Hambrecht, R., Holschermann, H., Yu, J., Corti, R., Mathey, D.G., Hamm, C.W., Suselbeck, T., Assmus, B., Tonn, T., Dimmeler, S., and Zeiher, A.M. (2006). Intracoronary bone marrow-derived progenitor cells in acute myocardial infarction. *N Engl J Med.* 355(12):1210-1221.

[76] Perin, E.C., Dohmann, H.F., Borojevic, R., Silva, S.A., Sousa, A.L., Mesquita, C.T., Rossi, M.I., Carvalho, A.C., Dutra, H.S., Dohmann, H.J., Silva, G.V., Belem, L., Vivacqua, R., Rangel, F.O., Esporcatte, R., Geng, Y.J., Vaughn, W.K., Assad, J.A., Mesquita, E.T., and Willerson, J.T. (2003). Transendocardial, autologous bone marrow cell transplantation for severe, chronic ischemic heart failure. *Circulation,* 107(18):2294-2302.

[77] Strauer, B.E., Brehm, M., Zeus, T., Kostering, M., Hernandez, A., Sorg, R.V., Kogler, G., and Wernet, P. (2002). Repair of infarcted myocardium by autologous intracoronary mononuclear bone marrow cell transplantation in humans. *Circulation,* 106(15):1913-1918.

[78] Chen, S.L., Fang, W.W., Ye, F., Liu, Y.H., Qian, J., Shan, S.J., Zhang, J.J., Chunhua, R.Z., Liao, L.M., Lin, S., and Sun, J.P. (2004). Effect on left ventricular function of intracoronary transplantation of autologous bone marrow mesenchymal stem cell in patients with acute myocardial infarction. *Am J Cardiol.* 94(1):92-95.

[79] Meyer, G.P., Wollert, K.C., Lotz, J., Steffens, J., Lippolt, P., Fichtner, S., Hecker, H., Schaefer, A., Arseniev, L., Hertenstein, B., Ganser, A., and Drexler, H. (2006). Intracoronary bone marrow cell transfer after myocardial infarction: eighteen months' follow-up data from the randomized, controlled BOOST (BOne marrOw transfer to enhance ST-elevation infarct regeneration) trial. *Circulation,* 113(10):1287-1294.

[80] Wollert, K.C., Meyer, G.P., Lotz, J., Ringes-Lichtenberg, S., Lippolt, P., Breidenbach, C., Fichtner, S., Korte, T., Hornig, B., Messinger, D., Arseniev, L., Hertenstein, B., Ganser, A., and Drexler, H. (2004). Intracoronary autologous bone-marrow cell transfer after myocardial infarction: the BOOST randomised controlled clinical trial. *Lancet,* 364(9429):141-148.

[81] Janssens, S., Dubois, C., Bogaert, J., Theunissen, K., Deroose, C., Desmet, W., Kalantzi, M., Herbots, L., Sinnaeve, P., Dens, J., Maertens, J., Rademakers, F., Dymarkowski, S., Gheysens, O., Van Cleemput, J., Bormans, G., Nuyts, J., Belmans, A., Mortelmans, L., Boogaerts, M., and Van de Werf, F. (2006). Autologous bone marrow-derived stem-cell transfer in patients with ST-segment elevation myocardial infarction: double-blind, randomised controlled trial. *Lancet,* 367(9505):113-121.

[82] Britten, M.B., Abolmaali, N.D., Assmus, B., Lehmann, R., Honold, J., Schmitt, J., Vogl, T.J., Martin, H., Schachinger, V., Dimmeler, S., and Zeiher, A.M. (2003). Infarct remodeling after intracoronary progenitor cell treatment in patients with acute myocardial infarction (TOPCARE-AMI): mechanistic insights from serial contrast-enhanced magnetic resonance imaging. *Circulation,* 108(18):2212-2218.

[83] Schachinger, V., Assmus, B., Britten, M.B., Honold, J., Lehmann, R., Teupe, C., Abolmaali, N.D., Vogl, T.J., Hofmann, W.K., Martin, H., Dimmeler, S., and Zeiher, A.M. (2004). Transplantation of progenitor cells and regeneration enhancement in acute myocardial infarction: final one-year results of the TOPCARE-AMI Trial. *J Am Coll Cardiol.* 44(8):1690-1699.

[84] Lunde, K., Solheim, S., Aakhus, S., Arnesen, H., Abdelnoor, M., Egeland, T., Endresen, K., Ilebekk, A., Mangschau, A., Fjeld, J.G., Smith, H.J., Taraldsrud, E., Grogaard, H.K., Bjornerheim, R., Brekke, M., Muller, C., Hopp, E., Ragnarsson, A., Brinchmann, J.E., and Forfang, K. (2006). Intracoronary injection of mononuclear bone marrow cells in acute myocardial infarction. *N Engl J Med.* 355(12):1199-1209.

[85] Menasche, P., Hagege, A.A., Vilquin, J.T., Desnos, M., Abergel, E., Pouzet, B., Bel, A., Sarateanu, S., Scorsin, M., Schwartz, K., Bruneval, P., Benbunan, M., Marolleau, J.P., and Duboc, D. (2003). Autologous skeletal myoblast transplantation for severe postinfarction left ventricular dysfunction. *J Am Coll Cardiol.* 41(7):1078-1083.

[86] Hagege, A.A., Marolleau, J.P., Vilquin, J.T., Alheritiere, A., Peyrard, S., Duboc, D., Abergel, E., Messas, E., Mousseaux, E., Schwartz, K., Desnos, M., and Menasche, P. (2006). Skeletal myoblast transplantation in ischemic heart failure: long-term follow-up of the first phase I cohort of patients. *Circulation,* 114(1 Suppl):I108-113.

[87] Herreros, J., Prosper, F., Perez, A., Gavira, J.J., Garcia-Velloso, M.J., Barba, J., Sanchez, P.L., Canizo, C., Rabago, G., Marti-Climent, J.M., Hernandez, M., Lopez-Holgado, N., Gonzalez-Santos, J.M., Martin-Luengo, C., and Alegria. E. (2003). Autologous intramyocardial injection of cultured skeletal muscle-derived stem cells in patients with non-acute myocardial infarction. *Eur Heart J.* 24(22):2012-2020.

[88] Siminiak, T., Kalawski, R., Fiszer, D., Jerzykowska, O., Rzezniczak, J., Rozwadowska, N., and Kurpisz, M. (2004). Autologous skeletal myoblast transplantation for the treatment of postinfarction myocardial injury: phase I clinical study with 12 months of follow-up. *Am Heart J.* 148(3):531-537.

[89] Siminiak, T., Fiszer, D., Jerzykowska, O., Grygielska, B., Rozwadowska, N., Kalmucki, P., and Kurpisz, M. (2005). Percutaneous trans-coronary-venous transplantation of autologous skeletal myoblasts in the treatment of post-infarction myocardial contractility impairment: the POZNAN trial. *Eur Heart J.* 26(12):1188-1195.

[90] Chachques, J.C., Herreros, J., Trainini, J., Juffe, A., Rendal, E., Prosper, F., and Genovese, J. (2004). Autologous human serum for cell culture avoids the implantation of cardioverter-defibrillators in cellular cardiomyoplasty. *Int J Cardiol.*95 Suppl 1:S29-33.

[91] Perin, E.C., Dohmann, H.F., Borojevic, R., Silva, S.A., Sousa, A.L., Silva, G.V., Mesquita, C.T., Belem, L., Vaughn, W.K., Rangel, F.O., Assad, J.A., Carvalho, A.C., Branco, R.V., Rossi, M.I., Dohmann, H.J., and Willerson, J.T. (2004). Improved exercise capacity and ischemia 6 and 12 months after transendocardial injection of autologous bone marrow mononuclear cells for ischemic cardiomyopathy. *Circulation,* 110(11 Suppl 1):II213-218.

[92] Fuchs, S., Satler, L.F., Kornowski, R., Okubagzi, P., Weisz, G., Baffour, R., Waksman, R., Weissman, N.J., Cerqueira, M., Leon, M.B., and Epstein, S.E. (2003). Catheter-

based autologous bone marrow myocardial injection in no-option patients with advanced coronary artery disease: a feasibility study. *J Am Coll Cardiol.* 41(10):1721-1724.

[93] Stamm, C., Kleine, H.D., Choi, Y.H., Dunkelmann, S., Lauffs, J.A., Lorenzen, B., David, A., Liebold, A., Nienaber, C., Zurakowski, D., Freund, M., and Steinhoff, G. (2007). Intramyocardial delivery of CD133+ bone marrow cells and coronary artery bypass grafting for chronic ischemic heart disease: safety and efficacy studies. *J Thorac Cardiovasc Surg.* 133(3):717-725.

[94] Archundia, A., Aceves, J.L., Lopez-Hernandez, M., Alvarado, M., Rodriguez, E., Diaz Quiroz, G., Paez, A., Rojas, F.M., and Montano, L.F. (2005). Direct cardiac injection of G-CSF mobilized bone-marrow stem-cells improves ventricular function in old myocardial infarction. *Life Sci.* 78(3):279-283.

[95] Hendrikx, M., Hensen, K., Clijsters, C., Jongen, H., Koninckx, R., Bijnens, E., Ingels, M., Jacobs, A., Geukens, R., Dendale, P., Vijgen, J., Dilling, D., Steels, P., Mees, U., and Rummens, J.L. (2006). Recovery of regional but not global contractile function by the direct intramyocardial autologous bone marrow transplantation: results from a randomized controlled clinical trial. *Circulation.* 114(1 Suppl):I101-107.

[96] Brehm, M. and Strauer, B.E. (2006). Stem cell therapy in postinfarction chronic coronary heart disease. *Nat Clin Pract Cardiovasc Med.* 3 Suppl 1:S101-104.

[97] Patel, A.N., Geffner, L., Vina, R.F., Saslavsky, J., Urschel, H.C., Jr., Kormos, R., and Benetti, F. (2005). Surgical treatment for congestive heart failure with autologous adult stem cell transplantation: a prospective randomized study. *J Thorac Cardiovasc Surg.* 130(6):1631-1638.

[98] Erbs, S., Linke, A., Adams, V., Lenk, K., Thiele, H., Diederich, K.W., Emmrich, F., Kluge, R., Kendziorra, K., Sabri, O., Schuler, G., and Hambrecht, R. (2005). Transplantation of blood-derived progenitor cells after recanalization of chronic coronary artery occlusion: first randomized and placebo-controlled study. *Circ Res.* 97(8):756-762.

[99] Assmus, B., Honold, J., Schachinger, V., Britten, M.B., Fischer-Rasokat, U., Lehmann, R., Teupe, C, Pistorius, K., Martin, H., Abolmaali, N.D., Tonn, T., Dimmeler, S., and Zeiher, A.M. (2006). Transcoronary transplantation of progenitor cells after myocardial infarction. *N Engl J Med.* 355(12):1222-1232.

In: Stem Cell Applications in Disease and Health
Editors: W. Burnsides and R. Ellsley
ISBN: 978-1-60456-255-2

Chapter IX

Overview of the Recent Research Progress in Parkinson's Disease: From Molecular Pathogenesis to Stem Cell-Based Therapeutic Strategy

Hiroshi Yamashita*[*] *and Masayasu Matsumoto
Department of Clinical Neuroscience and Therapeutics, Hiroshima University Graduate School of Biomedical Sciences, 1-2-3 Kasumi, Hiroshima 734-8551, Japan

Abstract

Parkinson's disease (PD) is a progressive neurodegenerative disorder characterized by rigidity, bradykinesia, postural instability, and resting tremor. The major symptoms are related to the progressive loss of dopaminergic neurons in the substantia nigra pars compacta (SNc). Although approximately 95% of PD cases are sporadic, the recent discovery of specific mutations in genes (PARK) that cause familial PD has led to a new approach in the study of the disease. Identification of PARK genes has allowed study of the pathogenesis of PD at the molecular level, and led to a new approach in the study of the disease. Numerous recent studies suggest that PARK gene products, such as α-synuclein, Parkin, DJ-1, PINK1, ATP13A2, UCH-L1, or LRRK2, interact with each other and that their loss of function or gain of toxic functions play an important role in the pathogenesis of both familial and sporadic PD. Nonetheless, the cause and pathogenesis of PD remains unknown; mitochondrial dysfunction, oxidative damage, endoplasmic reticulum stress, failure of the ubiquitin-proteasome system, environmental factors, and genetic predisposition might all be involved. To investigate the molecular pathogenesis and the novel treatment of the disease, toxin-induced PD animal models and genetic animal models that mimic familial PD have contributed. A mouse model, a Drosophila model, a cell model, and a stem cell-derived neuron model carrying the disease-causing PARK gene with genetic modification, have been generated. Using PD models, recent experimental studies have demonstrated that embryonic stem cells,

[*] E-mail address: yamashih@hiroshima-u.ac.jp, Tel.: +81-82-257-5201, Fax: +81-82-505-0490

neuronal stem cells, or bone marrow cells can be used as donor cells for transplantation to reverse symptoms. Moreover, neurogenesis in the SNc has been reported. Recently, molecular mechanisms for dopaminergic neuronal death in PD have been clarified and many advances in stem cell-based therapeutic approaches have been made. Nonetheless, PD involves multiple neuronal systems and the pathology is widespread: not only the SNc, but also the vagus, locus coeruleus, raphe, thalamus, amygdala, olfactory nuclei, and cerebral cortex are all involved. Progression of the disease is associated with pathological staging based on the topographical extent of the brain lesions. This review discusses molecular genetics and pathogenesis, experimental disease models, and the basis for the successful stem cell-based therapeutic approaches for PD.

Key words: Parkinson's disease, dopaminergic neuron, genetic mouse model, neurogenesis, neural precursor cell

Introduction

Parkinson’s disease (PD) represents the second most common neurodegenerative disorder, affecting about 1 percent of the population over the age of 60 years [1]. PD is clinically characterized by motor dysfunction such as rigidity, bradykinesia, postural instability, and resting tremor, which worsen over time; 95% of PD cases are sporadic [2]. In 1817 James Parkinson provided the first detailed description of the disease in “An Essay on the Shaking Palsy”. PD is characterized pathologically by selective degeneration of neurons in the substantia nigra (SN) and locus coeruleus, and the presence of Lewy bodies in the remaining neurons. The major symptoms of PD are related to the progressive loss of dopaminergic neurons in the SN pars compacta, resulting in depletion of striatal dopamine. It has long been believed that PD is caused by environmental factors, and factors such as rural living, pesticide use, and the drinking of well water were thought to be the main etiological hypotheses of PD. Recently, combinations of genetic and environmental factors are thought to be involved in the pathogenesis of the disease, although the cause of the disease remains unknown. However, a shift in the etiological concepts of PD has seen in the past 10 years, moving from an environmentally mediated disease towards a complex disorder with important genetic contributors [3]. PD is inherited in well-characterized kindreds, and PD patients have an affected family member more often than expected, suggesting that genetic predispositions play a role in PD. Genetic studies have identified genetic loci that may contribute to the development of PD or genetic variability that is associated with increased disease risk [4]. Furthermore, the discovery of single gene mutations responsible for causing familial PD that can be indistinguishable from sporadic PD strongly suggest that genetic causes play an important role in PD.

Molecular Pathogenesis

The discoveries of causative genes for familial forms of PD (PARK) have led to a new approach in the study of PD. Loss of function or gain of toxic function of the PARK gene products, such as α-synuclein, Parkin, DJ-1, PINK1, ATP13A2, UCH-L1 or LRRK2, are postulated to be involved in the pathogenesis of PD (Table). The discovery of missense mutations, A30P and A53T, in the *α-synuclein* gene in autosomal dominant PD (PARK1) provided new insights into the pathophysiology of this disease [5,6]. The identification of α-synuclein proteins as being one of the major components of Lewy bodies in sporadic PD further supported a role for this protein in the pathogenesis of PD [7,8]. In addition, several PARK genes have been identified to date: mutations in *parkin* (PARK2), *DJ-1* (PARK7), *PINK1* (PARK6), or *ATP13A2* (PARK9) in autosomal recessive PD and mutations in *UCH-L1* (PARK5) and *LRRK2* (PARK8) or genomic multiplication of *α-synuclein* (PARK4) in autosomal dominant PD [9-17] (Table).

Table. Familial forms of Parkinson's disease identified through genome-wide linkage analysis. Loci were cited from the recent gene map loci of PARK in the OMIM database. AD: autosomal dominant; AR: autosomal recessive.

PARK	Locus	Pattern	Gene	Function
PARK1	4q21	AD	α-synuclein	Presynaptic protein
PARK2	6q25.2-27	AR	parkin	Ubiquitin ligase
PARK4	4q21	AD	α-synuclein	Presynaptic protein
PARK5	4p14	AD	UCH-L1	Ubiquitin recycling enzyme
PARK6	1p36	AR	PINK1	Mitochondrial kinase
PARK7	1p36	AR	DJ1	Oxidative stress response
PARK8	12p12	AD	LRRK2	Kinase
PARK9	1p36	AR	ATP13A2	Lysosomal type 5 ATPase

α-Synuclein (PARK1 and PARK4) is a presynaptic protein that is involved in synaptic vesicle function and axonal transport [18,19]. Although α-synuclein is soluble and degraded by both autophagy and proteasomes [20,21], phosphorylated and ubiquitinated α-synuclein is present in Lewy bodies as aggregated and insoluble filaments [7,22]. Abnormally aggregated protein is postulated to inhibit cellular functions. Abnormal aggregation of α-synuclein into a toxic form is postulated to contribute to neuronal cell death [23]. Mutations in the α-synuclein protein, oxidative stress, nitration, phosphorylation, mitochondrial dysfunction, and proteasomal dysfunction can influence aggregation of α-synuclein into a variety of forms including protofibrils, fibrils, and filaments [4]. It appears that the protofibrils and fibrils are the most toxic forms and the creation and stabilization of these forms might represent a central pathologic mechanism [23] (Figure).

In familial forms of PD, *parkin* mutations (PARK2) are the most common cause of autosomal recessive PD [12]. Autopsies on patients with PARK2, except for one case, commonly show lack of Lewy bodies, suggesting that the normal function of Parkin is essential for the formation of Lewy bodies [24]. Also, the discovery that Parkin is an ubiquitin ligase suggests that the ubiquitin-proteasome system might have an important role in maintaining dopaminergic neurons. Parkin facilitates ubiquitination of numerous substrate proteins, such as O-glycosylated α-synuclein, synphilin-1, CDCrel-1 or Pael-R. Nitrosylation of Parkin, which is found in brains of PD patients, impairs its ubiquitin ligase activity. Loss of Parkin function is postulated to lead to the abnormal accumulation of toxic substrates and resultant cell death [4] (Figure). Parkin plays a role in degradation of misfolded proteins in the endoplasmic reticulum (ER); Parkin dysfunction causes accumulation of misfolded Pael-R in ER, leading to ER stress. Moreover, UCH-L1 (PARK5) is also involved in the ubiquitin-

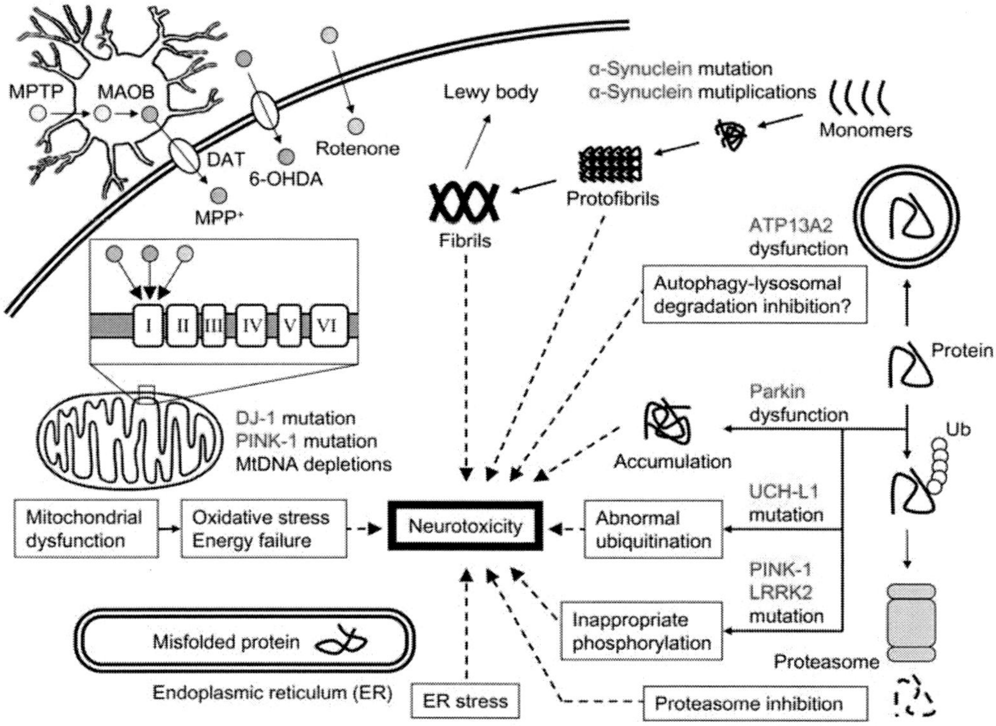

Figure. Model for pathogenesis of Parkinson's disease Mutations of causative genes for familial forms of PD lead to dopaminergic cell death. These gene products are involved in a variety of cellular processes. Parkin (PARK2) and UCH-L1 (PARK5) are postulated to be involved in protein ubiquitination and degradation via proteasomes. PINK1 (PARK6) and LRRK2 (PARK8) are postulated to be involved in protein phosphorylation; DJ-1 (PARK7) and PINK1 play a role in mitochondrial function. Deletion of mitochondrial DNA (mtDNA) has been reported in neurons in the substantia nigra from PD patients. ATP13A2 (PARK9) is postulated to be involved in lysosomal function. Parkin plays a role in degradation of misfolded proteins in the ER. Abnormal aggregation of α-synuclein (PARK1 and PARK4) into a misfolded toxic form is postulated to contribute to neuronal cell death. Toxins such as 1-methyl-4-phenyl-1,2,3,6- tetrahydropyridine (MPTP), 6-hydroxydopamine (6-OHDA), and rotenone, induce dopaminergic cell death via mitochondrial damage. In glial cells, monoamine oxidase B (MAO-B) metabolizes MPTP to an intermediate product that is converted to 1-methyl-4-phenylpyridinium (MPP^+) by spontaneous oxidation. MPP^+ and 6-OHDA are taken up by the dopamine transporter (DAT).

proteasome system through its ubiquitin hydrolase and ligase activities. These findings suggest a link between PD and dysfunction of the ubiquitin-proteasome system [25] (Figure). DJ-1 (PARK7) is an integral mitochondrial protein that is thought to act as an antioxidant, oxidation sensor, chaperone, or protease [26,27]. DJ-1 can interact with Parkin under conditions of oxidative stress [28]. In an oxidated state, DJ-1 interacts with α-synuclein to prevent fibril formation and acts as a chaperone that protects cells from α-synuclein toxicity [4,29]. PINK1 (PARK6) is a mitochondrial protein kinase whose substrates are unknown [30]. DJ-1 suppresses the activity of PTEN tumor suppressor, which induces PINK1 expression [31]. PINK1 mutation leads to mitochondrial dysfunction and increased sensitivity to cellular stress [32]. Loss of PINK1 leads to defects in mitochondrial function, which is rescued by Parkin overexpression, suggesting that Parkin and PINK1 function in a common biochemical pathway [33]. The presence of PINK1 and DJ-1 in the mitochondria underscores the role that this cellular organelle plays in PD pathogenesis [4]. Recently, the presence of a high level of deleted mitochondrial DNA (mtDNA) has been reported in neurons in the SN from both aged controls and individuals with PD [34], suggesting that somatic mtDNA deletions are important in the selective neuronal loss observed in brain aging and PD. Mitochondrial dysfunction might play a central role in the pathogenesis of PD (Figure). LRRK2 (PARK8) is a large cytoplasmic protein that is postulated to acts as a protein kinase [35]. Increased LRRK2 kinase activity is thought to be associated with PARK8 pathogenesis [36]. LRRK2 binding to Parkin leads to increased protein aggregation and ubiquitination [37]. Furthermore, LRRK2 participates in mitochondrial fission or interaction with the cytoskeleton, suggesting a role in membrane, vesicular, or protein trafficking [38]. Involvement of both LRRK2 and PINK1 kinases suggests that protein phosphorylation might play a role in PD pathogenesis [4]. ATP13A2 (PARK9) is a neuronal P-type ATPase that is located in the lysosome [14]. Lysosome activity is up-regulated in α-synuclein-overexpressing cells [39]. The autophagy-lysosomal degradation system might also be involved in the pathogenesis of PD (Figure).

Numerous recent studies suggest that PARK gene products interact with each other and that their loss of function or gain of toxic functions is involved in the pathogenesis of both familial and sporadic PD. Nonetheless, the cause and pathogenesis of PD remain unknown; mitochondrial dysfunction, oxidative damage, endoplasmic reticulum stress, failure of the ubiquitin-proteasome system, environmental factors, and genetic predisposition might all be involved [2,24] (Figure). The identification of responsible mutations in certain genes, and functional analysis of the gene products such as α-synuclein, Parkin, PINK1, DJ-1, LRRK2 and ATP13A2 have increased our understanding of the clinical and pathological changes underlying PD [3].

Toxin-Induced Models

A recent large cohort study demonstrated that exposure to pesticides or herbicides increases the risk for subsequently developing PD [40]. The discovery of toxins that induce parkinsonism in humans supports the involvement of environmental factors. 1-methyl-4-phenyl-1,2,3,6-tetrahydropyridine (MPTP) causes damage to the nigrostriatal dopaminergic neurons and results in clinical features of PD in humans [41]. MPTP has been demonstrated to cross the blood-brain barrier. In glial cells, monoamine oxidase B (MAO-B) metabolizes MPTP to an intermediate product that is converted to 1-methyl-4-phenylpyridinium (MPP^+)

by spontaneous oxidation [42] (Figure). MPP^+ is taken up by the dopamine transporter (DAT) and is concentrated in mitochondria [42]. MPP^+ impairs mitochondrial respiration by inhibiting complex I of the electron transport chain, resulting in an increased production of free radicals leading to oxidative stress [42] (Figure). Monkey and rodent MPTP models have been used as a standard experimental model for PD. The neurotoxic herbicide paraquat exhibits a structural similarity to MPP^+ and also exerts deleterious effects through oxidative stress. 6-hydroxydopamine (6-OHDA) can enter dopaminergic neurons through the DAT and destroy them by a combined effect of reactive oxygen species and quinines [43] (Figure). The rodent 6-OHDA model is one of the most popular experimental models for PD [43]. Rotenone freely crosses cellular membranes and impairs mitochondrial respiration by inhibiting complex I of the electron transport chain [43] (Figure). Rotenone has been extensively used in cell cultures, but less frequently in animal models. Unlike the 6-OHDA and MPTP models, cytoplasmic inclusions are observed in the remaining dopaminergic neurons in the SN of rotenone-infused rats [44,45]. Toxin-induced animal models have been widely used to test new symptomatic therapies, neuroprotective strategies, transplantation, and neurogenesis approaches. Nonetheless, these models do not fully reconstitute the pathology and disease progression seen in PD patients. Toxin models do not present the molecular pathogenesis of PD that was exclusively clarified within last 10 years.

Genetic Models

Animal models that harbor genetic lesions of familial forms of PD have been generated. α-Synuclein (PARK1 and PARK4) knockout mice demonstrate decreased dopamine content in the striatum [18] and have a resistance to MPTP [46]. Mice overexpressing the doubly mutated A30P and A53T α-synuclein under the tyrosine hydroxylase (TH) promoter demonstrate progressive reduction of striatal DAT density and dopamine content, age-related motor impairments, and a reduced response to dopamine agonists [47,48]. Similarly, mice overexpressing wild-type α-synuclein under the platelet-derived growth factor (PDGF) β promoter show reduced striatal dopamine content and motor impairment [49]. Mice overexpressing wild-type α-synuclein under the Thy1 promoter show a wide range of progressive sensorimotor impairments and an abnormal behavioral response to dopaminergic agonists [50]. However, several lines of α-synuclein transgenic mice show no pathology in the nigrostriatal pathway [51] and only a few show alterations in the nigrostriatal dopaminergic system [47,49,52]. Although α-synuclein transgenic mice lack a massive loss of nigrostriatal dopaminergic neurons, abnormal responses to dopamine agonists and behavioral impairments suggest that such mice may represent a model for dopaminergic dysfunction seen in PD [47,48]. Parkin (PARK2) knockout mice have been generated by deletion of exon 3, exon 7, or exon 2 in the *parkin* gene. Parkin knockout mice with an exon 3 deletion demonstrate increased extracellular striatal dopamine, progressive sensorimotor impairments [53], reduced mitochondrial respiration, and increased oxidative damage in the midbrain [54]. Another line of knockout mice with an exon 3 deletion demonstrate inhibition of dopamine release in fetal midbrain cultures, reduced DAT in the striatum, and motor and cognitive deficits [55]. Parkin knockout mice with an exon 7 deletion demonstrate a loss of catecholaminergic neurons in the locus coeruleus and a reduction in the norepinephrine-dependent startle response. [56]. Parkin knockout mice with an exon 2 deletion show no evidence for nigrostriatal, cognitive, or noradrenergic dysfunction [57]. Although several lines of Parkin knockout mice are limited by the absence of loss of nigrostriatal dopaminergic neurons, they demonstrate changes in dopamine release and synaptic dysfunction in the striatum. DJ-1 (PARK5) knockout mice with an exon 2 deletion [58] or insertion of a stop codon in exon 1 [59] have been generated. DJ-1 knockout mice exhibit increased susceptibility to MPTP. In culture, neurons from these mice were more sensitive to oxidative stress than controls [59]. Similar to α-synuclein transgenic and Parkin knockout mice, DJ-1 knockout mice fail to demonstrate loss of nigrostriatal dopaminergic neurons, although they do demonstrate reduced dopamine release in the striatum and decreased locomotor activity [58,59]. Homeobox transcription factor Pitx3 is involved in the development of nigrostriatal dopaminergic neurons. Pitx3-aphakia mice that lack Pitx3 demonstrate a loss of nigrostriatal dopaminergic neurons, a reduction in the dopamine content of the striatum, and sensorimotor impairments [60]. Transcription factor Nurr1 is essential for the development of ventral mesencephalon dopaminergic neurons [61]. Nurr1 heterozygous mice develop decreased dopamine levels in the striatum, decreased amount of nigrostriatal dopaminergic neurons, and decreased locomotor activity [62]. Although the dopaminergic phenotypes of the α-synuclein, Parkin or DJ-1 knockout mice are not as profound as the Nurr1 heterozygous and Pitx3-aphakia mice, they may provide insight into dopaminergic dysfunction and pathogenesis of the disease.

A Drosophila model of PD has been generated to investigate the molecular pathogenesis of the disease. A Drosophila model that expresses human wild-type or A30P or A53T α-synuclein exhibited adult-onset loss of dopaminergic neurons, filamentous intraneuronal inclusions containing α-synuclein, and locomotor dysfunction [63]. A Drosophila model that expresses PINK1 mutants exhibits mitochondrial dysfunction, indirect flight muscle and dopaminergic neuronal degeneration, accompanied by locomotive defects [33]. Loss of Drosophila Parkin results in phenotypes similar to those seen with loss of Pink1 function. [64]. The Drosophila model recapitulates the essential features of the human disease, yet may allow a powerful genetic approach to PD. We have established a novel PD cell model using mouse embryonic stem (ES) cells. The ES cell-derived PD model neurons expressing wild-type or mutant α-synuclein exhibited fundamental characteristics consistent with dopaminergic neurons in the SN and exhibited degenerative cell death accompanied by various features related to the neurodegenerative pathology of PD, including accumulation of α-synuclein [65]. In addition, recent developments in recombinant viral-vector technology have opened up an interesting alternative possibility, based on direct gene transfer to selected subregions in the brain. Using the highly efficient adeno-associated virus or lentivirus vectors, overexpression of mutant α-synuclein in nigrostriatal neurons induces progressive neuropathology and neurodegeneration, similar to that seen in PD. Targeted overexpression of disease-causing genes by viral vectors provides a new and highly flexible approach for in vivo modeling of neurodegenerative diseases, not only in rodents, but also in primates [66]. Although no perfect PD model exists yet, such models remain important tools, as they allow further investigation of disease pathogenesis, as well as identification of therapeutic targets. The discovery that neuronal degeneration can be caused by mutations in single proteins has resulted in new approaches to model the underlying disease processes by in vitro genetic modification of cells or by generation of transgenic animals carrying the disease-causing gene.

Stem Cell-Based Therapeutic Strategy

Treatment with L-dihydroxyphenylalanine (L-DOPA) and dopamine agonists can attenuate symptoms of PD patients at early stage. Nonetheless, the efficacy of such treatments gradually decreases over time, with the dose response decreasing with disease progression. The development of motor complications then requires additional treatment, such as deep brain stimulation. Since the major symptoms of PD result from loss of nigrostriatal dopaminergic neurons, it would appear reasonable that replenishment of lost neurons might reverse symptoms and cell transplantation therapy has been investigated. The transplantation of human adult adrenomedullary tissue was first examined and proven to have little efficacy or effect on mortality. A cell-based approach was next utilized for transplantation of human fetal mesencephalic tissue from aborted fetuses, rich in primary dopaminergic neurons in the putamen or caudate nucleus [67]. Clinical and experimental investigations have shown that transplantation of fetal dopaminergic neurons provides symptomatic relief [61,68-71]. Nonetheless, two controlled clinical trials aimed at using human fetal mesencephalic tissue transplanted into the striatum to treat PD failed to show a significant clinical benefit based on their primary endpoint variables in spite of substantial evidence of graft survival based on both autopsy and imaging studies [68,70,72]. Transplantation of fetal mesencephalic tissue has turned out to be less effective than deep brain stimulation [67]. Nevertheless, these transplantation studies were important as the first experimental cell-based treatment strategies, revolutionizing treatment strategy for neurodegenerative diseases [67]. Recent experimental studies have demonstrated that ES cells, neuronal stem cells, and bone marrow cells can be used as donor cells for transplantation in PD animal models to reverse symptoms. Such cells have advantages since they can be maintained, proliferated, and differentiated by use of in vitro cell biology and gene manipulation methods. Improvement of purification, differentiation, and implantation protocols might lead to efficient transplantation. For the moment, further developments are needed for cell-based approaches for them to become clinically effective treatments [52].

Neural stem/precursor cells exist not only in the developing mammalian nervous system, but also in the adult nervous system of all mammalian organisms, including humans [73]. In rodents, adult neural precursor cells reside in the subventricular zone (SVZ) of the lateral ventricles and the subgranular layer (SGL) of the dentate gyrus [67]. Neural precursor cells can differentiate into neurons, astroglia, and oligodendroglia. To investigate neural precursor cells in PD, MPTP or 6-OHDA models, as well as genetic mouse models and viral vector based models have been used [67]. In adult mice, a significant increase in 5-bromo-2'-deoxyuridine (BrdU)-labeled precursor cells is present in the striatum early after MPTP administration. Nonetheless, depletion of dopamine in MPTP-treated mice and 6-OHDA-lesioned rats and mice resulted in a decreased proliferation of precursor cells in the SVZ and SGL [74]. In animals, SVZ precursor cells can be recruited to the striatum after administration of several growth factors, such as transforming growth factor-α (TGF-α) and brain-derived neurotrophic factor (BDNF) [67]. Intrastriatal TGF-α infusion in 6-OHDA-lesioned rats induced a significant proliferation in SVZ precursor cells, as well as an increase in their migration [75,76]. Nonetheless, dopaminergic neurogenesis in the striatum and an improved behavioral response were not confirmed. Accordingly, correlation between newly born dopaminergic neurons and functional recovery remains unclear. In addition, PDGF and BDNF also recruit neural precursor cells to the striatum and the SN in 6-OHDA-lesioned rats, although no newborn neurons were detected in the SN [77]. Glial cell line-derived neurotrophic factor (GDNF) delivered via lentivirus increased the number of striatal TH-positive cells in MPTP-treated monkeys [78]. Such studies suggest the possibility of precursor cell recruitment to the striatum in PD models.

Not only growth factors, but dopamine itself has also been reported to play a role in the control of adult dopaminergic neurogenesis. Experimental depletion of dopamine in rodents decreases precursor cell proliferation in the SVZ and the SGL. A recent study in macaques revealed a topographically organized projection from the SN to the SVZ and a significant decrease in the number of proliferating cell nuclear antigen (PCNA)-positive precursor cells and polysialated-neural cell-adhesion molecule (PSA-NCAM)-positive immature neurons in the SVZ after MPTP treatment, suggesting that intact dopaminergic nigro-subventricular innervation is crucial for sustained neurogenesis [79]. Administration of the selective dopamine D2 receptor (D2L) agonist ropinirole increased the number of PCNA-expressing cells in both the 6-OHDA-lesioned and non-lesioned sides [74]. In addition, dopamine D3 receptor stimulation by 7-hydroxy-*N,N*-di-*n*propyl-2-aminotetralin (7-OH-DPAT) triggers neurogenesis in the SVZ and the SN of healthy adult rat brain. 7-OH-DPAT increased cell proliferation and dopaminergic differentiation in the SN of 6-OHDA-lesioned rats [80]. The results of such studies suggest that dopaminergic signals might regulate neurogenesis in adult brain and neural precursor cells might be impaired by dopaminergic denervation in PD. In the SN of PD patients, PSA-NCAM-positive immature neurons were detected. Such cells were sometimes TH-positive, suggesting the possibility of neurogenesis in the SN of patients with PD [81]. Successful endogenous stem cell-based therapy for PD will require efficient neural precursor cell proliferation and recruitment, dopaminergic differentiation, and survival of newly generated neurons. Nonetheless, reconstruction of nigrostriatal projection of newly generated dopaminergic neurons might prove a real challenge. Improvement of the local environment by viral vector-mediated overexpression of growth factors, as well as stimulation of dopamine D2L or D3 receptors might increase cell survival and help with dopaminergic differentiation. Further investigation is requisite for successful neural stem/precursor cell-based therapy for PD.

Conclusion

Molecular pathogenesis of PD including the mechanisms for dopaminergic neuronal death has been clarified in past 10 years, and many advances in cell-based therapeutic approaches have been made. Experimental PD models have been generated through treatment of animals with toxins that cause damage of dopaminergic neurons in the SN. Underlying disease processes have been also produced by in vitro genetic modification of cells or by generation of mice and flies that harbor the disease-causing PARK genes. Although no perfect PD model exists yet, the existing models are important tools that allow identification of therapeutic strategies utilizing stem cells.

On the other hand, PD involves multiple neuronal systems and the pathology is widespread: not only the SN, but also the dorsal motor nucleus of the vagus, locus coeruleus, raphe and reticular formation nuclei, thalamus, amygdala, olfactory nuclei, pediculopontine nucleus, and cerebral cortex are involved [4,82]. PD is a progressive disease and the course of PD patients is associated with pathological staging based on the topographical extent of the brain findings. Accordingly, PD is not a disease simply demonstrating loss of nigrostriatal dopaminergic neurons. Further investigation of the molecular pathogenesis is requisite to address the degeneration of multiple neuronal systems affected by PD. It is hoped that with the development of better experimental models for PD, the neurodegenerative pathogenesis and therapeutic approaches for PD can be better elucidated. The discovery of neurogenesis in the SN has opened attractive therapeutic perspectives for PD. Endogenous stem cell therapy for PD offers several potential advantages over other cell-based treatment strategies. However, many challenges still need to be overcome before this strategy can be brought into the clinic. There is still a great need of basic research on the mechanisms that control adult neurogenesis. The development of novel PD models that reproduce the pathophysiology of the human condition more accurately than the currently available models will also help to predict whether experimental stem cell-based therapeutic approaches have the potential for future clinical application. Further investigation is requisite for successful stem cell-based therapy for PD.

Acknowledgments

This work was supported by grants-in-aid for the Encouragement of Young Scientists (2007, 2008) and for Scientific Research (2006, 2007) from the Japanese Ministry of Education, Culture, Sports, Science, and Technology.

References

[1] Nutt, J.G. and Wooten, G.F. (2005) Clinical practice. Diagnosis and initial management of Parkinson's disease. *N Engl J Med* 353, 1021-7.

[2] Yamashita, H. and Matsumoto, M. (2007) Molecular pathogenesis, experimental models and new therapeutic strategies for Parkinson's disease. *Regen Med* 2, 447-55.

[3] Douglas, M.R., Lewthwaite, A.J. and Nicholl, D.J. (2007) Genetics of Parkinson's disease and parkinsonism. *Expert Rev Neurother* 7, 657-66.

[4] Savitt, J.M., Dawson, V.L. and Dawson, T.M. (2006) Diagnosis and treatment of Parkinson disease: molecules to medicine. *J Clin Invest* 116, 1744-54.

[5] Kruger, R., Kuhn, W., Muller, T., Woitalla, D., Graeber, M., Kosel, S., Przuntek, H., Epplen, J.T., Schols, L. and Riess, O. (1998) Ala30Pro mutation in the gene encoding alpha-synuclein in Parkinson's disease. *Nat Genet* 18, 106-8.

[6] Polymeropoulos, M.H., Lavedan, C., Leroy, E., Ide, S.E., Dehejia, A., Dutra, A., Pike, B., Root, H., Rubenstein, J., Boyer, R., Stenroos, E.S., Chandrasekharappa, S., Athanassiadou, A., Papapetropoulos, T., Johnson, W.G., Lazzarini, A.M., Duvoisin, R.C., Di Iorio, G., Golbe, L.I. and Nussbaum, R.L. (1997) Mutation in the alpha-synuclein gene identified in families with Parkinson's disease. Science 276, 2045-7.

[7] Spillantini, M.G., Crowther, R.A., Jakes, R., Hasegawa, M. and Goedert, M. (1998) alpha-Synuclein in filamentous inclusions of Lewy bodies from Parkinson's disease and dementia with lewy bodies. *Proc Natl Acad Sci U S A* 95, 6469-73.

[8] Spillantini, M.G., Schmidt, M.L., Lee, V.M., Trojanowski, J.Q., Jakes, R. and Goedert, M. (1997) Alpha-synuclein in Lewy bodies. *Nature* 388, 839-40.

[9] Bonifati, V., Rizzu, P., van Baren, M.J., Schaap, O., Breedveld, G.J., Krieger, E., Dekker, M.C., Squitieri, F., Ibanez, P., Joosse, M., van Dongen, J.W., Vanacore, N., van Swieten, J.C., Brice, A., Meco, G., van Duijn, C.M., Oostra, B.A. and Heutink, P. (2003) Mutations in the DJ-1 gene associated with autosomal recessive early-onset parkinsonism. *Science* 299, 256-9.

[10] Chartier-Harlin, M.C., Kachergus, J., Roumier, C., Mouroux, V., Douay, X., Lincoln, S., Levecque, C., Larvor, L., Andrieux, J., Hulihan, M., Waucquier, N., Defebvre, L., Amouyel, P., Farrer, M. and Destee, A. (2004) Alpha-synuclein locus duplication as a cause of familial Parkinson's disease. *Lancet* 364, 1167-9.

[11] Ibanez, P., Bonnet, A.M., Debarges, B., Lohmann, E., Tison, F., Pollak, P., Agid, Y., Durr, A. and Brice, A. (2004) Causal relation between alpha-synuclein gene duplication and familial Parkinson's disease. *Lancet* 364, 1169-71.

[12] Kitada, T., Asakawa, S., Hattori, N., Matsumine, H., Yamamura, Y., Minoshima, S., Yokochi, M., Mizuno, Y. and Shimizu, N. (1998) Mutations in the parkin gene cause autosomal recessive juvenile parkinsonism. *Nature* 392, 605-8.

[13] Paisan-Ruiz, C., Jain, S., Evans, E.W., Gilks, W.P., Simon, J., van der Brug, M., Lopez de Munain, A., Aparicio, S., Gil, A.M., Khan, N., Johnson, J., Martinez, J.R., Nicholl, D., Carrera, I.M., Pena, A.S., de Silva, R., Lees, A., Marti-Masso, J.F., Perez-Tur, J., Wood, N.W. and Singleton, A.B. (2004) Cloning of the gene containing mutations that cause PARK8-linked Parkinson's disease. *Neuron* 44, 595-600.

[14] Ramirez, A., Heimbach, A., Grundemann, J., Stiller, B., Hampshire, D., Cid, L.P., Goebel, I., Mubaidin, A.F., Wriekat, A.L., Roeper, J., Al-Din, A., Hillmer, A.M., Karsak, M., Liss, B., Woods, C.G., Behrens, M.I. and Kubisch, C. (2006) Hereditary parkinsonism with dementia is caused by mutations in ATP13A2, encoding a lysosomal type 5 P-type ATPase. *Nat Genet* 38, 1184-91.

[15] Singleton, A.B., Farrer, M., Johnson, J., Singleton, A., Hague, S., Kachergus, J., Hulihan, M., Peuralinna, T., Dutra, A., Nussbaum, R., Lincoln, S., Crawley, A., Hanson, M., Maraganore, D., Adler, C., Cookson, M.R., Muenter, M., Baptista, M., Miller, D., Blancato, J., Hardy, J. and Gwinn-Hardy, K. (2003) alpha-Synuclein locus triplication causes Parkinson's disease. *Science* 302, 841.

[16] Valente, E.M., Abou-Sleiman, P.M., Caputo, V., Muqit, M.M., Harvey, K., Gispert, S., Ali, Z., Del Turco, D., Bentivoglio, A.R., Healy, D.G., Albanese, A., Nussbaum, R., Gonzalez-Maldonado, R., Deller, T., Salvi, S., Cortelli, P., Gilks, W.P., Latchman,

D.S., Harvey, R.J., Dallapiccola, B., Auburger, G. and Wood, N.W. (2004) Hereditary early-onset Parkinson's disease caused by mutations in PINK1. *Science* 304, 1158-60.

[17] Wintermeyer, P., Kruger, R., Kuhn, W., Muller, T., Woitalla, D., Berg, D., Becker, G., Leroy, E., Polymeropoulos, M., Berger, K., Przuntek, H., Schols, L., Epplen, J.T. and Riess, O. (2000) Mutation analysis and association studies of the UCHL1 gene in German Parkinson's disease patients. *Neuroreport* 11, 2079-82.

[18] Abeliovich, A., Schmitz, Y., Farinas, I., Choi-Lundberg, D., Ho, W.H., Castillo, P.E., Shinsky, N., Verdugo, J.M., Armanini, M., Ryan, A., Hynes, M., Phillips, H., Sulzer, D. and Rosenthal, A. (2000) Mice lacking alpha-synuclein display functional deficits in the nigrostriatal dopamine system. *Neuron* 25, 239-52.

[19] Jensen, P.H., Li, J.Y., Dahlstrom, A. and Dotti, C.G. (1999) Axonal transport of synucleins is mediated by all rate components. *Eur J Neurosci* 11, 3369-76.

[20] Cuervo, A.M., Stefanis, L., Fredenburg, R., Lansbury, P.T. and Sulzer, D. (2004) Impaired degradation of mutant alpha-synuclein by chaperone-mediated autophagy. Science 305, 1292-5.

[21] Webb, J.L., Ravikumar, B., Atkins, J., Skepper, J.N. and Rubinsztein, D.C. (2003) Alpha-Synuclein is degraded by both autophagy and the proteasome. *J Biol Chem* 278 25009-13.

[22] Hasegawa, M., Fujiwara, H., Nonaka, T., Wakabayashi, K., Takahashi, H., Lee, V.M., Trojanowski, J.Q., Mann, D. and Iwatsubo, T. (2002) Phosphorylated alpha-synuclein is ubiquitinated in alpha-synucleinopathy lesions. *J Biol Chem* 277, 49071-6.

[23] Ross, C.A. and Poirier, M.A. (2004) Protein aggregation and neurodegenerative disease. *Nat Med* 10 Suppl, S10-7.

[24] Hattori, N. and Mizuno, Y. (2004) Pathogenetic mechanisms of parkin in Parkinson's disease. *Lancet* 364, 722-4.

[25] Chung, K.K., Thomas, B., Li, X., Pletnikova, O., Troncoso, J.C., Marsh, L., Dawson, V.L. and Dawson, T.M. (2004) S-nitrosylation of parkin regulates ubiquitination and compromises parkin's protective function. *Science* 304, 1328-31.

[26] Moore, D.J., West, A.B., Dawson, V.L. and Dawson, T.M. (2005) Molecular pathophysiology of Parkinson's disease. *Annu Rev Neurosci* 28, 57-87.

[27] Zhang, L., Shimoji, M., Thomas, B., Moore, D.J., Yu, S.W., Marupudi, N.I., Torp, R., Torgner, I.A., Ottersen, O.P., Dawson, T.M. and Dawson, V.L. (2005) Mitochondrial localization of the Parkinson's disease related protein DJ-1: implications for pathogenesis. *Hum Mol Genet* 14, 2063-73.

[28] Moore, D.J., Zhang, L., Troncoso, J., Lee, M.K., Hattori, N., Mizuno, Y., Dawson, T.M. and Dawson, V.L. (2005) Association of DJ-1 and parkin mediated by pathogenic DJ-1 mutations and oxidative stress. *Hum Mol Genet* 14, 71-84.

[29] Zhou, W., Zhu, M., Wilson, M.A., Petsko, G.A. and Fink, A.L. (2006) The oxidation state of DJ-1 regulates its chaperone activity toward alpha-synuclein. *J Mol Biol* 356, 1036-48.

[30] Silvestri, L., Caputo, V., Bellacchio, E., Atorino, L., Dallapiccola, B., Valente, E.M. and Casari, G. (2005) Mitochondrial import and enzymatic activity of PINK1 mutants associated to recessive parkinsonism. *Hum Mol Genet* 14, 3477-92.

[31] Kim, R.H., Peters, M., Jang, Y., Shi, W., Pintilie, M., Fletcher, G.C., DeLuca, C., Liepa, J., Zhou, L., Snow, B., Binari, R.C., Manoukian, A.S., Bray, M.R., Liu, F.F., Tsao, M.S. and Mak, T.W. (2005) DJ-1, a novel regulator of the tumor suppressor PTEN. *Cancer Cell* 7, 263-73.

[32] Petit, A., Kawarai, T., Paitel, E., Sanjo, N., Maj, M., Scheid, M., Chen, F., Gu, Y., Hasegawa, H., Salehi-Rad, S., Wang, L., Rogaeva, E., Fraser, P., Robinson, B., St George-Hyslop, P. and Tandon, A. (2005) Wild-type PINK1 prevents basal and induced neuronal apoptosis, a protective effect abrogated by Parkinson disease-related mutations. *J Biol Chem* 280, 34025-32.

[33] Park, J., Lee, S.B., Lee, S., Kim, Y., Song, S., Kim, S., Bae, E., Kim, J., Shong, M., Kim, J.M. and Chung, J. (2006) Mitochondrial dysfunction in Drosophila PINK1 mutants is complemented by parkin. *Nature* 441, 1157-61.

[34] Bender, A., Krishnan, K.J., Morris, C.M., Taylor, G.A., Reeve, A.K., Perry, R.H., Jaros, E., Hersheson, J.S., Betts, J., Klopstock, T., Taylor, R.W. and Turnbull, D.M. (2006) High levels of mitochondrial DNA deletions in substantia nigra neurons in aging and Parkinson disease. *Nat Genet* 38, 515-7.

[35] West, A.B., Moore, D.J., Biskup, S., Bugayenko, A., Smith, W.W., Ross, C.A., Dawson, V.L. and Dawson, T.M. (2005) Parkinson's disease-associated mutations in leucine-rich repeat kinase 2 augment kinase activity. *Proc Natl Acad Sci U S A* 102, 16842-7.

[36] Gloeckner, C.J., Kinkl, N., Schumacher, A., Braun, R.J., O'Neill, E., Meitinger, T., Kolch, W., Prokisch, H. and Ueffing, M. (2006) The Parkinson disease causing LRRK2 mutation I2020T is associated with increased kinase activity. *Hum Mol Genet* 15 223-32.

[37] Smith, W.W., Pei, Z., Jiang, H., Moore, D.J., Liang, Y., West, A.B., Dawson, V.L., Dawson, T.M. and Ross, C.A. (2005) Leucine-rich repeat kinase 2 (LRRK2) interacts with parkin, and mutant LRRK2 induces neuronal degeneration. *Proc Natl Acad Sci U S A* 102, 18676-81.

[38] Li, C. and Beal, M.F. (2005) Leucine-rich repeat kinase 2: a new player with a familiar theme for Parkinson's disease pathogenesis. *Proc Natl Acad Sci U S A* 102, 16535-6.

[39] Fujita, M., Sugama, S., Nakai, M., Takenouchi, T., Wei, J., Urano, T., Inoue, S. and Hashimoto, M. (2007) alpha-Synuclein Stimulates Differentiation of Osteosarcoma Cells: Relevance to Down-Regulation of Proteasome Activity. *J Biol Chem* 282, 5736-5748.

[40] Ascherio, A., Chen, H., Weisskopf, M.G., O'Reilly, E., McCullough, M.L., Calle, E.E., Schwarzschild, M.A. and Thun, M.J. (2006) Pesticide exposure and risk for Parkinson's disease. *Ann Neurol* 60, 197-203.

[41] Langston, J.W. and Ballard, P.A., Jr. (1983) Parkinson's disease in a chemist working with 1-methyl-4-phenyl-1,2,5,6-tetrahydropyridine. *N Engl J Med* 309, 310.

[42] Vila, M. and Przedborski, S. (2003) Targeting programmed cell death in neurodegenerative diseases. *Nat Rev Neurosci* 4, 365-75.

[43] Bove, J., Prou, D., Perier, C. and Przedborski, S. (2005) Toxin-induced models of Parkinson's disease. *NeuroRx* 2, 484-94.

[44] Betarbet, R., Sherer, T.B., MacKenzie, G., Garcia-Osuna, M., Panov, A.V. and Greenamyre, J.T. (2000) Chronic systemic pesticide exposure reproduces features of Parkinson's disease. *Nat Neurosci* 3, 1301-6.

[45] Hoglinger, G.U., Feger, J., Prigent, A., Michel, P.P., Parain, K., Champy, P., Ruberg, M., Oertel, W.H. and Hirsch, E.C. (2003) Chronic systemic complex I inhibition induces a hypokinetic multisystem degeneration in rats. *J Neurochem* 84, 491-502.

[46] Dauer, W., Kholodilov, N., Vila, M., Trillat, A.C., Goodchild, R., Larsen, K.E., Staal, R., Tieu, K., Schmitz, Y., Yuan, C.A., Rocha, M., Jackson-Lewis, V., Hersch, S., Sulzer, D., Przedborski, S., Burke, R. and Hen, R. (2002) Resistance of alpha - synuclein null mice to the parkinsonian neurotoxin *MPTP. Proc Natl Acad Sci U S A* 99, 14524-9.

[47] Richfield, E.K., Thiruchelvam, M.J., Cory-Slechta, D.A., Wuertzer, C., Gainetdinov, R.R., Caron, M.G., Di Monte, D.A. and Federoff, H.J. (2002) Behavioral and neurochemical effects of wild-type and mutated human alpha-synuclein in transgenic mice. *Exp Neurol* 175, 35-48.

[48] Thiruchelvam, M.J., Powers, J.M., Cory-Slechta, D.A. and Richfield, E.K. (2004) Risk factors for dopaminergic neuron loss in human alpha-synuclein transgenic mice. *Eur J Neurosci* 19, 845-54.

[49] Masliah, E., Rockenstein, E., Veinbergs, I., Mallory, M., Hashimoto, M., Takeda, A., Sagara, Y., Sisk, A. and Mucke, L. (2000) Dopaminergic loss and inclusion body formation in alpha-synuclein mice: implications for neurodegenerative disorders. *Science* 287, 1265-9.

[50] Fleming, S.M., Salcedo, J., Fernagut, P.O., Rockenstein, E., Masliah, E., Levine, M.S. and Chesselet, M.F. (2004) Early and progressive sensorimotor anomalies in mice overexpressing wild-type human alpha-synuclein. *J Neurosci* 24, 9434-40.

[51] Fleming, S.M., Salcedo, J., Hutson, C.B., Rockenstein, E., Masliah, E., Levine, M.S. and Chesselet, M.F. (2006) Behavioral effects of dopaminergic agonists in transgenic mice overexpressing human wildtype alpha-synuclein. *Neuroscience*.

[52] Rockenstein, E., Mallory, M., Hashimoto, M., Song, D., Shults, C.W., Lang, I. and Masliah, E. (2002) Differential neuropathological alterations in transgenic mice expressing alpha-synuclein from the platelet-derived growth factor and Thy-1 promoters. *J Neurosci Res* 68, 568-78.

[53] Goldberg, M.S., Fleming, S.M., Palacino, J.J., Cepeda, C., Lam, H.A., Bhatnagar, A., Meloni, E.G., Wu, N., Ackerson, L.C., Klapstein, G.J., Gajendiran, M., Roth, B.L., Chesselet, M.F., Maidment, N.T., Levine, M.S. and Shen, J. (2003) Parkin-deficient mice exhibit nigrostriatal deficits but not loss of dopaminergic neurons. *J Biol Chem* 278, 43628-35.

[54] Palacino, J.J., Sagi, D., Goldberg, M.S., Krauss, S., Motz, C., Wacker, M., Klose, J. and Shen, J. (2004) Mitochondrial dysfunction and oxidative damage in parkin-deficient mice. *J Biol Chem* 279, 18614-22.

[55] Itier, J.M., Ibanez, P., Mena, M.A., Abbas, N., Cohen-Salmon, C., Bohme, G.A., Laville, M., Pratt, J., Corti, O., Pradier, L., Ret, G., Joubert, C., Periquet, M., Araujo, F., Negroni, J., Casarejos, M.J., Canals, S., Solano, R., Serrano, A., Gallego, E., Sanchez, M., Denefle, P., Benavides, J., Tremp, G., Rooney, T.A., Brice, A. and Garcia

de Yebenes, J. (2003) Parkin gene inactivation alters behaviour and dopamine neurotransmission in the mouse. *Hum Mol Genet* 12, 2277-91.

[56] Von Coelln, R., Thomas, B., Savitt, J.M., Lim, K.L., Sasaki, M., Hess, E.J., Dawson, V.L. and Dawson, T.M. (2004) Loss of locus coeruleus neurons and reduced startle in parkin null mice. *Proc Natl Acad Sci U S A* 101, 10744-9.

[57] Perez, F.A. and Palmiter, R.D. (2005) Parkin-deficient mice are not a robust model of parkinsonism. Proc Natl Acad Sci U S A 102, 2174-9.

[58] Goldberg, M.S., Pisani, A., Haburcak, M., Vortherms, T.A., Kitada, T., Costa, C., Tong, Y., Martella, G., Tscherter, A., Martins, A., Bernardi, G., Roth, B.L., Pothos, E.N., Calabresi, P. and Shen, J. (2005) Nigrostriatal dopaminergic deficits and hypokinesia caused by inactivation of the familial Parkinsonism-linked gene DJ-1. *Neuron* 45 489-96.

[59] Kim, R.H., Smith, P.D., Aleyasin, H., Hayley, S., Mount, M.P., Pownall, S., Wakeham, A., You-Ten, A.J., Kalia, S.K., Horne, P., Westaway, D., Lozano, A.M., Anisman, H., Park, D.S. and Mak, T.W. (2005) Hypersensitivity of DJ-1-deficient mice to 1-methyl-4-phenyl-1,2,3,6-tetrahydropyrindine (MPTP) and oxidative stress. *Proc Natl Acad Sci U S A* 102, 5215-20.

[60] Semina, E.V., Murray, J.C., Reiter, R., Hrstka, R.F. and Graw, J. (2000) Deletion in the promoter region and altered expression of Pitx3 homeobox gene in aphakia mice. *Hum Mol Genet* 9, 1575-85.

[61] Zetterstrom, R.H., Solomin, L., Jansson, L., Hoffer, B.J., Olson, L. and Perlmann, T. (1997) Dopamine neuron agenesis in Nurr1-deficient mice. *Science* 276, 248-50.

[62] Jiang, C., Wan, X., He, Y., Pan, T., Jankovic, J. and Le, W. (2005) Age-dependent dopaminergic dysfunction in Nurr1 knockout mice. *Exp Neurol* 191, 154-62.

[63] Feany, M.B. and Bender, W.W. (2000) A Drosophila model of Parkinson's disease. *Nature* 404, 394-8.

[64] Clark, I.E., Dodson, M.W., Jiang, C., Cao, J.H., Huh, J.R., Seol, J.H., Yoo, S.J., Hay, B.A. and Guo, M. (2006) Drosophila pink1 is required for mitochondrial function and interacts genetically with parkin. *Nature* 441, 1162-6.

[65] Yamashita, H., Nakamura, T., Takahashi, T., Nagano, Y., Hiji, M., Hirabayashi, T., Amano, T., Yagi, T., Sakai, N., Kohriyama, T. and Matsumoto, M. (2006) Embryonic stem cell-derived neuron models of Parkinson's disease exhibit delayed neuronal death. *J Neurochem* 98, 45-56.

[66] Kirik, D. and Bjorklund, A. (2003) Modeling CNS neurodegeneration by overexpression of disease-causing proteins using viral vectors. *Trends Neurosci* 26, 386-92.

[67] Geraerts, M., Krylyshkina, O., Debyser, Z. and Baekelandt, V. (2007) Concise review: therapeutic strategies for Parkinson disease based on the modulation of adult neurogenesis. *Stem Cells* 25, 263-70.

[68] Freed, C.R., Greene, P.E., Breeze, R.E., Tsai, W.Y., DuMouchel, W., Kao, R., Dillon, S., Winfield, H., Culver, S., Trojanowski, J.Q., Eidelberg, D. and Fahn, S. (2001) Transplantation of embryonic dopamine neurons for severe Parkinson's disease. *N Engl J Med* 344, 710-9.

[69] Lindvall, O., Sawle, G., Widner, H., Rothwell, J.C., Bjorklund, A., Brooks, D., Brundin, P., Frackowiak, R., Marsden, C.D., Odin, P. and et al. (1994) Evidence for long-term survival and function of dopaminergic grafts in progressive Parkinson's disease. *Ann Neurol* 35, 172-80.

[70] Olanow, C.W., Goetz, C.G., Kordower, J.H., Stoessl, A.J., Sossi, V., Brin, M.F., Shannon, K.M., Nauert, G.M., Perl, D.P., Godbold, J. and Freeman, T.B. (2003) A double-blind controlled trial of bilateral fetal nigral transplantation in Parkinson's disease. *Ann Neurol* 54, 403-14.

[71] Widner, H., Tetrud, J., Rehncrona, S., Snow, B., Brundin, P., Gustavii, B., Bjorklund, A., Lindvall, O. and Langston, J.W. (1992) Bilateral fetal mesencephalic grafting in two patients with parkinsonism induced by 1-methyl-4-phenyl-1,2,3,6-tetrahydropyridine (MPTP). *N Engl J Med* 327, 1556-63.

[72] Kordower, J.H. and Sortwell, C.E. (2000) Neuropathology of fetal nigra transplants for Parkinson's disease. *Prog Brain Res* 127, 333-44.

[73] Gage, F.H. (2000) Mammalian neural stem cells. *Science* 287, 1433-8.

[74] Hoglinger, G.U., Rizk, P., Muriel, M.P., Duyckaerts, C., Oertel, W.H., Caille, I. and Hirsch, E.C. (2004) Dopamine depletion impairs precursor cell proliferation in Parkinson disease. *Nat Neurosci* 7, 726-35.

[75] Cooper, O. and Isacson, O. (2004) Intrastriatal transforming growth factor alpha delivery to a model of Parkinson's disease induces proliferation and migration of endogenous adult neural progenitor cells without differentiation into dopaminergic neurons. *J Neurosci* 24, 8924-31.

[76] Fallon, J., Reid, S., Kinyamu, R., Opole, I., Opole, R., Baratta, J., Korc, M., Endo, T.L., Duong, A., Nguyen, G., Karkehabadhi, M., Twardzik, D., Patel, S. and Loughlin, S. (2000) In vivo induction of massive proliferation, directed migration, and differentiation of neural cells in the adult mammalian brain. *Proc Natl Acad Sci U S A* 97, 14686-91.

[77] Mohapel, P., Frielingsdorf, H., Haggblad, J., Zachrisson, O. and Brundin, P. (2005) Platelet-derived growth factor (PDGF-BB) and brain-derived neurotrophic factor (BDNF) induce striatal neurogenesis in adult rats with 6-hydroxydopamine lesions. *Neuroscience* 132, 767-76.

[78] Palfi, S., Leventhal, L., Chu, Y., Ma, S.Y., Emborg, M., Bakay, R., Deglon, N., Hantraye, P., Aebischer, P. and Kordower, J.H. (2002) Lentivirally delivered glial cell line-derived neurotrophic factor increases the number of striatal dopaminergic neurons in primate models of nigrostriatal degeneration. *J Neurosci* 22, 4942-54.

[79] Freundlieb, N., Francois, C., Tande, D., Oertel, W.H., Hirsch, E.C. and Hoglinger, G.U. (2006) Dopaminergic substantia nigra neurons project topographically organized to the subventricular zone and stimulate precursor cell proliferation in aged primates. *J Neurosci* 26, 2321-5.

[80] Van Kampen, J.M. and Eckman, C.B. (2006) Dopamine D3 receptor agonist delivery to a model of Parkinson's disease restores the nigrostriatal pathway and improves locomotor behavior. *J Neurosci* 26, 7272-80.

[81] Yoshimi, K., Ren, Y.R., Seki, T., Yamada, M., Ooizumi, H., Onodera, M., Saito, Y., Murayama, S., Okano, H., Mizuno, Y. and Mochizuki, H. (2005) Possibility for neurogenesis in substantia nigra of parkinsonian brain. *Ann Neurol* 58, 31-40.

[82] Braak, H., Del Tredici, K., Rub, U., de Vos, R.A., Jansen Steur, E.N. and Braak, E. (2003) Staging of brain pathology related to sporadic Parkinson's disease. *Neurobiol Aging* 24, 197-211.

In: Stem Cell Applications in Disease and Health
Editors: W. Burnsides and R. Ellsley

ISBN: 978-1-60456-255-2

Chapter X

ES Cell-Derived Endoderm Stem Cells for Clinical Application

Masahiro Yasunaga*

Investigative Treatment Division, Research Center for Innovative Oncology
National Cancer Center Hospital East, 6-5-1 Kashiwanoha
Kashiwa, Chiba, 277-8577, Japan

Abstract

Recently, stem cell technology has been developed and be exploited for clinical use. Among stem cells, ES cells have a high application potential as a source to produce therapeutic cells in vitro because of their high growth capacity and differentiation competence. Endoderm cells compose respiratory, digestive and urological organs such as thyroid, lung, liver, pancreas, intestine, prostrate, and bladder. In-vitro manufactured hepatocytes and insulin-producing cells via endoderm cells were expected for use in the treatment of liver diseases and diabetes mellitus. There are many protocols to produce ES cell-derived endoderm. The basic technologies can be categorized as: 1. embryoid body, 2. transducing key gene, 3. coculture with stromal cells, 4. chemically defined medium, 5. linage-tracing with specific markers. However, many of methods are focused on an examination of the final differentiation step only. There are problems such as low reproducibility, low differentiation efficiency and lower function. Our strategy is to monitor intermediate stages of in-vivo endoderm development. Increasing of intermediate stage cells growth and differentiation efficiency can improve the therapeutic cell differentiation efficiency. A higher function is expected using a differentiation system according to in vivo development. We have developed the endoderm-specific induction and monitoring system to identify the Goosecoid+ Sox17+ definitive endoderm stem cells with cell surface markers CXCR4 and E-cadherin. Moreover, purified definitive endoderm stem cells can differentiate into various visceral organ cells in each secondary induction assay.

---

* E-mail address:mayasuna@east.ncc.go.jp

Introduction

Recently stem cell technology was developed and it is anticipated to use regenerative therapy [1-19]. A stem cell is a cell that has an ability of self-renewal and multipotency, which can give rise to mature somatic cells such as heart, neuron, liver and pancreas[1-19]. Therefore, transplanted stem cells or stem cell-derived somatic cells can replace and recover the function lost in the various human diseases and traumas [4-19]. As many types of stem cells have been reported at this time, it is important to consider what type of stem cell is most useful for appropriate treatment for each disease [4-19]. Embryonic stem cells (ES cells) [1-11], hematopoietic stem cells (HSC) [12, 13], mesenchymal stem cells (MSC) [14-17] and neural stem cells (NSC) [18, 19] are popular and well-researched. Among them, MSC have been the focus of much investigation for clinic use, because they can be obtained in quantities from adults including patients themselves and easily manipulated in vitro for growth and differentiation. Cardio-myocites in cardiac infarction, chondrocytes in joint injury, and muscular myocytes in degenerative dystrophy are good candidates for clinical application of MSCs [14-17]. However, MSCs exhibit limited differentiation-potential regarding some endoderm-associated cells such as pancreatic beta cells, unlike ES cells [4-6]. Efficacy of MSC therapy in a phase-I clinical trial with a few patients suffering from chronic liver diseases has been reported. But MSC therapy for liver injury is still controversial because the study was done in a non-controlled, non-randomized method. And it might be that some cytokines secreted from MSC rather than transdifferentiated hepatocytes contribute to regeneration in the injury, as in the case of cardio-myocites and endothelial regeneration by MSCs [10,13-16]. Thus characterization of MSC-derived hepatocytes is not fully understood at this time. There is a strong argument to be made that MSCs can become hepatocytes beyond endoderm-differentiation lineage [7, 10]. Otherwise, ES cells, which are originated from inner mass cells within a blastocyst, have pluripotency, and can naturally become every tissue including endoderm-derived cells in the body [1-11]. However, their clinical application was not favored because of ethical problems using destroyed human embryos [6]. In vitro producing of pancreatic beta cells is expected for the effective treatment of insulin-dependent diabetes mellitus [4-8]. There are two sources for beta cells: one is somatic stem cells; another is ES cells [4-8]. With regard to somatic pancreatic stem cells, growth activity is limited. Moreover, existence of adult pancreatic stem cells themselves is debated [4]. It certain that ES cells are currently the only type of stem cell that is unquestionably capable of differentiating into beta cells [4-8]. The problem is how to get differentiated cells, how to proceed efficienctly, and how to make them functional. ES cells have to go through multiple differentiation steps, from egg to organogenesis. Therefore, it is important to understand each developmental step for exploiting the manipulation techniques. In this report, ES cell-derived endoderm cells and related technologies are reviewed, and our induction and monitoring system toward clinical application is introduced.

Endoderm Stem Cells

In higher animals, including mammals, there are developmental periods called three germ layers consisting of the ectoderm, mesoderm and endoderm [20, 21]. Endoderm is generally understood as the cell population, which gives rise to internal organs such as thyroid, parathyroid, thymus, lung, liver, pancreas, bladder, prostate and gut [22, 23]. However, in the early development of mammals, there are two types of endoderm: extra-embryonic primitive endoderm (PE) and embryonic definitive endoderm (DE) [23-26] (Figure 1). PE, consisting of parietal endoderm (PE) and visceral endoderm (VE), which is specific to mammals, can become placenta and yolk sac [24-27]. PE is easily distinguished from VE and DE in their distinct scattered growing [2, 3, 24-27]. We can also distinguish between VE and DE according to the difference in their anatomical location [22-27] (Figure 3C, D). However, as in in-vitro culture condition, such anatomical information is absent and these two endoderm show similarity regarding cell-morphology (Figure 3A, B) and expression of molecular-markers such as Foxa2, GATA4, GATA6 and Sox17. Therefore, it is not easy to distinguish between them [28-34]. ES cells from inner cell mass can differentiate into both types of endoderm cells. Moreover, VE cells can be directly differentiated from ES cells and expand earlier than DE cells. Surprisingly, VE can easily grow in serum-free culture compared to other types of cells [29]. The reason that they can survive nutrient-poor conditions is that VE and VE-derivatives (yolk sac and placenta) have a role primarily to protect embryo development against the detrimental effects of external environments. Therefore, they have multiple functions in nutrient supply, protein (albumin, AFP, etc.) synthesis, hormone synthesis, toxic-protection and metabolic control, which is likely to be functionally combined of gut, liver and pancreas [24-26]. This acquired phenotype is explained as the result of mammal evolution from egg-laying animals. Actually, it is acceptable to use them as a source of cell therapy depending on their functional similarity to hepatocytes. However, some critical issues including quality of VE-derived hepatocyte-like cells should be reconsidered for their clinical use. Similarly, VE-derived insulin-producing cells, which show poor glucose-response and low insulin secretion, are not suitable for cell-based therapy as an alternative to the transplantation-donor derived islet cells [4-8]. Process and quality control is absolutely necessary in manufacturing therapeutic cells [28-34]. Exploiting of the endoderm differentiation system using ES cells for clinical application should be conducted from this viewpoint.

It is certain that endoderm stem cells exist in DE, and clearly have multiple differentiation potential and self renewal activity. And the role and function of DE in embryogenesis is evolutionary conserved in higher animals, unlike to VE [20-26]. Many of the key genes, which are involved in DE development, are commonly seen in various animal models such as the zebrafish, xenopus, chicken, and mouse [20-23, 35]. In many differentiation factors, activin and nodal are well known for their important role in DE specification in both in-vitro and in-vivo situations [36, 37]. Actually, activin is generally used in the DE differentiation system [28-30, 32-34]. In late DE development, they are

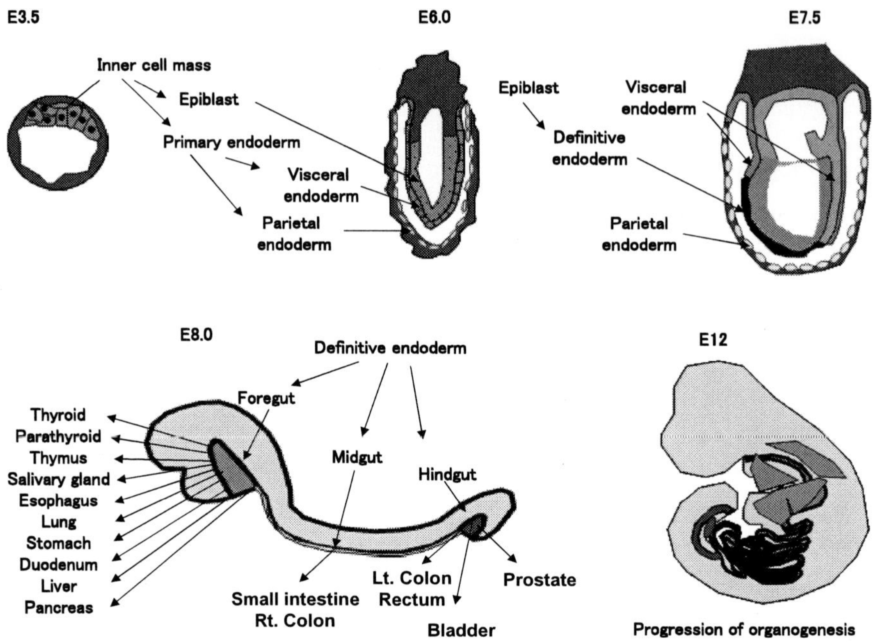

Figure 1. Endoderm development in vivo. Events of mouse endoderm development in vivo are depicted from embryonic day 3.5 (E3.5). The cells of inner cell mass (= ES cells origin) differentiate into primary endoderm cells (which consists of parietal endoderm and visceral endoderm cells) and definitive endoderm cells via epibast-cells from E6.0 to E7.5. Definitive endoderm cells convert to primitive gut tube (=foregut, midgut and hindgut) and give rise to organs of the gastrointestinal, respiratory and urological organs from E8.0 to E12.

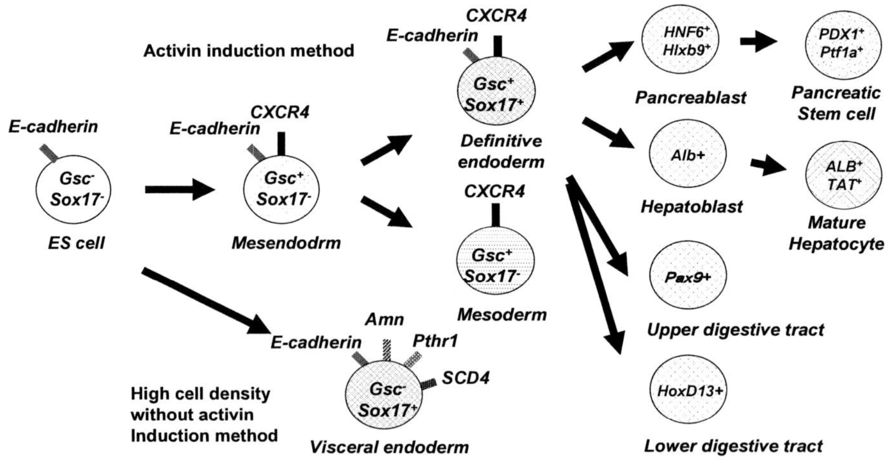

Figure 2. In vitro endoderm differentiation from ES cells. ES cells can differentiate into definitive endoderm cells via mesendoderm (=common precursor of definitive endoderm and mesoderm) cells [28] with activin induction method and visceral endoderm cells at a high cell density without activin method, respectively. CXCR4 is a useful marker in both identifying and separating definitive endoderm cells from visceral endoderm cells. In addition, a combination of CXCR4 and E-cadherin was used to distinguish endoderm cells and mesoderm cells in activin-inducted cells.

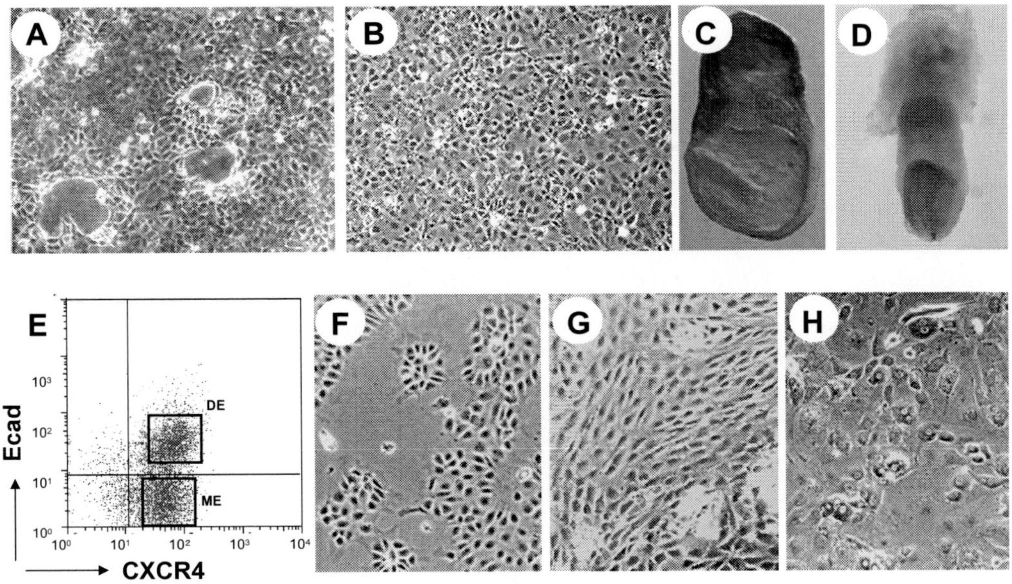

Figure 3. Visceral endoderm, definitive endoderm and derivatives. There are two type of endoderm stem cells: visceral (A) and definitive endoderm (B) cells. E7.5 mouse embyo was histologically stained with visceral endoderm specific marker SCD4 (C) or definitive endoderm specific marker CXCR4 (D). CXCR4+ and E-Cadhern+ definitive endoderm (DE) cells and CXCR4+ and E-Cadhern- mesoderm (ME) cells on activin induction day 6, were monitored in FACS (E), and purified (F). The cells differentiated into gut-associated squamous epithelium cells (G) and cuboidal hepatocytes (H) at each induction assay.

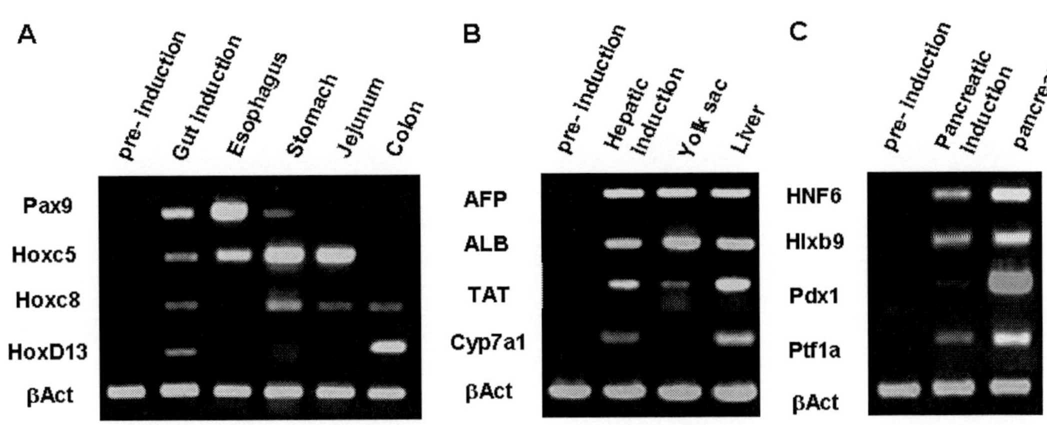

Figure. 4. Various induction assays using definitive endoderm.In purified Ecad+ CXCR4+ definitive endoderm (DE) cells, which were recultured with EGF (gut induction assay), all anteroposterior region-specific markers were induced (A). Sorted DE cells, were recultured with BMP4, FGF (hepatocyte induction assay), and differentiated into TAT+ Cyp7a1+ mature hepatocytes (B). In DE cells, which were recultured with Retinoic acid (pancreas induction assay), strong HNF6 and Hlxb9 expression and weak PDX1 and Ptf1a expressions were seen (C).

organized into a primitive gut tube along the anterior-posterior axis to form the foregut, midgut and hindgut [23, 35] (Figure 1). The foregut gives rise to the gastro-intestinal structures from the esophagus to the duodenum. In addition, many visceral organs such as the thyroid, parathyroid, thymus, salivary gland, lung, liver and pancreas come out from foregut. The midgut gives rise to the small intestine and right colon. The hindgut gives rise to the left colon, rectum, bladder and prostate [23, 35]. Their well-organized antero-posterior axisis formation is more understandable according to the evidence that site-specification was seen in various pathological ectopic mucosa and aberrant tissue [38-41]. Gastric ectopic mucosa is seen in the upper digestive tract, mainly the esophagus [38]. The occurrence of aberrant pancreatic tissue is seen in the lower foregut~midgut area in early endoderm development [38]. Otherwise, although few aberrant hepatocytes are reported, most of the occurrence was seen in the gallbladder [40]. Most of the aberrant endoderm cells in Meckel are gastric cells and some pancreatic cells, but not lung, esophageal or colorectal cells [41]. This site specificity could very likely be associated with the Hox code specificity. The Hox expression pattern is an early indicator of endoderm differentiation [29, 42]. Actually, the Hox pattern of the esophagus and colon is quite different (Figure 4A).

Production of Endoderm Cells from ES Cells

There are many reports regarding successful production of endoderm cells from ES cells. Basic technologies utilized in each method can be categorized into 1. using embryoid body [43-46], 2. transducing key gene [3, 47-50], 3. coculture with stromal cells [51-54], 4. chemically defined medium using soluble factors in various combinations [55-59], and 5. monitoring system using linage-specific markers [28-30,32-34,60-62].

Embryoid body is generally used in various induction systems. It is stably and effectively functional in some induction systems, such as the neuron [18, 19]. However, it is difficult to control their size distribution and understand their internal molecular mechanism. In addition, extra-embryonic VE cells can arise earlier from the embryoid body and grow more easily than DE cells [29]. Application of the embryoid body is expected for its physiological environment as model of embryonic development in the ES cell differentiation system. Definitive endoderm tissues are well organized along antero-posterior (AP) axis in vivo. Anterior induction seems to be clearly different from posterior [22,23,35,42]. Embryoid body is most likely to be the body plan of radial animals at the point of lacking body axis (AP axis, dorsal-ventral axis and left-right axis) [63]. Evolutionary body axis is thought to have involved endoderm-visceral organ induction from primitive gut tube in higher animals [22, 23, 35, 42, 64]. Therefore, the usefulness of the embryoid body for endoderm differentiation should be examined extensively and reconsidered carefully.

Gene recombination method to transduce key genes such as Foxa2 [47, 48], GATA6 [3], PAX4 [49], and PDX1 [50] for endoderm or endoderm-derivatives is used to manipulate ES cell differentiation in vitro. However, none have yet reached a clinical trial. There are some problems with each method. Foxa2 and GATA6 are well known early endoderm markers [3, 47, 48]. With regard to Foxa2, it is involved in not only endoderm differentiation, but also other germ layers associated with cell differentiation, such as the neural plate [22, 23, 35].

Therefore, there is the possibility that over-expression of Foxa2 can not only induce endoderm differentiation. There is another possibility that induced hepatocyte-like cells exist in yolk sac or placental cells derived from VE cells, as described in an earlier paragraph. GATA6 over-expressed ES cells can become AFP+TTR+ApoE+ VE cells very similar to hepatocytes [3]. Although Foxa2 and GATA6 are expressed in both DE and VE, they induce ES cells into VE lineage rather than DE lineage. Moreover, it is still unknown that VE-derived hepatocyte-like cells can act the same as hepatocytes in cell therapy for various liver diseases. PDX1 and PAX4 are well known as transcriptional factor which regulate insulin promoter [49, 50, 64]. They also play role in organogenesis of pancreas [49, 50, 64]. Therefore, it seems that they can act as master genes in pancreas differentiation from ES cells. Actually, PDX1 or PAX4 over-expressed ES cells become insulin-producing cells in vitro [49, 50]. However, it is not sufficient to treat diabetes mellitus because of poor glucose responsibility and low insulin-releasing activity. It seems that PDX1 or PAX are necessary genes, but not sufficient for mature beta cell differentiation. Some other genes must be introduced for complete beta cell function to be acquired, but it seems to be difficult to create and maintain the glucose-sensibility in vitro in their stable extracellular glucose level in contrast to in-vivo variables and stimulus situation (probably an in vitro adaptation mechanism is dominant in the cultured cells). Another possibility is that PDX1- or PAX4-overexpressed ES cells fail to differentiate into DE. As PDX1 and PAX4 are also expressed in neurons [65], there is a possibility that these genes over-expressing insulin-producing cells are not DE-derived cells but some type of neuronal cells. Actually, insulin-producing cells exist in some fetal neurons [66]. Their glucose-responsibility and insulin secretion activity is weak and cannot replace the function of pancreatic beta cells in vivo. This is also true of VE-derived insulin-producing cells. It is very important to understand the difference between sporadically-existing insulin producing cells and pancreas-specific beta cells regarding cell origin and endocrinological activity. Introduction of PDX1 or PAX4 into DE or foregut cells from ES cells could drive them to become proper pancreatic beta cells. There is a a general critical issue in therapeutic cell manufacturing through gene recombination, which would be deteriorating to human health. To settle this problem, some technologies such as the Cre/loxP system, artificial chromosome and protein-transduction are being researched and developed [67-71]. The Cre/loxP system is used to eliminate incorporated genes, which would adversely affect the host genome [67]. The human artificial chromosome that they are replicating during many cell divisions but not incorporating into host genome is more excellent [68, 69]. The protein transduction technique has been introduced recently [70, 71]. HIV1-derived Tat is a well known protein transduction domain. Tat-fused protein is easily membrane transducible. PDX1 has a Drosophila Antennapedia domain. It can be introduced into cells and function as a transcription factor. PDX1 can bind with PDX1 enhancer resulting in the activation of PDX1. Therefore, transient PDX1 protein activation can cause natural PDX1 expression [70]. Although the protein transduction system, which is nontoxic to the human body, is most suitable for clinical application, there are some technological problems such as short half-life and necessary high dose [70,71].

As gene-recombination, modification and transfer technology is powerful and effective to manipulate growth and differentiation of stem cells, much progress will facilitate their clinical application in the field of stem cell based therapy.

Embryonic stromal cells play an important role in organ morphogenesis because they secrete various signals that stimulate cell proliferation and differentiation in the DE development [23]. Actually, there has been a long history of using them to regulate growth and differentiation of various stem cells, including ES cells in vitro [72-74]. Coculture with OP9 cells can dramatically enhance hematopoiesis and angiogenesis/vasculogenesis from ES cells in vitro [72, 73]. Stromal cell line PA6 can strongly induce neuronal cell differentiation via ectoderm from ES cells [74]. Some coculture systems are applied to regulate hepato-pancreatic cells via endoderm from ES cells [51-53]. This technology is so practical that the culture system will be able to be developed even if the molecular mechanism of morphogenesis is unknown. However, it is not so easy to select and establish an appropriate stromal cell line. And, the culture system is somewhat unstable, as it is sometimes difficult to keep, bioactivity of sromal cells stable, which is dependent on various factors such as cell-stock status, culture conditions and individual manipulation techniques.

Development of chemically defined media is quite important as a basic technology in using stem cells for therapy [9, 28, 29]. Rigorous culture conditions including selection and dosage of soluble factors and optimizing of animal protein-free supplement should be prepared according to GLP (Good Laboratory Practice)/GMP (Good Manufacturing Practices). This technology is preferred because of its regulated procedure and guaranteed safety, close to physiological conditions. Among various soluble factors, Activin is well used in the DE induction system. In pancreatic specification, Activin, TGFb, retinoid, cyclopamine (shh antagonist), and FGF are applied [32, 34, 51, 56, 58, 61]. In hepatocyte specification, BMP, FGF, HGF and Oncostatin M are applied [28, 29, 43, 52, 53, 62]. It is important to add them at necessary concentration and at the precise time.

Although insulin-producing cells and hepatocytes via DE from human ES cells were induced with multiple steps, in most cases lineage tracing was not always conducted [5-8,10]. Observation of the differentiation process is required for the guidance and improvement of the beta cell and hepatocyte manufacturing. Several groups, including ours, reported the monitoring system of DE, using Brachyury [32, 62], Goosecoid [28-30], Mixl1[60] and CXCR4 [29, 33, 34] as ME~DE marker. Effectiveness of this method has already demonstrated that induction and monitoring of Flk1+/VE-cadherin+ mesoderm stem cells from ES cells specifically contribute to the production of blood vessels in vitro [73]. In addition, Flow cytometry, which allows targeted cells to be simultaneously identified and isolated in live in real time, is an indispensable tool for this system. With this machine, a few cells are detectable and dead cells, which sometimes become differentiation-marker pseudo-positive cells in PCR, immunohistochemistry and Flow cytometry itself, are excluded. Our practical method is described in the next paragraph.

Induction and Monitoring System

From the beginning, we planned to develop an in-vitro manipulation technique without gene recombination using ES cells as a source of therapeutic cells for clinical application. We focused on 2 technologies: 1. chemically defined medium for cell growth and differentiation as an induction system, and 2. cell sorting using antibodies against cell surface markers as a monitoring system.

ES cells can differentiate into every lineage derivative, with the result that a heterogeneous population of various organ cells and their intermediates are easily formed. Most current methods seem to be focused on an examination of the final differentiation step only. There are problems such as low reproducibility, low differentiation efficiency and low biological function. Our strategy is to monitor the intermediate stages of in-vivo endoderm development. Clarification of soluble factors which regulate the growth and differentiation of the intermediate stage cells leads to a stable culture condition.

Increasing of intermediate stage cell growth and differentiation efficiency can improve the endoderm cell differentiation efficiency. A higher function is expected using a differentiation system according to in vivo development.

Using intermediate stage marker knock-in ES cells and endoderm differentiation system, it is possible to purify the intermediate stage cells. Gene expression can be examined by DNA microarray and Bioinformatics. We can then screen the human homologs of cell surface markers, including cytokine receptors specific to mouse endoderm stem cells. According to this concept, we developed Goosecoid(Gsc)^{gfp} Sox17^{huCD25} ES cells and succeeded in visualizing goosecoid and Sox17 genes to monitor definitive endoderm in vitro [28, 29]. We also established two types of culture conditions: one for definitive endoderm differentiation with activin (activin induction method) [28, 29] and a second for visceral endoderm without activin and with a high ES cell density (high cell density without activin method) [29]. By using a combination of Gsc Sox17 double knock-in ES cells and these culture conditions, we were able to purify the Gsc+Sox17+ DE cells and Gsc-Sox17+ VE cells, respectively. A comparison of gene expressions of 2 types of endoderm cells was made using the DNA microarray system and software. Seven surface molecules that are expressed differentially in the two populations were identified [29]. The changing pattern of cell surface markers in the activin induction method and the non-activin, high cell density method are shown in Figure 2. Although both anatomical location and the cell fate of DE and VE are different (Figure 3C, D), the morphology is too similar to distinguish each other in the in-vitro culture system (Figure 3A, B). Our identified DE specific marker CXCR4 was useful in separating from VE cells (Figure 3D, E). In addition, a combination of CXCR4 and E cadherin was used to distinguish endoderm cells and mesoderm cells in the activin induction method. Purified Ecad+CXCR4+ DE cells (Figure 3F), which were recultured with EGF (gut induction assay), were able to differentiate into gut-tube associated cells, including squamous cells, (Figure 3G) and expressed all specific antero-posterior gastrointestinal markers such as Pax9, HoxC5, HoxC8 and HoxD13 (Figure 4A). Moreover, they could differentiate into AFP+ALB+TAT+Cyp7a1+mature cuboidal hepatocytes (Figure 3H), which were different from VE derived AFP+ALB+TAT^{low}Cyp7a1- yolk sac cells, when exposed to BMP4 and FGF (hepatic induction assay) (Figure 4B) [28, 29, 62, 75]. Furthermore, they were able to differentiate into pancreatic blast/pancreatic stem cells expressing strong HNF6 and Hlxb9 and weak PDX1 and Ptf1a under the culture condition containing Retinoic acid as a central factor (pancreatic induction assay) (Figure 4C).

We want to emphasize the importance of DE purification to eliminate some unfavorable cells: 1. undifferentiated ES cells are true origin of in vivo tumor teratoma; and 2. some mesoderm cells, which sometimes interfere with growth and differentiation of endoderm cells like a lateral inhibition in neuronal differentiation. A monitoring system using antibodies is a good application for clinical use depending on convenience and easy technological transfer. Actually, we finished the technology transfer from mouse ES cell system to human ES cell system in a very short period of time. From this outcome, our strategy for clinical application was re confirmed as the right way forward.

Conclusion

The number of people suffering from liver disease and diabetes mellitus is increasing throughout the world. Although progress in medical science contributes to developing drugs and transplantation procedures to treat these diseases, the discovery of the new treatment is eagerly anticipated for many patients resistant to current protocols. In this circumstance, stem-cell based therapy is a new hope. There are several candidates as source of stem cell for therapy. Among them, ES cells are most attractive, because they have a great growth and differentiation potential compared to other stem cells. Moreover, ES cell-based disease and toxicity models will contribute to new drug discovery and development [76]. Rightly, there should be clear ethical and legal guidelines for conducting research. We focused on safety and physiological conditions in developing a new system and completed the induction and monitoring system in a fully chemically defined medium without gene recombination. It would be the right way forward to clinical application at this time. However, advanced innovational technology might bring new methodology in the near future.

I hope that the day that ES cell-derived endoderm stem cells and related technologies are applied to disease treatment is here as early as possible.

Acknowledgment

The work of "induction and monitoring system" was performed in collaboration with both the Laboratory for Stem Cell Biology in RIKEN Center for Development Biology (Kobe, Japan) and the Basic Research Laboratory in Stem Cell Science KK (Kobe, Japan). I would like to thank Dr. Shin-Ichi Nishikawa, Dr. Takumi Era, Dr. Hitoshi Niwa, Dr. Shinsuke Tada, Dr. Mitsuhiro Okada, Dr. Lars Martin Jakt, Dr. Yoshiaki Kanai, Dr. Naoki Izumi, Yasuhiro Tosaka, Satomi Torikai-Nishikawa, Yoko Nakano, Kaori Tsujino, Muneaki Miyata, Satomi Nishikawa and other collaborators for assistance and advice during experiments.

References

[1] Zernicka-Goetz M, Pines J, McLean Hunter S, Dixon JP, Siemering KR, Haseloff J, Evans MJ. Following cell fate in the living mouse embryo. *Development*. 1997. 124. 1133-7.

[2] Nichols J, Zevnik B, Anastassiadis K, Niwa H, Klewe-Nebenius D, Chambers I, Scholer H, Smith A. Formation of pluripotent stem cells in the mammalian embryo depends on the POU transcription factor Oct4. *Cell*. 1998. 95. 379-91.
[3] Fujikura J, Yamato E, Yonemura S, Hosoda K, Masui S, Nakao K, Miyazaki Ji J, Niwa H. Differentiation of embryonic stem cells is induced by GATA factors. *Genes Dev*. 2002. 16. 784-9.
[4] Dor Y, Brown J, Martinez OI, Melton DA. Adult pancreatic beta-cells are formed by self-duplication rather than stem-cell differentiation. *Nature*. 2004. 429. 41-6.
[5] Bonner-Weir S, Weir GC. New sources of pancreatic beta-cells. *Nat Biotechnol*. 2005. 23. 857-61.
[6] Trucco M. Regeneration of the pancreatic beta cell. *J Clin Invest*. 2005. 115. 5-12.
[7] Fujikawa T, Oh SH, Shupe T, Petersen BE. Stem-cell therapy for hepatobiliary pancreatic disease. *J Hepatobiliary Pancreat Surg*. 2005. 12. 190-5.
[8] Madsen OD, Serup P. Towards cell therapy for diabetes. *Nat Biotechnol*. 2006. 24. 1481-3.
[9] Ludwig TE, Levenstein ME, Jones JM, Berggren WT, Mitchen ER, Frane JL, Crandall LJ, Daigh CA, Conard KR, Piekarczyk MS, Llanas RA, Thomson JA. Derivation of human embryonic stem cells in defined conditions. *Nat Biotechnol*. 2006. 24. 185-7.
[10] Cantz T, Manns MP, Ott M. Stem cells in liver regeneration and therapy Cell Tissue Res. 2007. Sep. 28. [Epub ahead of print]
[11] Mimeault M, Hauke R, Batra SK. Stem cells: a revolution in herapeutics-recent advances in stem cell biology and their therapeutic applications in regenerative medicine and cancer therapies. *Clin Pharmacol Ther*. 2007. 82. 252-64.
[12] Appelbaum FR. Hematopoietic-cell transplantation at 50. *N Engl J Med*. 2007. 357. 1472-5.
[13] Ueno H, Weissman IL. Stem cells: blood lines from embryo to adult. *Nature*. 2007. 446. 996-7.
[14] Devine SM. Mesenchymal stem cells: will they have a role in the clinic? *J Cell Biochem Suppl*. 2002. 38. 73-9.
[15] Miyahara Y, Nagaya N, Kataoka M, Yanagawa B, Tanaka K, Hao H, Ishino K, Ishida H, Shimizu T, Kangawa K, Sano S, Okano T, Kitamura S, Mori H. Monolayered mesenchymal stem cells repair scarred myocardium after myocardial infarction. Nat Med. 2006. 12. 459-65.
[16] Giordano A, Galderisi U, Marino IR. From the laboratory bench to the patient's bedside: an update on clinical trials with mesenchymal stem cells. *J Cell Physiol*. 2007. 211. 27-35.
[17] Valfre di Bonzo L, Ferrero I, Cravanzola C, Mareschi K, Rustichelli D, Novo E, Sanavio F, Cannito S, Zamara E, Bertero M, Davit A, Francica S, Novelli F, Colombatto S, Fagioli F, Parola M.Human mesenchymal stem cells as a two-edged sword in hepatic regenerative medicine: Engraftment and hepatocyte differentiation versus profibrogenic potential. Gut. 2007. Jul. 16. {Epub ahead of print]
[18] Shin S, Sun Y, Liu Y, Khaner H, Svant S, Cai J, Xu QX, Davidson BP, Stice SL, Smith AK, Goldman SA, Reubinoff BE, Zhan M, Rao MS, Chesnut JD. Whole genome

analysis of human neural stem cells derived from embryonic stem cells and stem and progenitor cells isolated from fetal tissue. *Stem Cells.* 2007. 25. 1298-306.

[19] Redmond DE Jr, Bjugstad KB, Teng YD, Ourednik V, Ourednik J, Wakeman DR, Parsons XH, Gonzalez R, Blanchard BC, Kim SU, Gu Z, Lipton SA, Markakis EA, Roth RH, Elsworth JD, Sladek JR Jr, Sidman RL, Snyder EY. Behavioral improvement in a primate Parkinson's model is associated with multiple homeostatic effects of human neural stem cells. *Proc Natl Acad Sci U S A.* 2007. 104. 12175-80.

[20] Beddington, R.S. & Robertson, E.J. Axis development and early asymmetry in mammals. 1999. *Cell* 96. 195–209.

[21] Rossant, J. Lineage development and polar asymmetries in the peri-implantation mouse blastocyst. Semin. 2004. *Cell Dev. Biol.* 15. 573–581.

[22] Tam, P.P., Kanai-Azuma, M. & Kanai, Y. Early endoderm development in vertebrates:lineage differentiation and morphogenetic function. *Curr. Opin. Genet. Dev.* 2003. 13. 393–400.

[23] Wells, J.M. & Melton, D.A. Vertebrate endoderm development. *Annu. Rev. Cell Dev.Biol.* 1999. 15. 393–410.

[24] Gardner, R.L. Investigation of cell lineage and differentiation in the extraembryonic endoderm of the mouse embryo. *J. Embryol. Exp. Morphol.* 1982. 68. 175–198.

[25] Meehan RR, Barlow DP, Hill RE, Hogan BL, Hastie ND Pattern of serum protein gene expression in mouse visceral yolk sac and foetal liver. *EMBO J.* 1984. 3. 1881-5

[26] Bielinska M, Narita N, Wilson DB. Distinct roles for visceral endoderm during embryonic mouse development. *Int J Dev Biol.* 1999. 43. 183-205.

[27] Kunath T, Arnaud D, Uy GD, Okamoto I, Chureau C, Yamanaka Y, Heard E, Gardner RL, Avner P, Rossant J. Imprinted X-inactivation in extra-embryonic endoderm cell lines from mouse blastocysts. *Development.* 2005. 132. 1649-61.

[28] Tada S, Era T, Furusawa C, Sakurai H, Nishikawa S, Kinoshita M, Nakao K, Chiba T, Nishikawa S.Characterization of mesendoderm: a diverging point of the definitive endoderm and mesoderm in embryonic stem cell differentiation culture. *Development.* 2005. 132. 4363-74.

[29] Yasunaga M, Tada S, Torikai-Nishikawa S, Nakano Y, Okada M, Jakt LM, Nishikawa S, Chiba T, Era T, Nishikawa S. Induction and monitoring of definitive and visceral endoderm differentiation of mouse ES cells. *Nat Biotechnol.* 2005. 23. 1542-50.

[30] Izumi N, Era T, Akimaru H, Yasunaga M, Nishikawa S. Dissecting the molecular hierarchy for mesendoderm differentiation through a combination of embryonic stem cell culture and RNA interference. *Stem Cells.* 2007. 25. 1664-74.

[31] Milne HM, Burns CJ, Kitsou-Mylona I, Luther MJ, Minger SL, Persaud SJ, Jones PM. Generation of insulin-expressing cells from mouse embryonic stem cells. *Biochem Biophys Res Commun.* 2005. 328. 399-403.

[32] Kubo A, Shinozaki K, Shannon JM, Kouskoff V, Kennedy M, Woo S, Fehling HJ, Keller G. evelopment of definitive endoderm from embryonic stem cells in culture. *Development.* 2004. 131. 1651-62.

[33] D'Amour KA, Agulnick AD, Eliazer S, Kelly OG, Kroon E, Baetge EE. Efficient differentiation of human embryonic stem cells to definitive endoderm. *Nat Biotechnol.* 2005. 23. 1534-41.

[34] D'Amour KA, Bang AG, Eliazer S, Kelly OG, Agulnick AD, Smart NG, Moorman MA, Kroon E, Carpenter MK, Baetge EE. Production of pancreatic hormone-expressing endocrine cells from human embryonic stem cells. *Nat Biotechnol*. 2006. 24. 1392-401.

[35] Hogan, B. L. M. and Zaret, K. S. Development of the endoderm and its tissue derivatives. J. Rosant and P. Tam, editors Mouse Development: Patterning, Morphogenesis, and Organogenesis. *Academic Press* 2002. 301-330.

[36] Conlon FL, Lyons KM, Takaesu N, Barth KS, Kispert A, Herrmann B, Robertson EJ. A primary requirement for nodal in the formation and maintenance of the primitive streak in the mouse. *Development*. 1994. 120. 1919-28.

[37] Huelsken J, Vogel R, Brinkmann V, Erdmann B, Birchmeier C, Birchmeier W.Requirement for beta-catenin in anterior-posterior axis formation in mice. *J Cell Biol*. 2000.148. 567-78.

[38] Ueno J, Davis SW, Tanakami A, Seo K, Yoshida S, Nishitani H, Irie H, Lu CC. Ectopic gastric mucosa in the upper esophagus: detection and radiographic findings. *Radiology*. 1994.191. 751-3.

[39] Fukuda A, Kawaguchi Y, Furuyama K, Kodama S, Horiguchi M, Kuhara T, Koizumi M, Boyer DF, Fujimoto K, Doi R, Kageyama R, Wright CV, Chiba T. Ectopic pancreas formation in Hes1 -knockout mice reveals plasticity of endodermal progenitors of the gut, bile duct, and pancreas.*J Clin Invest*. 2006. 116. 1484-93.

[40] Leone N, Saettone S, De Paolis P, Carucci P, Brunello F, De Angelis C, Menozzi G, Rizzetto M. Ectopic livers and related pathology: report of three cases of benign lesions.*Dig Dis Sci*. 2005. 50. 1818-22.

[41] Artigas V, Calabuig R, Badia F, Rius X, Allende L, Jover J. Meckel's diverticulum: value of ectopic tissue.*Am J Surg*. 1986. 151. 631-4.

[42] Beck F, Tata F, Chawengsaksophak K. Homeobox genes and gut development. *Bioessays*. 2000. 22. 431-41.

[43] Chinzei R, Tanaka Y, Shimizu-Saito K, Hara Y, Kakinuma S, Watanabe M, Teramoto K, Arii S, Takase K, Sato C, Terada N, Teraoka H. Embryoid-body cells derived from a mouse embryonic stem cell line show differentiation into functional hepatocytes. *Hepatology*. 2002. 36. 22-9.

[44] Yamada T, Yoshikawa M, Kanda S, Kato Y, Nakajima Y, Ishizaka S, Tsunoda Y. In vitro differentiation of embryonic stem cells into hepatocyte-like cells identified by cellular uptake of indocyanine green. *Stem Cells*. 2002. 20. 146-54.

[45] Asahina K, Fujimori H, Shimizu-Saito K, Kumashiro Y, Okamura K, Tanaka Y, Teramoto K, Arii S, Teraoka H. Expression of the liver-specific gene Cyp7a1 reveals hepatic differentiation in embryoid bodies derived from mouse embryonic stem cells. *Genes Cells*. 2004. 9. 1297-308.

[46] Choi D, Lee HJ, Jee S, Jin S, Koo SK, Paik SS, Jung SC, Hwang SY, Lee KS, Oh B. In vitro differentiation of mouse embryonic stem cells: enrichment of endodermal cells in the embryoid body. *Stem Cells*. 2005. 23. 817-27.

[47] Levinson-Dushnik M, Benvenisty N. Involvement of hepatocyte nuclear factor 3 in endoderm differentiation of embryonic stem cells. *Mol Cell Biol*. 1997. 17. 3817-22.

[48] Ishizaka S, Shiroi A, Kanda S, Yoshikawa M, Tsujinoue H, Kuriyama S, Hasuma T, Nakatani K, Takahashi K. Development of hepatocytes from ES cells after transfection with the HNF-3beta gene.*FASEB J.* 2002. 16. 1444-6.

[49] Blyszczuk P, Czyz J, Kania G, Wagner M, Roll U, St-Onge L, Wobus AM. Expression of Pax4 in embryonic stem cells promotes differentiation of nestin-positive progenitor and insulin-producing cells. *Proc Natl Acad Sci U S A*. 2003. 100. 998-1003.

[50] Miyazaki S, Yamato E, Miyazaki J. Regulated expression of pdx-1 promotes in vitro differentiation of insulin-producing cells from embryonic stem cells. *Diabetes*. 2004. 53. 1030-7.

[51] Shiraki N, Lai CJ, Hishikari Y, Kume S. TGF-beta signaling potentiates differentiation of embryonic stem cells to Pdx-1 expressing endodermal cells. *Genes Cells*. 2005. 10. 503-16.

[52] Ishii T, Yasuchika K, Fujii H, Hoppo T, Baba S, Naito M, Machimoto T, Kamo N, Suemori H, Nakatsuji N, Ikai I. In vitro differentiation and maturation of mouse embryonic stem cells into hepatocytes. *Exp Cell Res*. 2005. 309. 68-77.

[53] Teratani T, Yamamoto H, Aoyagi K, Sasaki H, Asari A, Quinn G, Sasaki H, Terada M, Ochiya T. Direct hepatic fate specification from mouse embryonic stem cells. *Hepatology*. 2005. 41. 836-46.

[54] Soto-Gutierrez A, Kobayashi N, Rivas-Carrillo JD, Navarro-Alvarez N, Zhao D, Okitsu T, Noguchi H, Basma H, Tabata Y, Chen Y, Tanaka K, Narushima M, Miki A, Ueda T, Jun HS, Yoon JW, Lebkowski J, Tanaka N, Fox IJ. Reversal of mouse hepatic failure using an implanted liver-assist device containing ES cell-derived hepatocytes. *Nat Biotechnol*. 2006. 24. 1412-9.

[55] Schuldiner M, Yanuka O, Itskovitz-Eldor J, Melton DA, Benvenisty N. Effects of eight growth factors on the differentiation of cells derived from human embryonic stem cells.*Proc Natl Acad Sci U S A*. 2000. 97. 11307-12.

[56] Lumelsky N, Blondel O, Laeng P, Velasco I, Ravin R, McKay R. Differentiation of embryonic stem cells to insulin-secreting structures similar to pancreatic islets. *Science*. 2001. 292. 1389-94.

[57] Hamazaki T, Iiboshi Y, Oka M, Papst PJ, Meacham AM, Zon LI, Terada N. Hepatic maturation in differentiating embryonic stem cells in vitro. *FEBS Lett*. 2001. 497. 15-9.

[58] Hori Y, Rulifson IC, Tsai BC, Heit JJ, Cahoy JD, Kim SK. Growth inhibitors promote differentiation of insulin-producing tissue from embryonic stem cells. *Proc Natl Acad Sci U S A*. 2002. 99. 16105-10.

[59] Kim D, Gu Y, Ishii M, Fujimiya M, Qi M, Nakamura N, Yoshikawa T, Sumi S, Inoue K. In vivo functioning and transplantable mature pancreatic islet-like cell clusters differentiated from embryonic stem cell. *Pancreas*. 2003. 27. e34-41.

[60] Ng ES, Azzola L, Sourris K, Robb L, Stanley EG, Elefanty AG.The primitive streak gene Mixl1 is required for efficient haematopoiesis and BMP4-induced ventral mesoderm patterning in differentiating ES cells. *Development*. 2005. 132. 873-84.

[61] Micallef SJ, Janes ME, Knezevic K, Davis RP, Elefanty AG, Stanley EG.Retinoic acid induces Pdx1-positive endoderm in differentiating mouse embryonic stem cells. *Diabetes*. 2005. 54. 301-5.

[62] Gouon-Evans V, Boussemart L, Gadue P, Nierhoff D, Koehler CI, Kubo A, Shafritz DA, Keller G. BMP-4 is required for hepatic specification of mouse embryonic stem cell-derived definitive endoderm. *Nat Biotechnol.* 2006. 24. 1402-11.

[63] Matus DQ, Pang K, Marlow H, Dunn CW, Thomsen GH, Martindale MQ.Molecular evidence for deep evolutionary roots of bilaterality in animal development. *Proc Natl Acad Sci U S A.* 2006. 103. 11195-200.

[64] Jensen J. Gene regulatory factors in pancreatic development. *Dev Dyn.* 2004. 229. 176-200.

[65] Kemp DM, Lin JC, Habener JF. Regulation of Pax4 paired homeodomain gene by neuron-restrictive silencer factor. *J Biol Chem.* 2003. 278. 35057-62.

[66] Hori Y, Gu X, Xie X, Kim SK. Differentiation of insulin-producing cells from human neural progenitor cells. *PLoS Med.* 2005. Apr;2. e103.

[67] Narushima M, Kobayashi N, Okitsu T, Tanaka Y, Li SA, Chen Y, Miki A, Tanaka K, Nakaji S, Takei K, Gutierrez AS, Rivas-Carrillo JD, Navarro-Alvarez N, Jun HS, Westerman KA, Noguchi H, Lakey JR, Leboulch P, Tanaka N, Yoon JW. A human beta-cell line for transplantation therapy to control type 1 diabetes. *Nat Biotechnol.* 2005. 23. 1274-82.

[68] Suzuki N, Nishii K, Okazaki T, Ikeno M. Human artificial chromosomes constructed using the bottom-up strategy are stably maintained in mitosis and efficiently transmissible to progeny mice.*J Biol Chem.* 2006. 281. 26615-23.

[69] Ren X, Tahimic CG, Katoh M, Kurimasa A, Inoue T, Oshimura M. Human artificial chromosome vectors meet stem cells: new prospects for gene delivery. *Stem Cell Rev.* 2006. 2. 43-50.

[70] Noguchi H, Matsushita M, Matsumoto S, Lu YF, Matsui H, Bonner-Weir S.noguchi Mechanism of PDX-1 protein transduction. *Biochem Biophys Res Commun.* 2005. 332. 68-74.

[71] Schorderet DF, Manzi V, Canola K, Bonny C, Arsenijevic Y, Munier FL, Maurer F. D-TAT transporter as an ocular peptide delivery system. *Clin Experiment Ophthalmol.* 2005. 33. 628-35.

[72] Nakano T, Kodama H, Honjo T. Generation of lymphohematopoietic cells from embryonic stem cells in culture. *Science.* 1994. 265. 1098-101.

[73] Yamashita J, Itoh H, Hirashima M, Ogawa M, Nishikawa S, Yurugi T, Naito M, Nakao K, Nishikawa S. Flk1-positive cells derived from embryonic stem cells serve as vascular progenitors. *Nature.* 2000. 408. 92-6.

[74] Kawasaki H, Suemori H, Mizuseki K, Watanabe K, Urano F, Ichinose H, Haruta M, Takahashi M, Yoshikawa K, Nishikawa S, Nakatsuji N, Sasai Y. Generation of dopaminergic neurons and pigmented epithelia from primate ES cells by stromal cell-derived inducing activity. *Proc Natl Acad Sci U S A.* 2002. 99. 1580-5.

[75] Rossi JM, Dunn NR, Hogan BL, Zaret KS. Distinct mesodermal signals, including BMPs from the septum transversum mesenchyme, are required in combination for hepatogenesis from the endoderm. *Genes Dev.* 2001. 15. 1998-2009.

[76] Pouton CW, Haynes JM. Embryonic stem cells as a source of models for drug discovery. *Nat Rev Drug Discov.* 2007 6. 605-16.

In: Stem Cell Applications in Disease and Health
Editors: W. Burnsides and R. Ellsley
ISBN: 978-1-60456-255-2

Chapter XI

Hematopoietic Stem/Progenitor Cell Migration in Health and Disease

Susannah H. Kassmer, Jeanette Seidel, Bernd Niggemann, Kurt S. Zänker and Thomas Dittmar *
Institute of Immunology, Witten/ Herdecke University, Stockumer Str. 10
58448 Witten, Germany

Abstract

Hematopoietic stem cells (HSC) are a rare population of self renewing cells within the bone marrow, which constantly reproduce all hematopoietic lineages. The homing and migration capacities of HSC are of fundamental importance for successful transplantation during treatment of hematological diseases. Transplantation of HSC collected from peripheral blood has become more and more significant, requiring directed homing to the bone marrow after intravenous infusion. The multi-step process of HSC homing, involving adhesion, transendothelial migration and migration across the bone marrow stroma, depends on active migration along a gradient of the chemokine SDF-1α. The potential of HSC to undergo self-renewing cell divisions has stimulated investigations of *ex vivo* stem cell expansion, which would be valuable to improve the quality of stem cell transplants by increasing the number of stem cells. Hematopoietic stem and progenitor cells (HSPC) can be expanded *in vitro* by supplementation with hematopoietic cytokines, for instance, Flt3-Ligand and stem cell factor. However, analyses of migratory activity using a collagen matrix migration assay recently revealed a dramatic impact of various hematopoietic cytokines on the SDF-1α-directed migration of these *ex vivo* expanded HSPC.

Although most hematopoietic progenitors reside in the bone marrow, a small fraction of these motile cells constantly circulate via the blood vessel system during steady state homeostasis. In stress situations, like bleeding, organ injury and inflammation, the number of circulating HSPC increases dramatically. During inflammation and tissue damage, SDF-1α as well as other chemokines are released, attracting migrating HSPC to

* E-mail address: thomasd@uni-wh.de,; Phone: +49 2302 926 165; FAX: +49 2302 926 158

sites of injury, which thereupon participate in tissue repair. Due to their plasticity and pluripotency as well as their capacity to fuse with other cells, HSPC develop into hepatocytes, muscle and neuronal cell types. Thus, the question arises regarding what the outcome of these events of trans-differentiation and fusion will be, if a hematopoietic stem cell enters chronically inflamed or even tumor tissue. This directed migration and homing of HSPC to their target tissues underlies the influence of start and stop signals, exerted, apart from CXCR4/SDF-1α interaction, by various factors. Recent observations indicate a modulating effect of the chemokine IL-8 and even neurotransmitters on the migratory activity of hematopoietic stem cells. Further elucidation of these mechanisms may facilitate the selective manipulation of HSPC homing to their target tissues.

Thus, it appears that understanding the mechanisms influencing and regulating the migratory processes of HSPC, comprising homing, retention, release and mobilization, may allow the improvement of their clinical applications.

Introduction

Hematopoietic stem cells (HSC) are a rare population of immature cells within the bone marrow, possessing the ability to self-renew and continuously give rise to all hematopoietic lineages [1]. The application of HSC for the treatment of hematologic malignancies and myeloproliferative disorders requires their inherent migratory capacity, which is fundamental during homing and mobilization. In the 1950s, Dr. E. Donnall Thomas discovered that bone marrow cells infused intravenously repopulated the bone marrow and produced new blood cells. Besides transplantation of HSC from bone marrow, transplantation of HSC collected from peripheral blood after mobilization with granulocyte-colony stimulating factor (G-CSF) or granulocyte macrophage-colony stimulating factor (GM-CSF) has become more and more significant since the early 1990s. After intravenous infusion, successful engraftment depends on directed homing of the transplanted HSC to the bone marrow niche. During the homing process, HSC have to actively migrate along a chemotactic gradient secreted by the bone marrow stroma and endothelium. The engrafted HSC repopulate the bone marrow, leading to generation of maturing blood cells of all lineages as well as maintenance of a small stem cell pool within the host bone marrow. Under healthy conditions, very low levels of these motile cells constantly migrate throughout the bloodstream. The migratory capacity of HSC becomes significant during inflammation and tissue injury, when migrating HSC are attracted to injured or inflamed tissues, and participate in their repair. Under several circumstances, trans-differentiation or fusion of HSC with other cells was observed in injured or inflamed tissues. Within this chapter, we will describe the conditions and factors influencing HSC migration, and briefly discuss the question as to the outcome of trans-differentiation or fusion events when a hematopoietic stem cell is recruited to malignant tissue.

HSC Migration during Mobilization and Homing to Bone Marrow

Under steady state conditions, hematopoietic cells at various stages of differentiation reside within specialized bone marrow niches, whereas terminally differentiated blood cells migrate into the blood. However, a very small proportion of primitive HSC constantly leaves the bone marrow and circulates through the bloodstream. HSC numbers in the peripheral blood are extremely low, ranging from 0.05-0.1% within the peripheral blood mononuclear cell fraction [2]. The circulating pool of HSC can be significantly increased by their enforced migration from the bone marrow via administration of cytokines or chemokines. Mostly, mobilization of HSC from the bone marrow into the bloodstream is induced by frequent injection of G-CSF, and involves HSC migration. Following G-CSF mobilization, activated neutrophils release a set of proteases from their granules, which then cleave multiple target molecules on hematopoietic and stromal cells as well as their matrix [3]. These include adhesion receptors, chemokines, and their receptors. Serine proteases, especially neutrophil elastase, accumulate within the bone marrow, cleaving substrates comprising VCAM-1, vascular endothelial cadherin, CXCR4 and SDF-1α [4, 5]. G-CSF mediated mobilization likely involves CD26, a membrane bound extracellular peptidase expressed by hematopoietic stem and progenitor cells, which cleaves some of the same target molecules as serine proteases, including SDF-1α, yielding its truncated from. This results in a reduced migratory response to normal SDF-1α, thereby reducing retention of HSC within the bone marrow and facilitating mobilization into the periphery [6]. Inhibition of CD26 on HSC increases their SDF-1α directed homing and chemotactic migration [7]. Another substrate of proteases is membrane bound urokinase receptor (uPAR), which plays an important role in short term homing and engraftment, and constitutes a cell-autonomous retention signal for HSC. In response to G-CSF stimulation, uPAR is proteolytically cleaved by the protease plasmin, leading to formation of soluble uPAR. Soluble uPAR further amplifies mobilization and also promotes HSC migration [8]. In general, disruption of the interaction of SDF-1α with its receptor CXCR4 is one of the most important mechanisms causing HSC mobilization [9], whereas its activation is essential for homing, as will be described below. For instance, the CXCR4 antagonist AMD3100 induced rapid mobilization of HSC, alone or in addition to G-CSF [10, 11]. Mobilized cells exhibit decreased expression of very late antigen-4 (VLA-4). More important, however, is the reduced functional state of VLA receptors. As migratory responses require the participation of integrins, functional down-regulation of integrins is an important step during the final stages of transmigration through the endothelial sinuses [9].

Homing is defined as the active migration of circulating HSC across the blood/bone endothelial barrier into the bone marrow compartment within a fairly short time span. The multi-step process of HSC homing involving recruitment of circulating HSC by the bone marrow microvasculature, adhesion, transendothelial migration and migration across the bone marrow stroma, depends on active migration along a gradient of the chemokine SDF-1α [12-15]. SDF-1α plays a major role in migration, proliferation, differentiation and survival of various cell types including hematopoietic stem and progenitor cells (HSPC). The CXC chemokine SDF-1α is constitutively expressed in many organs, including bone marrow endothelium and the stem cell rich endosteum region, as well as multiple stromal cell types. SDF forms a decreasing gradient from the marrow extravascular compartment toward the lumen of infiltrating vessels [12, 16, 17], thus promoting homing of transplanted HSC to bone marrow. The G-protein-coupled chemokine receptor CXCR4, which is expressed by HSC, binds SDF-1α and thus mediates chemotaxis and Integrin-mediated firm adhesion and transendothelial migration [15]. Integrin activation is mediated by chemokine and cytokine receptors via 'inside out' signaling [18]. Cytokines and chemokines, like, for instance, GM-CSF, IL-3, SCF and SDF-1α, activate integrin function, thereby increasing adhesiveness of human CD34$^+$ bone marrow cells to fibronectin and collagen [19, 20]. Thus, SDF-1α stimulates firm adhesion of HSC to endothelium and transendothelial migration, which depends on lymphocyte function associated antigen-1 (LFA-1) / ICAM-1 (intracellular adhesion molecule-1) and VLA-4 / VCAM-1 (vascular adhesion molecule-1) interactions [9]. For instance, SDF-1α induced polarization and extravasation of HSC has been shown to depend on VLA-4 and VLA-5 [13, 14, 21]. P-selectin glycoprotein ligand, the sialomucin receptor for P-selectin, likewise seems to play a key role in homing of HSC [22, 23]. Polarization is also induced when SDF stimulated cells bind to immobilized hyaloronic acid (HA), and the HA-receptor CD44 is involved in adhesion and immobilization of cells [24]. Transendothelial migration requires the degradation of the basement membrane and thus the production of matrix degrading enzymes, like matrix-metalloproteinases (MMP) [25]. Their expression and secretion by cells crossing the endothelial layer are stimulated by various cytokines and chemokines. Additionally, MMP-9 expressed by bone marrow stromal cells cleaves membrane bound SCF, which then enhances Integrin-mediated binding of HSC to extracellular matrix components. Exposure of HSC to SCF upregulates the levels of the homing related molecules CD49e, CD49f, CD54, CXCR4, MMP-4 and MMP-2, leading to increased transmigration and homing potential [26]. After firm adhesion to the bone marrow endostelium, HSC transmigrate within the marrow to specific sites know as stem cell niches, where their lodgment, proliferation and self-renewal is regulated by a complex interplay of cells, growth factors and adhesion molecules. HSC predominantly transmigrate to the endosteal region, whereas lineage committed or differentiated cells move away from the endosteal region and into the central marrow region [27, 28]. Our current understanding about the molecules involved in this process is still fragmentary, however, hyaluronic acid, the transmembrane isoform of SCF and osteopontin seem to play a role in transendothelial migration and lodgment of transplanted HSC within their niche [29]. On human CD34$^+$ HSPC, RHAMM, a receptor for hyaluronic acid, leads to an increase in cell motility, whereby CD44 leads to an increase in adhesion.

HSC Migration and CXCR4 Signal Transduction

The SDF-1α receptor CXCR4 belongs to the group of 7-transmembrane helix receptors, as SDF-1α signaling via CXCR4 involves a pertussis toxin sensitive $G_{\alpha i}$ protein [30]. Upon SDF binding, PI3-kinase (PI3K), phospholipase C-γ (PLC-γ), protein kinase C (PKC) and MAPK$^{p42/44}$ are activated, finally leading to tyrosine phosphorylation of focal adhesion components such as paxillin, related adhesion focal tyrosine kinase (RAFTK/ pyk2), p130^{CAS}, Crk-II, and Crk-L [31, 32]. Furthermore, cell polarization, adhesion to bone marrow stromal cells, chemotaxis and homing of human CD34$^+$ HSPC were shown to depend upon PKC-ζ, representing an atypical PKC subtype independent of calcium or diacylglycerole [33].

SDF also induces transient activation of molecules implicated in cell survival, like ribosomal S6 kinase and Akt [34], and nuclear factor-κB (NF-κB) [31, 35]. SDF binding to CXCR4 triggers receptor dimerization and activates the JAK/STAT pathway, and thus seems to influence gene regulation [35]. Association of CXCR4 with lipid rafts and colocalization with Rac-1 primes the SDF-directed homing response of HSC and facilitates activation of Rac-1 [36]. Rac-1 expression is crucial during the engraftment phase of HSC homing, while lack of Rac-1 and Rac-2 expression leads to massive mobilization of HSC and impaired hematopoiesis. Thus, Rac proteins seem to regulate homing as well as mobilization processes [37].

Activation of PLC-γ by SDF binding to CXCR4 leads to release of calcium from IP_3 sensitive endoplasmatic reticulum stores. Intracellular calcium ions regulate the structure and function of cytoskeletal proteins like actin. Thus, CXCR4 signaling finally culminates in the highly synchronized processes of actin polymerization, and cytoskeleton reorganization.

Migration Capacity and Engraftment of *Ex Vivo* Expanded HSC

The ability of HSC to self-renew allows them to undergo symmetric cell divisions, yielding two daughter cells retaining the characteristics of the originating cell. This has stimulated investigations of *ex vivo* stem cell expansion, which would be valuable to increase the number of stem cells and thus improve the quality of stem cell transplants. For instance, umbilical cord blood transplants, which are suitable for transplantations of small children, contain only a very limited number of stem cells, and thus, *ex vivo* expansion may allow cord blood transplants to be used in adults [38]. To investigate *ex vivo* expansion and engraftment of HSC, murine models are frequently used. HSPC can be isolated from murine bone marrow and expanded *in vitro* by supplementation with a variety of hematopoietic cytokines. Recent studies have identified thrombopoietin (TPO), Flt3-ligand (FL), stem cell factor (SCF) and interleukin (IL)-11 as cytokines able to stimulate amplification of primitive murine hematopoietic stem cells *in vitro* [39-44]. Human HSC and HSPC have been successfully expanded *in vitro* by supplementation with combinations of FL, SCF, IL-1β, IL-6, IL-3, G-CSF, GM-CSF and SCF [45, 46]. Megakaryocyte growth and differentiation factor (MGDF) has also been used [38].

HSC expansion *in vitro* needs to induce self renewing cell divisions, however, proliferation itself may silence the ability of HSC to home [47]. As stated above, the homing and migration capacities of HSC cells are of fundamental importance for successful transplantation during treatment of hematological diseases. A number of recent observations suggest a specific influence of cytokines and cytokine combinations on the engraftment potential as well as on the homing ability of HSPC. Initially, Cerny et al. reported a cytokine induced defect in homing to the bone marrow and engraftment potential of murine Lin^- $Sca\text{-}1^+$ HSPC after cultivation in IL-3, IL-6, IL-11 and SCF for 48 hours [48]. Likewise, Ahmed et al. showed an irreversibly impaired homing of human $CD34^+$ HSPC to the bone marrow following 48 hour exposure to FL, SCF, IL-6 and IL-3 [49]. Von Drygalski and colleagues performed transplantation studies using murine bone marrow cells cultivated with three different cytokine combinations (FL, SCF, TPO; FL, SCF, TPO + IL-11 and FL, SCF, TPO, IL-11 + IL-3), and observed decreased 7 months survival rates and poorer donor engraftment of cultured compared to freshly isolated cells [50]. In line with these observations, one recent study analyzing the migratory activity of *ex vivo* expanded HSPC revealed a dramatic impact of various hematopoietic cytokines on SDF-1α-directed migration [51]. Murine $Lin^{-}c\text{-}kit^{+}$ (HSPC) were cultivated in the presence of twelve combinations of the hematopoietic cytokines TPO, FL, SCF and IL-11. As SDF-1α is the most potent migratory stimulant for HSC [52], spontaneous as well as SDF-1α stimulated locomotory activity were measured in a 3D collagen matrix using time lapse video microscopy. Each cytokine combination had a distinct effect on the expansion as well as the SDF-1α-stimulated migratory activity of murine Lin^{-} $c\text{-}kit^{+}$ HSPC. For instance, FL and SCF cultivated cells showed a high SDF-response, but only a moderate fold expansion of long term repopulating HSC. By contrast, cultivation with SCF, TPO and IL-11 resulted in a relatively high fold expansion of HSC, but absence of an elevated migratory activity in response to SDF stimulation. The differences in the migratory SDF-response were not related to different CXCR4 expression levels, but rather to the differential engagement of the CXCR4 dependent $MAPK^{p42/44}$ and PI3K signal transduction pathways, demonstrating a strong influence of different cytokines and cytokine combinations on signal processing downstream of CXCR4. In cells exhibiting an intermediate to low increase of migratory activity in response to SDF-1α, PI3K and $MAPK^{p42/44}$ pathways are not involved in migration in the presence of SDF-1α, while in cells exhibiting a high response to SDF-stimulation, PI3K and $MAPK^{p42/44}$ are crucial in SDF-1α directed migration [51]. This is in accordance with a study by Fukuda and colleagues, who previously demonstrated differential phosphorylation of key signal transduction proteins induced by FL treatment in human $CD34^{+}$ HSPC. Flt3-ligand and SDF-1α were observed to synergistically enhance the phosphorylation of $MAPK^{p42/44}$, PI3K and Akt, as well as migratory activity. In cells pretreated with FL for two days, weaker and delayed phosphorylation of $MAPK^{p42/44}$ was observed after SDF-1α stimulation [53]. In another study on *ex vivo* expanded human $CD133^{+}$ HSPC, analyses of the migratory activity revealed an enhancing effect of IL-6 on the SDF-1α-stimulated migratory activity of these cells [54]. This was due to recruitment of nonmoving cells, and an increase in the speed of actively migrating cells. One of the possible mechanisms underlying the altered migratory and homing capacity of *ex vivo* cultivated HSPC is a dramatically altered adhesion receptor expression and reduced functional binding to the extracellular matrix [55], or delayed deactivation of integrins and sustained attachment. This may result in either impaired engraftment within the bone marrow niche or decreased migratory activity, respectively. The involved mechanisms were further elucidated by Glimm et al. who observed a cell cycle dependent change in *in vivo* stem cell transplantability of $CD34^{+}$ human cord blood HSPC after 5 days of *ex vivo* expansion. [45]. After cultivation in FL, SCF, IL-3, IL-6 and G-CSF, cells in the G_1 phase of the cell cycle contained at least 85% of the transplantable stem cell activity, suggesting a cell cycle associated change in *in vivo* stem cell homing. A recent study by Yang et al. provides insight into the cell cycle associated processes resulting in homing and engraftment deficiencies of HSC [56]. In Cdc42 deficient mice, HSC within the bone marrow enter the active cell cycle, resulting in impaired adhesion, homing, lodging and retention of HSC within the bone marrow, which also causes a massive egress of HSC to distal organs and peripheral blood and an engraftment failure. These effects are associated with deregulated Integrin-β1 and N-cadherin expression and defective actin organization.

In summary, cytokines commonly used in *ex vivo* expansion protocols seem to influence homing and migration of hematopoietic cells in many different ways, depending on combination with other cytokines, cell type, incubation time, etc. Furthermore, hematopoietic cytokines strongly affect the intracellular signaling machinery of *ex vivo* expanded HSPC and their aptitude to react to SDF-1α stimulation with increased locomotory activity. As therapeutic success is fundamentally dependent on the migration capacity and homing of transplanted HSC to bone marrow, analysis of SDF-1α-stimulated migratory activity and migration parameters may prove valuable to investigate the effects of *ex vivo* expansion supplements, as these may enhance or even compromise the efficacy of therapy involving HSPC migration.

Migration of HSPC in Steady State Homeostasis and Tissue Repair

Although most hematopoietic progenitors reside in the bone marrow, a small fraction of these motile cells constantly circulates via the blood system during steady state homeostasis. Continuous remodeling within the bone marrow results in their constant and random release into the circulation. These constitutively migrating cells play a physiological role in the functional reengraftment of unconditioned bone marrow by filling open HSC niches [57]. The fact that HSC have been isolated from many other tissues besides bone marrow would indicate that circulating cells are in dynamic equilibrium with these other tissues [9]. For instance, HSPC present in substantial numbers in muscle tissue are of bone marrow origin [58, 59]. Krause et al. showed that after transplantation of single cells into irradiated mice, long term repopulating HSC initiated differentiation of bone marrow derived cells into epithelial cells of the liver, lung, gastrointestinal tract and skin in the absence of tissue injury [60]. An important player in this scenario is SDF-1α, which is a rather unique cytokine, as it is the product of resting cells in multiple organs, like bone, lymph node, muscle, lung, skin and liver [61-64]. Thus, transplanted HSC have been found to home, besides bone marrow, to spleen, liver and lung [65], which likely applies also to circulating HSC during homeostasis. For instance, human liver bile duct epithelium expresses SDF-1α under steady state conditions, attracting CD34$^+$ hematopoietic progenitors [66]. SDF-1α expressed in human heart and skeletal muscle constantly attracts CXCR4 positive CD34$^+$ HSPC to niches within the muscle tissue [61, 67].

In stress situations, like bleeding, organ injury and inflammation, the number of circulating HSPC increases dramatically. In response to tissue injury, multiple chemical signals in concert with cytokines, growth factors and chemokines like SDF-1α initiate the host response, involving activation and directed migration of leukocytes from the venous system to sites of tissue damage. Activated endothelial cells facilitate rolling of leukocytes on activated members of the selectin family of adhesion molecules (L-P-E-selectin). Cytokine-mediated activation of leukocyte-integrins leads to immobilization and firm adhesion by means of $\alpha_4\beta_1$- and $\alpha_4\beta_7$-integrins binding to VCAM-1, followed by transmigration through the endothelium to sites of injury, facilitated by matrix metalloproteinases.

During inflammation and tissue damage, SDF-1α as well as other chemokines are released from cells within the lesioned areas, attracting migrating HSPC to sites of injury, which thereupon participate in tissue repair. For HSC, it has been observed that cytokine exposure, which is likely to occur during an inflammatory process, seems to be able to change the homing behavior of circulating HSC, and to increase homing to tissues other than bone marrow, as murine HSPC exposed to a combination of cytokines showed an increased homing to lung tissue and decreased homing to marrow [65]. The complement anaphylatoxin C3a, which is generated during inflammatory responses, sensitizes the response of human $CD34^+$ HSPC to SDF-1α by potentiating SDF-1α dependent chemotaxis and stimulates MMP-9 secretion and VLA-4 mediated adhesion to VCAM-1 [68].

Various animal models for tissue damage have been used to detect migration and homing of bone marrow derived hematopoietic stem cells to sites of tissue injury. For instance, systemically injected HSC are attracted to the subretinal space in a murine model of retinal pigment epithelium damage by SDF-1α, C3a, hepatic growth factor (HGF), and leukemia inhibitory factor (LIF) which are released from the injured tissue. HSC were found to express the tissue specific marker RPE-65 after homing to the subretinal space [69, 70]. In the liver, irradiation or inflammation increases SDF-1α and HGF expression, promoting recruitment and directional migration of $CD34^+$ hematopoietic progenitors, which differentiate into albumin producing cells [66]. In a murine model for hepatic tissue damage, only transplantation of purified HSC gave rise to donor derived hepatic regeneration [71]. Another important mediator of stem cell homing is hyaluronic acid, which is increasingly expressed in injured tissue, as was recently demonstrated by Herrera et al. in a murine model of tubular injury in the kidney [72]. Expression of hyaluronic acid was crucial in the homing of transplanted stem cells, which subsequently increased renal healing. This finding is interesting in the light of a previous study, in which homing of human $CD34^+$ HSPC to bone marrow was found to depend on binding of hyaluronic acid to its receptor CD44 on the cell surface, whereby a crosstalk between the SDF-receptor CXCR4 and CD44 seems to be present [24]. Another stress situation leading to increased expression of chemokines is ischemic injury. In ischemic tissues, hypoxic stimuli lead to hypoxia inducible factor-1α (HIF) dependent expression of SDF-1α by endothelial cells, mediating ischemia specific recruitment of circulating progenitor cells [73, 74]. After ischemic injury, SDF-1α expression increases in the kidney and decreases in the bone marrow, causing mobilization of $CD34^+$ HSPC and their subsequent homing to damaged kidney tissue [75]. Bone marrow derived stem cells (BMDC) comprise HSC and mesenchymal stem cells, which both exhibit high plasticity and a high migratory potential. After ischemia induced by coronary heart occlusion, highly purified BMDC transplanted into mice differentiate into cardiomyocytes and contribute to the formation of functional tissue [76]. Myocardial infarction leads to increased expression of SDF-1α, vascular endothelial growth factor (VEGF), MMP-9, ICAM-1 and VCAM-1 [77], all of which may be involved in promoting recruitment of bone marrow derived cells including HSC to the injured heart. In the brains of mice recieving focal cerebral ischemia, SDF-1α expression by endothelial cells within the lesioned areas is profoundly increased and leads to the attraction of CXCR4 expressing leukocytes [78], including HSC. Mobilization of BMDC in mice by treatment with G-CSF and SCF significantly increased the number of bone marrow derived cells expressing neuronal markers within several brain areas, which may become a new strategy for neuronal repair in neurodegenerative diseases [79]. In mice within the subacute phase of cerebral infarction, the same treatment induced attraction of BMDC to the infracted area, associated with improved motor performance and brain function [80]. Thus, it appears that stimulated migration and homing of hematopoietic and other bone marrow derived stem cells may be a valuable process during healing and treatment of tissue injury.

In all the situations described so far, HSC entering the damaged or inflamed tissue seem to contribute to tissue repair, either by trans-differentiation, or via indirect mechanisms, like secretion of cytokines and attraction of other cells. Trans-differentiation may be due either to the local tissue milieu, which serves as a blueprint for the differentiation of the invading BMDC [81], to cell-fusion events [82], or to a combination of both. HSC show a high plasticity, as they differentiate into tissue cells even when derived from other embryonic origin, like ectoderm and endoderm, for instance, hepatocytes and neuronal cell types. For instance, bone marrow derived cells were demonstrated to fuse *in vivo* with hepatocytes, purkinje neurons, and cardiomyocytes [83]. However, in these cells, polyploidy and multiple nuclei are well tolerated, as hepatocytes and purkinje neurons tend to be polyploid, and muscle fibers form multinucleated syncytia. Interestingly, a recent study has shown that transplanted BMDC fuse with normal intestinal stem cells in gamma-irradiated mice. Additionally, they report fusion of BMDC with transformed intestinal stem cells, providing *in vivo* evidence that strongly supports the hypothesis that fusion between circulating hematopoietic cells and transformed intestinal epithelium is involved in intestinal tumorigenesis [84].

The concept that regeneration of liver after bone-marrow transplantation in mice occurs through cell fusion of hepatocytes with HSC progeny has received substantial support by two studies, in which male donor cells derived from bone marrow were injected into female recipients, and fusion of HSPC with liver cells was proven by sex-mismatch of repopulated liver cells [85, 86]. Other groups favor the theory that HSPC trans-differentiate into liver cells. Jang et al. could show via RT-PCR that HSPC exposed to supernatant of damaged or injured liver tissue, up-regulated liver marker expression [87]. Another study by Oh et al. also provides evidence that BMDC plasticity involves trans-differentiation in the absence of fusion. BMDC were induced *in vitro* to trans-differentiate into insulin producing cells, which showed an architecture and protein expression pattern characteristic of the pancreatic lineage after transplantation into mice [88].

Recruitment of HSC to Malignant Tissue – Contribution to Carcinogenesis?

Thus, the question arises what will be the outcome of these events of trans-differentiation and fusion, if a HSC enters chronically inflamed or even tumor tissue. Tumor cells produce various cytokines and chemokines, which act in concert to attract leukocytes comprising neutrophils, dendritic cells, macrophages, eosinophils and mast cells, as well as lymphocytes, which in turn secrete an array of cytokines, inflammatory mediators and chemokines. The enhanced proliferation within a tumor tissue results in hypoxic areas and the release of reactive oxygen species, promoting the loss of self-limitation of the mechanisms involved in normal inflammation. The inflammatory reaction is mediated by tumor associated macrophages, which are attracted to tumor tissue via hypoxia, and in turn secrete proinflammatory cytokines, such as TNF-α and chemokines such as monocyte chemoattractant protein 1, -2, -3 (MCP-1, -2, -3) [89, 90]. But not only cells of the immune system have been shown to enter sites of inflammation. More recently, also hematopoietic stem cells were found at inflammatory sites, in all likelihood recruited by a hypoxia induced SDF-1α-release of the inflamed or tumor tissue [73]. TGF-β and SDF-1α dependent migration of HSPC and homing towards gliomas *in vivo* are induced by hypoxia and irradiation, and, like in ischemic tissue, depend on HIF-1α [91]. However, the fate of HSPC recruited to neoplastic tissue still remains unclear.

Bone marrow-derived stem cells, including HSC, which are recruited to tumor tissue, encounter this malignant environment as instructing milieu. Thus, stem cells may adopt cancer cells characteristics by trans-differentiation, as the blueprint for this process is given by the local tissue environment, which is defined by a specific pattern of cytokines, chemokines and adhesion molecules, etc. [81]. The second possible mechanism leading to (trans-)differentiation is fusion of a BMDC with a tissue or even a tumor cell. As stated above, HSC possess the ability to fuse with tissue cells during processes of tissue repair. Fusion of a BMDC with a tumor cell might increase tumor cell diversity due to chromosomal aberrations and unbalanced chromosome distribution during mitosis [86, 92-94]. Cancer stem cells are certain cells within a tumor tissue sharing similarities to stem cells such as immortality, drug resistance, and responsiveness to chemokines, an undifferentiated phenotype and migratory activity leading to metastasis. It would be conceivable that cancer stem cells result from fusion events of BMDC or tissue stem cells and transformed tissue cells [95-98].

Several studies indicate that a chronically inflamed microenvironment may lead to a faulty set of instructions received by a tissue stem cell, which may then adopt a neoplastic phenotype [81]. For instance, Houghton et al. showed that chronic infection of mice with *helicobacter felis* induced the attraction of BMDC to the stomach, which progress to form intraepithelial cancer [99]. The chronically inflamed tissue within a tumor is characterized by the presence of chemokines, cytokines, reactive oxygen and nitrogen species and proteases, which have an impact on cell-cell communication via gap-junctions and adhesion molecules. Furthermore, reactive oxygen and nitrogen species can cause DNA damage, leading to multiple genetic aberrations. Thus, HSC and other BMDC entering a chronically inflamed tissue may encounter an altered set of instructions, which then results in a faulty (trans-) differentiation process.

Identification of Factors Terminating HSPC Migration

The directed migration and homing of HSPC to their target tissues, including inflamed and malignant tissue, underlies the influence of start and stop signals, exerted, apart from CXCR4/SDF-1α interaction, by various factors. In inflammatory responses, the chemokine

IL-8 is released by macrophages encountering an antigen, and is therefore present at sites of inflammation. Recent observations indicate a modulating effect of the chemokine IL-8 and even neurotransmitters on the migratory activity of human mobilized CD133^{+} HSPC (Figure 2A). Human HSPC express the receptors for IL-8, CXCR1 and CXCR2. Co-stimulation of HSPC with IL-8 revealed an inhibitory effect of IL-8 on their SDF-1α induced migratory activity. Thus, IL-8 might be one of the factors responsible for the retention of HSC at inflammatory sites within a tissue or a tumor. For instance, during development of melanoma, melanoma cells have been found to express IL-8 in response to TGF-β, TNF-α and IL-1α secretion by activated macrophages [100].

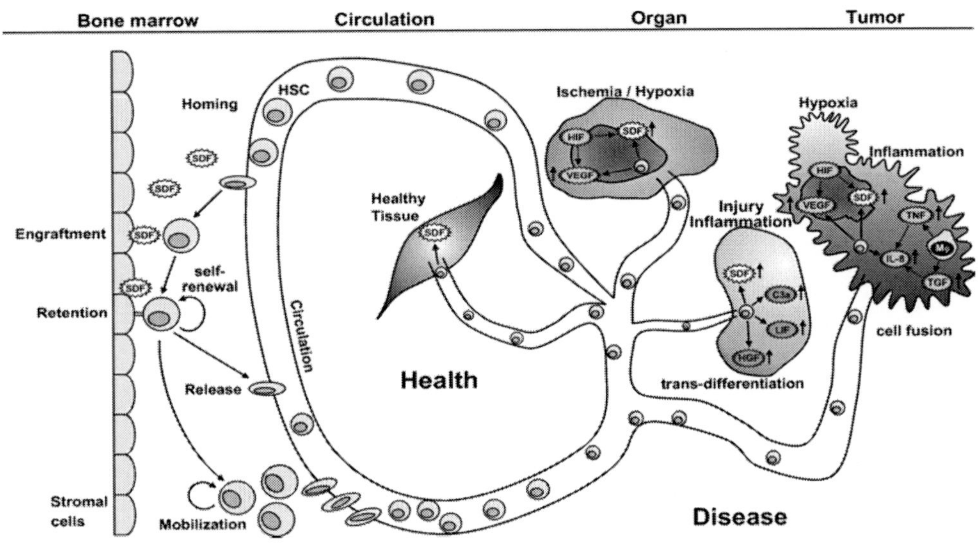

Figure 1. Hematopoietic stem cell migration. Under steady state conditions, HSC are retained within specialized bone marrow niches supported by stromal cells. HSC homing to the bone marrow depends on active migration along a gradient of the chemokine SDF-1α, which is secreted by the bone marrow stroma. A very small proportion of primitive HSC constantly leaves the bone marrow and circulates through the bloodstream. These circulating HSC enter other tissues, forming a dynamic equilibrium. Mobilization results in proliferation of HSC and their enforced migration from the bone marrow. During inflammation and tissue damage, SDF-1α as well as other chemokines are released from cells within the lesioned areas, attracting migrating HSPC to sites of injury. These include HGF, C3a and LIF. In ischemic tissues, including tumor tissue, hypoxic stimuli lead to HIF dependent expression of SDF-1α and VEGF. HSC recruited to injured, inflamed or hypoxic tissue thereupon participate in tissue repair by trans-differentiation. The inflammatory reaction within tumor tissue is mediated by tumor-associated macrophages (Mϕ) secreting TNF-α, IL-8 and TGF-β. Recruitment of HSPC to tumor tissue may result in faulty trans-differentiation, or cell-fusion with transformed cells.

Remarkably, the neurotransmitter γ-aminobutyric acid (GABA), which has formerly been mainly associated with neuronal tissues, was recently shown to exert an inhibitory influence on human HSPC-migration (Figure 2B). Therefore, functional characterization of distinct migration parameters revealed the inhibitory effect of GABA on spontaneous and SDF-1α-stimulated migration of HSPC to be primarily attributed to an increase in the average pause length and a decrease in speed and velocity. GABA-signaling on HSPC was mediated by the

$GABA_B$-receptor. Time dependent calcium measurements revealed a decrease in intracellular calcium level after GABA application. Inhibition of calcium release activated calcium (CRAC) channels with BTP-2 indicates that GABA-mediated reduction of HSPC-migration is exerted by inhibitions of CRAC channels, resulting in a decrease of intracellular calcium, which is crucial for cell migration [101].

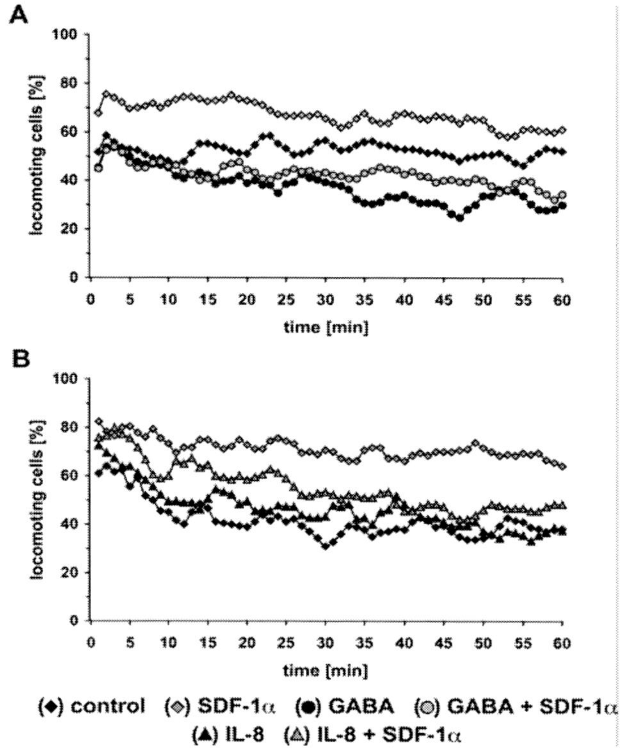

Figure 2. Termination of SDF-1α induced HSPC migration by IL-8 and GABA. Cell migration studies were performed using the 3-dimensional collagen matrix migration assay combined with computer-assisted cell-tracking. Addition of SDF-1α to human $CD133^+$ HSPC from mobilized peripheral blood resulted in an increased locomotory activity (black diamonds) as compared to untreated $CD133^+$ HSPC (grey diamonds). A) The neurotransmitter γ-aminobutyric acid (GABA) efficiently blocks both the spontaneous and SDF-1α induced migration of human $CD133^+$ HSPC. Thereby, GABA signals via the $GABA_B$-receptor, which causes the blockade of calcium release activated calcium (CRAC) channels, thereby inhibiting cell migration. B) The chemokine IL-8 potently inhibits the SDF-1α induced migration of human $CD133^+$ HSPC. Analyses of the appropriate SDF-1α and IL-8 signal transduction cascades indicate that IL-8 likely interrupts the intracellular calcium cycling. Shown are the mean of at least three independent experiments.

Conclusions

As cytokines used for *ex vivo* expansion extensively influence the engraftment and homing properties of HSC, the application of in vitro assays allowing the close analysis of the migratory capacity of *ex vivo* cultivated HSC may prove useful to predict their homing abilities during transplantation procedures. So far, *in vivo* mechanisms of recruitment and mobilization of *in vitro* propagated HSPCs have been extensively investigated, but very little is known about mechanisms terminating HSPC-migration. Further elucidation of these mechanisms may therefore allow the selective manipulation of HSPC homing to their target tissues. This may improve the future use of HSPC in regenerative medicine and may also prove valuable to prevent HSC from entering malignant tissue. Thus it appears that understanding the mechanisms influencing and regulating the migratory processes of HSPC, comprising homing, retention, release and mobilization, may allow the improvement of their clinical applications.

Acknowledgments

This work was supported by the Fritz Bender Foundation, Munich, Germany.

References

[1] Morrison SJ; Uchida N; Weissman IL. The biology of hematopoietic stem cells. *Annu Rev Cell Dev Biol*, 1995 11, 35-71.

[2] de Wynter EA; Buck D; Hart C; Heywood R; Coutinho LH; Clayton A; Rafferty JA; Burt D; Guenechea G; Bueren JA; Gagen D; Fairbairn LJ; Lord BI; Testa NG. CD34+AC133+ cells isolated from cord blood are highly enriched in long-term culture-initiating cells, NOD/SCID-repopulating cells and dendritic cell progenitors. *Stem Cells*, 1998 16, 387-396.

[3] Weidt C; Niggemann B; Kasenda B; Drell TL; Zänker KS; Dittmar T. Stem cell migration: a quintessential stepping stone to successful therapy. *Curr Stem Cell Res Treat*, 2007 2, 89-103.

[4] Levesque JP; Takamatsu Y; Nilsson SK; Haylock DN; Simmons PJ. Vascular cell adhesion molecule-1 (CD106) is cleaved by neutrophil proteases in the bone marrow following hematopoietic progenitor cell mobilization by granulocyte colony-stimulating factor. *Blood*, 2001 98, 1289-1297.

[5] Hermant B; Bibert S; Concord E; Dublet B; Weidenhaupt M; Vernet T; Gulino-Debrac D. Identification of proteases involved in the proteolysis of vascular endothelium cadherin during neutrophil transmigration. *J Biol Chem*, 2003 278, 14002-14012.

[6] Christopherson KW, 2nd; Cooper S; Broxmeyer HE. Cell surface peptidase CD26/DPPIV mediates G-CSF mobilization of mouse progenitor cells. *Blood*, 2003 101, 4680-4686.

[7] Christopherson KW, 2nd; Hangoc G; Mantel CR; Broxmeyer HE. Modulation of hematopoietic stem cell homing and engraftment by CD26. *Science*, 2004 305 1000-1003.

[8] Tjwa M, Sidenius N, Theunissen K, Moura R, Moons L, Janssens S, Blasi F, Collen D, Carmeliet P. Novel role of plasmin in mobilization and of uPAR in retention of hematopoietic stem/progenitor cells in the bone marrow niche: therapeutic implications. *Blood*, 2004 11, 37a.

[9] Papayannopoulou T. Current mechanistic scenarios in hematopoietic stem/progenitor cell mobilization. *Blood*, 2004 103, 1580-1585.

[10] Larochelle A; Krouse A; Metzger M; Orlic D; Donahue RE; Fricker S; Bridger G; Dunbar CE; Hematti P. AMD3100 mobilizes hematopoietic stem cells with long-term repopulating capacity in nonhuman primates. *Blood*, 2006 107, 3772-3778.

[11] Flomenberg N; DiPersio J; Calandra G. Role of CXCR4 chemokine receptor blockade using AMD3100 for mobilization of autologous hematopoietic progenitor cells. *Acta Haematol*, 2005 114, 198-205.

[12] Lapidot T; Kollet O. The essential roles of the chemokine SDF-1 and its receptor CXCR4 in human stem cell homing and repopulation of transplanted immune-deficient NOD/SCID and NOD/SCID/B2m(null) mice. *Leukemia*, 2002 16, 1992-2003.

[13] Peled A; Petit I; Kollet O; Magid M; Ponomaryov T; Byk T; Nagler A; Ben-Hur H; Many A; Shultz L; Lider O; Alon R; Zipori D; Lapidot T. Dependence of human stem cell engraftment and repopulation of NOD/SCID mice on CXCR4. *Science*, 1999 283, 845-848.

[14] Peled A; Kollet O; Ponomaryov T; Petit I; Franitza S; Grabovsky V; Slav MM; Nagler A; Lider O; Alon R; Zipori D; Lapidot T. The chemokine SDF-1 activates the integrins LFA-1, VLA-4, and VLA-5 on immature human CD34(+) cells: role in transendothelial/stromal migration and engraftment of NOD/SCID mice. *Blood*, 2000 95, 3289-3296.

[15] Aiuti A; Webb IJ; Bleul C; Springer T; Gutierrez-Ramos JC. The chemokine SDF-1 is a chemoattractant for human CD34+ hematopoietic progenitor cells and provides a new mechanism to explain the mobilization of CD34+ progenitors to peripheral blood. *J Exp Med*, 1997 185, 111-120.

[16] Ponomaryov T; Peled A; Petit I; Taichman RS; Habler L; Sandbank J; Arenzana-Seisdedos F; Magerus A; Caruz A; Fujii N; Nagler A; Lahav M; Szyper-Kravitz M; Zipori D; Lapidot T. Induction of the chemokine stromal-derived factor-1 following DNA damage improves human stem cell function. *J Clin Invest*, 2000 106, 1331-1339.

[17] Imai K; Kobayashi M; Wang J; Shinobu N; Yoshida H; Hamada J; Shindo M; Higashino F; Tanaka J; Asaka M; Hosokawa M. Selective secretion of chemoattractants for haemopoietic progenitor cells by bone marrow endothelial cells: a possible role in homing of haemopoietic progenitor cells to bone marrow. *Br J Haematol*, 1999 106, 905-911.

[18] Clark EA; Brugge JS. Integrins and signal transduction pathways: the road taken. *Science*, 1995 268, 233-239.

[19] Levesque JP; Leavesley DI; Niutta S; Vadas M; Simmons PJ. Cytokines increase human hemopoietic cell adhesiveness by activation of very late antigen (VLA)-4 and VLA-5 integrins. *J Exp Med*, 1995 181, 1805-1815.

[20] Hidalgo A; Sanz-Rodriguez F; Rodriguez-Fernandez JL; Albella B; Blaya C; Wright N; Cabanas C; Prosper F; Gutierrez-Ramos JC; Teixido J. Chemokine stromal cell-derived factor-1alpha modulates VLA-4 integrin-dependent adhesion to fibronectin and VCAM-1 on bone marrow hematopoietic progenitor cells. *Exp Hematol*, 2001 29, 345-355.

[21] Kollet O; Spiegel A; Peled A; Petit I; Byk T; Hershkoviz R; Guetta E; Barkai G; Nagler A; Lapidot T. Rapid and efficient homing of human CD34(+)CD38(-/low)CXCR4(+) stem and progenitor cells to the bone marrow and spleen of NOD/SCID and NOD/SCID/B2m(null) mice. *Blood*, 2001 97, 3283-3291.

[22] Frenette PS; Subbarao S; Mazo IB; von Andrian UH; Wagner DD. Endothelial selectins and vascular cell adhesion molecule-1 promote hematopoietic progenitor homing to bone marrow. *Proc Natl Acad Sci U S A*, 1998 95, 14423-14428.

[23] Ma Q; Jones D; Borghesani PR; Segal RA; Nagasawa T; Kishimoto T; Bronson RT; Springer TA. Impaired B-lymphopoiesis, myelopoiesis, and derailed cerebellar neuron migration in CXCR4- and SDF-1-deficient mice. *Proc Natl Acad Sci U S A*, 1998 95, 9448-9453.

[24] Avigdor A; Goichberg P; Shivtiel S; Dar A; Peled A; Samira S; Kollet O; Hershkoviz R; Alon R; Hardan I; Ben-Hur H; Naor D; Nagler A; Lapidot T. CD44 and hyaluronic acid cooperate with SDF-1 in the trafficking of human CD34+ stem/progenitor cells to bone marrow. *Blood*, 2004 103, 2981-2989.

[25] Chang C; Werb Z. The many faces of metalloproteases: cell growth, invasion, angiogenesis and metastasis. *Trends Cell Biol*, 2001 11, S37-43.

[26] Zheng Y; Watanabe N; Nagamura-Inoue T; Igura K; Nagayama H; Tojo A; Tanosaki R; Takaue Y; Okamoto S; Takahashi TA. Ex vivo manipulation of umbilical cord blood-derived hematopoietic stem/progenitor cells with recombinant human stem cell factor can up-regulate levels of homing-essential molecules to increase their transmigratory potential. *Exp Hematol*, 2003 31, 1237-1246.

[27] Nilsson SK; Dooner MS; Tiarks CY; Weier HU; Quesenberry PJ. Potential and distribution of transplanted hematopoietic stem cells in a nonablated mouse model. *Blood*, 1997 89, 4013-4020.

[28] Nilsson SK; Johnston HM; Coverdale JA. Spatial localization of transplanted hemopoietic stem cells: inferences for the localization of stem cell niches. *Blood*, 2001 97, 2293-2299.

[29] Nilsson SK; Simmons PJ; Bertoncello I. Hemopoietic stem cell engraftment. *Exp Hematol*, 2006 34, 123-129.

[30] Dutt P; Wang JF; Groopman JE. Stromal cell-derived factor-1 alpha and stem cell factor/kit ligand share signaling pathways in hemopoietic progenitors: a potential mechanism for cooperative induction of chemotaxis. *J Immunol*, 1998 161, 3652-3658.

[31] Ganju RK; Brubaker SA; Meyer J; Dutt P; Yang Y; Qin S; Newman W; Groopman JE. The alpha-chemokine, stromal cell-derived factor-1alpha, binds to the transmembrane G-protein-coupled CXCR-4 receptor and activates multiple signal transduction pathways. *J Biol Chem*, 1998 273, 23169-23175.

[32] Wang JF; Park IW; Groopman JE. Stromal cell-derived factor-1alpha stimulates tyrosine phosphorylation of multiple focal adhesion proteins and induces migration of hematopoietic progenitor cells: roles of phosphoinositide-3 kinase and protein kinase C. *Blood*, 2000 95, 2505-2513.

[33] Petit I; Goichberg P; Spiegel A; Peled A; Brodie C; Seger R; Nagler A; Alon R; Lapidot T. Atypical PKC-zeta regulates SDF-1-mediated migration and development of human CD34+ progenitor cells. *J Clin Invest*, 2005 115, 168-176.

[34] Lee Y; Gotoh A; Kwon HJ; You M; Kohli L; Mantel C; Cooper S; Hangoc G; Miyazawa K; Ohyashiki K; Broxmeyer HE. Enhancement of intracellular signaling associated with hematopoietic progenitor cell survival in response to SDF-1/CXCL12 in synergy with other cytokines. *Blood*, 2002 99, 4307-4317.

[35] Vila-Coro AJ; Rodriguez-Frade JM; Martin De Ana A; Moreno-Ortiz MC; Martinez AC; Mellado M. The chemokine SDF-1alpha triggers CXCR4 receptor dimerization and activates the JAK/STAT pathway. *Faseb J*, 1999 13, 1699-1710.

[36] Wysoczynski M; Reca R; Ratajczak J; Kucia M; Shirvaikar N; Honczarenko M; Mills M; Wanzeck J; Janowska-Wieczorek A; Ratajczak MZ. Incorporation of CXCR4 into membrane lipid rafts primes homing-related responses of hematopoietic stem/progenitor cells to an SDF-1 gradient. *Blood*, 2005 105, 40-48.

[37] Cancelas JA; Lee AW; Prabhakar R; Stringer KF; Zheng Y; Williams DA. Rac GTPases differentially integrate signals regulating hematopoietic stem cell localization. *Nat Med*, 2005 11, 886-891.

[38] Hofmeister CC; Zhang J; Knight KL; Le P; Stiff PJ. Ex vivo expansion of umbilical cord blood stem cells for transplantation: growing knowledge from the hematopoietic niche. *Bone Marrow Transplant*, 2007 39, 11-23.

[39] Audet J; Miller CL; Eaves CJ; Piret JM. Common and distinct features of cytokine effects on hematopoietic stem and progenitor cells revealed by dose-response surface analysis. *Biotechnol Bioeng*, 2002 80, 393-404.

[40] Ema H; Takano H; Sudo K; Nakauchi H. In vitro self-renewal division of hematopoietic stem cells. *J Exp Med*, 2000 192, 1281-1288.

[41] Miller CL; Eaves CJ. Expansion in vitro of adult murine hematopoietic stem cells with transplantable lympho-myeloid reconstituting ability. *Proc Natl Acad Sci U S A*, 1997 94, 13648-13653.

[42] Ramsfjell V; Bryder D; Bjorgvinsdottir H; Kornfalt S; Nilsson L; Borge OJ; Jacobsen SE. Distinct requirements for optimal growth and In vitro expansion of human CD34(+)CD38(-) bone marrow long-term culture-initiating cells (LTC-IC), extended LTC-IC, and murine in vivo long-term reconstituting stem cells. *Blood*, 1999 94, 4093-4102.

[43] Piacibello W; Sanavio F; Severino A; Dane A; Gammaitoni L; Fagioli F; Perissinotto E; Cavalloni G; Kollet O; Lapidot T; Aglietta M. Engraftment in nonobese diabetic severe combined immunodeficient mice of human CD34(+) cord blood cells after ex vivo expansion: evidence for the amplification and self-renewal of repopulating stem cells. *Blood*, 1999 93, 3736-3749.

[44] Albella B; Segovia JC; Guenechea G; Pragnell IB; Bueren JA. Preserved long-term repopulation and differentiation properties of hematopoietic grafts subjected to ex vivo expansion with stem cell factor and interleukin 11. *Transplantation*, 1999 67, 1348-1357.

[45] Glimm H; Oh IH; Eaves CJ. Human hematopoietic stem cells stimulated to proliferate in vitro lose engraftment potential during their S/G(2)/M transit and do not reenter G(0). *Blood*, 2000 96, 4185-4193.

[46] Haylock DN; To LB; Dowse TL; Juttner CA; Simmons PJ. Ex vivo expansion and maturation of peripheral blood CD34+ cells into the myeloid lineage. *Blood*, 1992 80, 1405-1412.

[47] Verfaillie CM. Hematopoietic stem cells for transplantation. *Nat Immunol*, 2002 3, 314-317.

[48] Cerny J; Dooner M; McAuliffe C; Habibian H; Stencil K; Berrios V; Reilly J; Carlson J; Cerny AM; d'Hondt L; Benoit B; Lambert JF; Colvin G; Nilsson S; Becker P; Quesenberry P. Homing of purified murine lymphohematopoietic stem cells: a cytokine-induced defect. *J Hematother Stem Cell Res*, 2002 11, 913-922.

[49] Ahmed F; Ings SJ; Pizzey AR; Blundell MP; Thrasher AJ; Ye HT; Fahey A; Linch DC; Yong KL. Impaired bone marrow homing of cytokine-activated CD34+ cells in the NOD/SCID model. *Blood*, 2004 103, 2079-2087.

[50] Von Drygalski A; Alespeiti G; Ren L; Adamson JW. Murine bone marrow cells cultured ex vivo in the presence of multiple cytokine combinations lose radioprotective and long-term engraftment potential. *Stem Cells Dev*, 2004 13, 101-111.

[51] Kassmer SH; Niggemann B; Zaenker KS; Dittmar T. Cytokine Combinations differentially influence the SDF-1α dependent migratory activity of cultured murine hematopoietic stem and progenitor cells *Biological Chemistry*, 2007 submitted,

[52] Wright DE; Bowman EP; Wagers AJ; Butcher EC; Weissman IL. Hematopoietic stem cells are uniquely selective in their migratory response to chemokines. *J Exp Med*, 2002 195, 1145-1154.

[53] Fukuda S; Broxmeyer HE; Pelus LM. Flt3 ligand and the Flt3 receptor regulate hematopoietic cell migration by modulating the SDF-1alpha(CXCL12)/CXCR4 axis. *Blood*, 2005 105, 3117-3126.

[54] Weidt C; Niggemann B; Hatzmann W; Zanker KS; Dittmar T. Differential effects of culture conditions on the migration pattern of stromal cell-derived factor-stimulated hematopoietic stem cells. *Stem Cells*, 2004 22, 890-896.

[55] Berrios VM; Dooner GJ; Nowakowski G; Frimberger A; Valinski H; Quesenberry PJ; Becker PS. The molecular basis for the cytokine-induced defect in homing and engraftment of hematopoietic stem cells. *Exp Hematol*, 2001 29, 1326-1335.

[56] Yang L; Wang L; Geiger H; Cancelas JA; Mo J; Zheng Y. Rho GTPase Cdc42 coordinates hematopoietic stem cell quiescence and niche interaction in the bone marrow. *Proc Natl Acad Sci U S A*, 2007 104, 5091-5096.

[57] Wright DE; Wagers AJ; Gulati AP; Johnson FL; Weissman IL. Physiological migration of hematopoietic stem and progenitor cells. *Science*, 2001 294, 1933-1936.

[58] McKinney-Freeman SL; Jackson KA; Camargo FD; Ferrari G; Mavilio F; Goodell MA. Muscle-derived hematopoietic stem cells are hematopoietic in origin. *Proc Natl Acad Sci U S A*, 2002 99, 1341-1346.

[59] Kawada H; Ogawa M. Bone marrow origin of hematopoietic progenitors and stem cells in murine muscle. *Blood*, 2001 98, 2008-2013.

[60] Krause DS; Theise ND; Collector MI; Henegariu O; Hwang S; Gardner R; Neutzel S; Sharkis SJ. Multi-organ, multi-lineage engraftment by a single bone marrow-derived stem cell. *Cell*, 2001 105, 369-377.

[61] Ratajczak MZ; Majka M; Kucia M; Drukala J; Pietrzkowski Z; Peiper S; Janowska-Wieczorek A. Expression of functional CXCR4 by muscle satellite cells and secretion of SDF-1 by muscle-derived fibroblasts is associated with the presence of both muscle progenitors in bone marrow and hematopoietic stem/progenitor cells in muscles. *Stem Cells*, 2003 21, 363-371.

[62] Kucia M; Jankowski K; Reca R; Wysoczynski M; Bandura L; Allendorf DJ; Zhang J; Ratajczak J; Ratajczak MZ. CXCR4-SDF-1 signalling, locomotion, chemotaxis and adhesion. *J Mol Histol*, 2004 35, 233-245.

[63] Pablos JL; Amara A; Bouloc A; Santiago B; Caruz A; Galindo M; Delaunay T; Virelizier JL; Arenzana-Seisdedos F. Stromal-cell derived factor is expressed by dendritic cells and endothelium in human skin. *Am J Pathol*, 1999 155, 1577-1586.

[64] Rossi D; Zlotnik A. The biology of chemokines and their receptors. *Annu Rev Immunol*, 2000 18, 217-242.

[65] Dooner M; Cerny J; Colvin G; Demers D; Pimentel J; Greer D; Abedi M; McAuliffe C; Quesenberry P. Homing and conversion of murine hematopoietic stem cells to lung. *Blood Cells Mol Dis*, 2004 32, 47-51.

[66] Kollet O; Shivtiel S; Chen YQ; Suriawinata J; Thung SN; Dabeva MD; Kahn J; Spiegel A; Dar A; Samira S; Goichberg P; Kalinkovich A; Arenzana-Seisdedos F; Nagler A; Hardan I; Revel M; Shafritz DA; Lapidot T. HGF, SDF-1, and MMP-9 are involved in stress-induced human CD34+ stem cell recruitment to the liver. *J Clin Invest*, 2003 112, 160-169.

[67] Pituch-Noworolska A; Majka M; Janowska-Wieczorek A; Baj-Krzyworzeka M; Urbanowicz B; Malec E; Ratajczak MZ. Circulating CXCR4-positive stem/progenitor cells compete for SDF-1-positive niches in bone marrow, muscle and neural tissues: an alternative hypothesis to stem cell plasticity. *Folia Histochem Cytobiol*, 2003 41, 13-21.

[68] Reca R; Mastellos D; Majka M; Marquez L; Ratajczak J; Franchini S; Glodek A; Honczarenko M; Spruce LA; Janowska-Wieczorek A; Lambris JD; Ratajczak MZ. Functional receptor for C3a anaphylatoxin is expressed by normal hematopoietic stem/progenitor cells, and C3a enhances their homing-related responses to SDF-1. *Blood*, 2003 101, 3784-3793.

[69] Li Y; Reca RG; Atmaca-Sonmez P; Ratajczak MZ; Ildstad ST; Kaplan HJ; Enzmann V. Retinal pigment epithelium damage enhances expression of chemoattractants and migration of bone marrow-derived stem cells. *Invest Ophthalmol Vis Sci*, 2006 47, 1646-1652.

[70] Atmaca-Sonmez P; Li Y; Yamauchi Y; Schanie CL; Ildstad ST; Kaplan HJ; Enzmann V. Systemically transferred hematopoietic stem cells home to the subretinal space and express RPE-65 in a mouse model of retinal pigment epithelium damage. *Exp Eye Res*, 2006 83, 1295-1302.

[71] Lagasse E; Connors H; Al-Dhalimy M; Reitsma M; Dohse M; Osborne L; Wang X; Finegold M; Weissman IL; Grompe M. Purified hematopoietic stem cells can differentiate into hepatocytes in vivo. *Nat Med*, 2000 6, 1229-1234.

[72] Herrera MB; Bussolati B; Bruno S; Morando L; Mauriello-Romanazzi G; Sanavio F; Stamenkovic I; Biancone L; Camussi G. Exogenous mesenchymal stem cells localize to

the kidney by means of CD44 following acute tubular injury. *Kidney Int*, 2007 72, 430-441.

[73] Ceradini DJ; Kulkarni AR; Callaghan MJ; Tepper OM; Bastidas N; Kleinman ME; Capla JM; Galiano RD; Levine JP; Gurtner GC. Progenitor cell trafficking is regulated by hypoxic gradients through HIF-1 induction of SDF-1. *Nat Med*, 2004 10, 858-864.

[74] Ceradini DJ; Gurtner GC. Homing to hypoxia: HIF-1 as a mediator of progenitor cell recruitment to injured tissue. *Trends Cardiovasc Med*, 2005 15, 57-63.

[75] Togel F; Isaac J; Hu Z; Weiss K; Westenfelder C. Renal SDF-1 signals mobilization and homing of CXCR4-positive cells to the kidney after ischemic injury. *Kidney Int*, 2005 67, 1772-1784.

[76] Goodell MA; Jackson KA; Majka SM; Mi T; Wang H; Pocius J; Hartley CJ; Majesky MW; Entman ML; Michael LH; Hirschi KK. Stem cell plasticity in muscle and bone marrow. *Ann N Y Acad Sci*, 2001 938, 208-218; discussion 218-220.

[77] Abbott JD; Huang Y; Liu D; Hickey R; Krause DS; Giordano FJ. Stromal cell-derived factor-1alpha plays a critical role in stem cell recruitment to the heart after myocardial infarction but is not sufficient to induce homing in the absence of injury. *Circulation*, 2004 110, 3300-3305.

[78] Stumm RK; Rummel J; Junker V; Culmsee C; Pfeiffer M; Krieglstein J; Hollt V; Schulz S. A dual role for the SDF-1/CXCR4 chemokine receptor system in adult brain: isoform-selective regulation of SDF-1 expression modulates CXCR4-dependent neuronal plasticity and cerebral leukocyte recruitment after focal ischemia. *J Neurosci*, 2002 22, 5865-5878.

[79] Corti S; Locatelli F; Strazzer S; Salani S; Del Bo R; Soligo D; Bossolasco P; Bresolin N; Scarlato G; Comi GP. Modulated generation of neuronal cells from bone marrow by expansion and mobilization of circulating stem cells with in vivo cytokine treatment. *Exp Neurol*, 2002 177, 443-452.

[80] Kawada H; Takizawa S; Takanashi T; Morita Y; Fujita J; Fukuda K; Takagi S; Okano H; Ando K; Hotta T. Administration of hematopoietic cytokines in the subacute phase after cerebral infarction is effective for functional recovery facilitating proliferation of intrinsic neural stem/progenitor cells and transition of bone marrow-derived neuronal cells. *Circulation*, 2006 113, 701-710.

[81] Dittmar T; Seidel J; Zaenker KS; Niggemann B. Carcinogenesis driven by bone marrow-derived stem cells. *Contrib Microbiol*, 2006 13, 156-169.

[82] Terada N; Hamazaki T; Oka M; Hoki M; Mastalerz DM; Nakano Y; Meyer EM; Morel L; Petersen BE; Scott EW. Bone marrow cells adopt the phenotype of other cells by spontaneous cell fusion. *Nature*, 2002 416, 542-545.

[83] Alvarez-Dolado M; Pardal R; Garcia-Verdugo JM; Fike JR; Lee HO; Pfeffer K; Lois C; Morrison SJ; Alvarez-Buylla A. Fusion of bone-marrow-derived cells with Purkinje neurons, cardiomyocytes and hepatocytes. *Nature*, 2003 425, 968-973.

[84] Rizvi AZ; Swain JR; Davies PS; Bailey AS; Decker AD; Willenbring H; Grompe M; Fleming WH; Wong MH. Bone marrow-derived cells fuse with normal and transformed intestinal stem cells. *Proc Natl Acad Sci U S A*, 2006 103, 6321-6325.

[85] Quintana-Bustamante O; Alvarez-Barrientos A; Kofman AV; Fabregat I; Bueren JA; Theise ND; Segovia JC. Hematopoietic mobilization in mice increases the presence of

bone marrow-derived hepatocytes via in vivo cell fusion. *Hepatology*, 2006 43, 108-116.

[86] Wang X; Willenbring H; Akkari Y; Torimaru Y; Foster M; Al-Dhalimy M; Lagasse E; Finegold M; Olson S; Grompe M. Cell fusion is the principal source of bone-marrow-derived hepatocytes. *Nature*, 2003 422, 897-901.

[87] Jang YY; Collector MI; Baylin SB; Diehl AM; Sharkis SJ. Hematopoietic stem cells convert into liver cells within days without fusion. *Nat Cell Biol*, 2004 6, 532-539.

[88] Oh SH; Muzzonigro TM; Bae SH; LaPlante JM; Hatch HM; Petersen BE. Adult bone marrow-derived cells trans-differentiating into insulin-producing cells for the treatment of type I diabetes. *Lab Invest*, 2004 84, 607-617.

[89] Coussens LM; Werb Z. Inflammation and cancer. *Nature*, 2002 420, 860-867.

[90] Balkwill F; Mantovani A. Inflammation and cancer: back to Virchow? *Lancet*, 2001 357, 539-545.

[91] Tabatabai G; Frank B; Mohle R; Weller M; Wick W. Irradiation and hypoxia promote homing of haematopoietic progenitor cells towards gliomas by TGF-beta-dependent HIF-1alpha-mediated induction of CXCL12. *Brain*, 2006 129, 2426-2435.

[92] Sotillo R; Hernando E; Diaz-Rodriguez E; Teruya-Feldstein J; Cordon-Cardo C; Lowe SW; Benezra R. Mad2 overexpression promotes aneuploidy and tumorigenesis in mice. *Cancer Cell*, 2007 11, 9-23.

[93] Weaver BA; Silk AD; Montagna C; Verdier-Pinard P; Cleveland DW. Aneuploidy acts both oncogenically and as a tumor suppressor. *Cancer Cell*, 2007 11, 25-36.

[94] Vassilopoulos G; Wang PR; Russell DW. Transplanted bone marrow regenerates liver by cell fusion. *Nature*, 2003 422, 901-904.

[95] Jamieson CH; Ailles LE; Dylla SJ; Muijtjens M; Jones C; Zehnder JL; Gotlib J; Li K; Manz MG; Keating A; Sawyers CL; Weissman IL. Granulocyte-macrophage progenitors as candidate leukemic stem cells in blast-crisis CML. *N Engl J Med*, 2004 351, 657-667.

[96] Seidel J; Batistin E; Schwitalla S; Niggemann B; Zänker KS; Dittmar T. Cancer Cell + Stem Cell = Cancer Stem Cell? In: Saitama H. *New Cell Differentiation Research Topics*. Hauppauge, NY: Nova Science Publishers; 2007; in press.

[97] Eisenberg LM; Eisenberg CA. Stem cell plasticity, cell fusion, and transdifferentiation. *Birth Defects Res C Embryo Today*, 2003 69, 209-218.

[98] Ying QL; Nichols J; Evans EP; Smith AG. Changing potency by spontaneous fusion. *Nature*, 2002 416, 545-548.

[99] Houghton J; Stoicov C; Nomura S; Rogers AB; Carlson J; Li H; Cai X; Fox JG; Goldenring JR; Wang TC. Gastric cancer originating from bone marrow-derived cells. *Science*, 2004 306, 1568-1571.

[100] 100. Torisu H; Ono M; Kiryu H; Furue M; Ohmoto Y; Nakayama J; Nishioka Y; Sone S; Kuwano M. Macrophage infiltration correlates with tumor stage and angiogenesis in human malignant melanoma: possible involvement of TNFalpha and IL-1alpha. *Int J Cancer*, 2000 85, 182-188.

[101] Seidel J; Niggemann B; Punzel M; Fischer J; Zänker KS; Dittmar T. The neurotransmitter gamma-aminobutyric-acid (GABA) is a potent inhibitor of the stromal cell-derived factor-1. *Stem Cells Dev*, 2007 16, in press.

In: Stem Cell Applications in Disease and Health
Editors: W. Burnsides and R. Ellsley
ISBN: 978-1-60456-255-2

Chapter XII

Ordered Commitment of Hematopoietic Stem Cells to Lineage Fates

Geoffrey Brown[1] *, Philip J. Hughes[1], Robert H. Michell[2], Antonius G. Rolink[3] and Rod Ceredig[3]

[1]Division of Immunity and Infection, The Medical School, Edgbaston Birmingham B15 2TT, UK

[2] School of Biosciences, University of Birmingham, Edgbaston Birmingham B15 2TT, UK

[3]Developmental and Molecular Immunology, Centre for Biomedicine Department of Clinical and Biological Sciences, University of Basel Basel, Switzerland

Abstract

There is considerable interest worldwide in being able to expand stem cells *ex vivo* in order to develop therapies for blood cancers and to repair tissues such as the heart and central nervous system. Such efforts rely on having a sound conceptual framework of the nature of stem cells and how they adopt terminal developmental pathways. Lessons learned from studies of blood cell development in the bone marrow, in particular how self-renewing hematopoietic stem cells give rise to lineage committed progenitors, have laid important foundations for our understanding of how stem cells generate differentiated progeny. However, several models have been proposed to describe hematopoiesis. Many models consider development of blood cell lineages to be an ordered process and a longstanding view has been that there is an early and irrevocable choice between progenitors developing towards either the lymphoid or myeloid families of cells. Recent studies have identified new subsets of progenitor cells, each possessing particular combinations of lineage potentials – for example, Early Progenitors with Lymphoid and Myeloid potential (EPLM) and a myeloid/lymphoid progenitor lacking erythroid and megakaryocyte lineage potential. These and other progenitors having both

* E-mail address: g.brown@bham.ac.uk

lymphoid and myeloid potential would seem to negate a strict lympho/myeloid dichotomy. The notion that progenitors possessing certain combinations of differentiation options indicate which fates are developmentally closely related led us to propose some time ago a Sequential Determination model whereby stem cells undergo stepwise progression through binary options all the way from megakaryocytes to T lymphocytes. In this article, we review recent developments in the field of hematopoiesis with particular emphasis on how they can be incorporated into the Sequential Determination model.

In adult mammals, hematopoietic stem cells (HSCs) in the bone marrow generate neutrophils, basophils, eosinophils, macrophages, at least two types of dendritic cells, several types of B and T lymphocytes, natural killer cells, megakaryocytes/platelets, erythrocytes and maybe others. Except for T lymphocytes, all these cell types are generated in large numbers throughout life. As to unravelling how stem cells in general can give rise to a number of terminal differentiated progeny, Don Metcalf wrote: "If paradise exists for an experimental biologist, then hematopoiesis has surely got to be it" [1].

Exploration of hematopoiesis has provided important insights into the nature of stem cells and how they ultimately give rise to a wide variety of functional end-cells. HSCs are generally viewed as tissue-specific stem cells, and their definitive features are two-fold. Thus, they possess in an as-yet-unresolved manner the capacity to give rise to all members of the hematopoietic family with developmental progression being normally restricted to the generation of cells of the blood and immune system. Importantly, HSCs can replicate substantially without any diminution of their intrinsic developmental options. This ability to self-renew ensures that blood cells are replenished throughout life. HSCs initially give rise to a heterogeneous mixture of hematopoietic progenitor cells (HPCs) possessing various combinations of lineage potentials, but as development progresses, differentiation potentials become increasingly restricted. Binary choices between certain hematopoietic options undoubtedly occur, and Holtzer has argued that cellular fate decisions throughout development are essentially binary ([2], and see later). HPCs are substantially limited as to their capacity to self-renew - as they divide they progressively age to become functional end cells.

An overall objective of many current research programmes in hematopoiesis is to achieve the *in vitro* expansion of HSCs without any loss of their multi-lineage differentiation potential, that is, without compromising their 'stemness'. These efforts have regularly been frustrated by the fact that HSCs, when cultured in any of the tested combinations of growth factors, cannot be prevented from differentiating into lineage-committed progenitors that then mature. Overcoming this tendency of cultured HSC to lose their 'stemness' would lead to an immediate clinical benefit since a limitation in the number of available HSCs frequently compromises treatments. HSC expansion without loss of 'stemness' would, for example, allow the wider application of umbilical cord blood-derived HSCs, because an individual sample of cord blood only provides sufficient cells for transplanting a young child. The ongoing establishment of large public and private cord blood banks could then make stored cord blood stem cells the ideal choice for prompt bone marrow transplantation. Achieving expansion of HSC would also be of strategic benefit to stem cell research, particularly for the use of adult stem cells in a variety of regenerative medical applications. In addition, a corollary would be that a better understanding of how HSC maintain their "stemness" would provide information regarding how development along a particular pathway might be controlled. The possibility of enriching for progenitors retaining certain combinations of developmental potential, for example enriched in T lymphocyte-reconstituting potential, would have immediate therapeutic applications.

Lineage Selections – Random or Systematic?

Understanding the identity of stem cells, of whatever nature, and of their progeny progenitor cells is a matter of understanding how multiple lineage options are made available and then selected. An early view of hematopoiesis focused on the so-called "priming" of HSC for differentiation. It was argued that the genes characteristically expressed by alternative lineages are simultaneously accessible for transcription and that in early progenitors there is low-level expression of genes characteristic of many lineages [3, 4]. The pattern of this "promiscuous" gene expression appears to be complex. For their transcription, genes require DNA-binding protein complexes called transcription factors to bind to their promotors. Transcription factors can be classified as having either a "general" or "specific" nature. Presumably, there is no limitation in the availability of general transcription factors. In contrast, changes in the concentrations of "specific" transcription factors are the ultimate arbitrators of cell fate decisions. One view has been that the various transcription factors that drive lineage specification (e.g. Ikaros, PU.1) fluctuate continuously, with random fluctuations governing fate adoption [5]. Evidence cited in favour of this random/stochastic process of lineage commitment of HSC is that these cells appear to assort commitment to various lineages within colony-forming cells in complex patterns ([6, 7], reviewed in [8]).

The obvious argument against adopting a completely random commitment model is that important circumstances that restrict developmental possibilities may simply not have been defined. Variable and regulated rates of replenishment of various blood cell types by a completely random commitment process could only be achieved by combining over-production of all cell types with regulated cell loss. This would be grossly wasteful for the production of bulk populations of erythrocytes or neutrophils, for example. More importantly, lineage options are not sorted randomly in committed colony-forming cells. An experiment of Ray Bradley's that transformed experimental haematology showed that bone marrow cell suspensions could form lineage-committed colonies in semi-solid medium [9]. Moreover, the early identification of colony-forming units (CFU) capable of forming colonies containing only granulocytes (neutrophils) and macrophages (monocytes) (GM-CFU) led to the view that these two lineages are developmentally particularly related.

Identification of additional close relationships between options for different lineages (see later) has led to the stochastic model of lineage commitment becoming less favoured. More recent models of hematopoiesis have focused on drawing maps for developmental relationships between lineage options. As a result there are numerous lineage maps for the blood cell system, and these vary in their views of the proposed degrees of relatedness between cells of the 'myeloid' and 'lymphoid' subdivisions of the hematopoietic system.

A 'Conventional' View of Hematopoiesis Is Inadequate

Lineage maps have arisen largely from the use *in vivo* and *in vitro* of clonal assays that identify the differentiation potentials of various populations of progenitor cells. From such studies, a longstanding conclusion has been that the progeny of HSCs first become committed either to the myeloid or lymphoid group of lineages (for review see [10, 11]). Cells that appear to be restricted to these lineage spectra have been purified from the bone marrow of adult mice [12, 13], and also appear to be present in human bone marrow [14]. Figure 1 shows one of the resulting models of hematopoiesis that is often used in text books. In addition to an early lymphoid/myeloid dichotomy, it features a common progenitor for granulocytes/monocytes (see above). In an updated version of this model, erythroid and megakaryocyte are alternative fates of another bipotent intermediate progenitor [11].

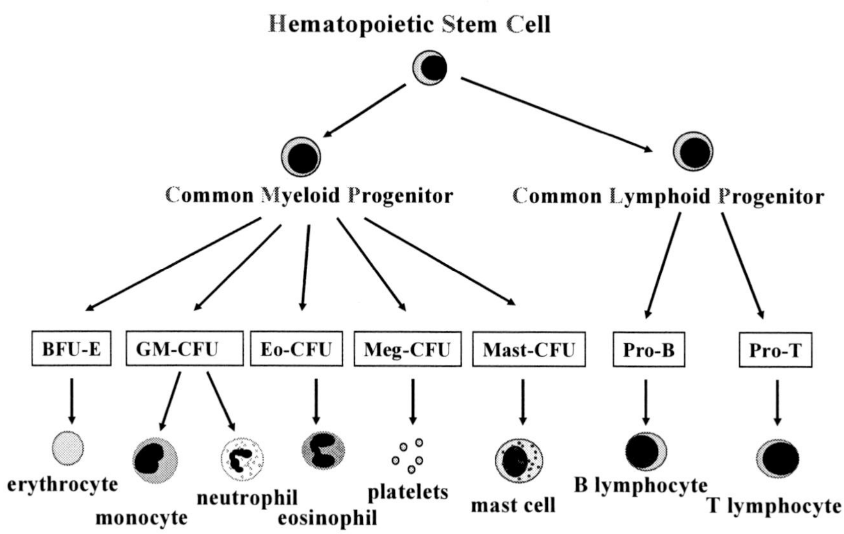

Figure 1. A 'conventional' model of hematopoiesis.

Central tenets of this model are an early dichotomy of the potentials for development along the myeloid and lymphoid pathways, and that the potentials for neutrophil and monocyte differentiation are closely related. BFU-E is burst forming unit-erythroid. CFU-X are colony forming unit for granulocytes/monocytes (GM), eosinophils (Eo), megakaryocytes (Meg) or mast cells and basophils (Mast).

A key test of the validity of the conventional model is whether the myeloid and lymphoid families of cells are really distant relatives developmentally. Relatedness of cell lineages is best proven by identifying sub-sets of progenitors that are each restricted to two (or more) potentials. A common bi-potent progenitor might relate to the two end-cells having similar functions. This, in turn, might relate to one cell type having evolved from the other. That monocytes and B cells are closely related is supported by evidence pertaining to all three of these notions.

The core of the 'conventional' model was first negated by evidence that the potentials for monocyte and B lymphocyte differentiation are not always segregated early. Progenitors with restricted potential that can differentiate either to monocytes or to B lymphocytes are generated in both adult mouse bone marrow and foetal liver [15, 16]. Transgenic mice that over-express IL-7 display a lymphoproliferative disorder and frequently, tumours arise that involve this progenitor [17, 18]. Moreover, there are several cell lines in which the close association of monocytes and B cells has been demonstrated, namely: i) the 5-azacytidine-induced macrophage-lineage switching of a pre-B lymphoma line [19]; ii) a B lymphocyte progenitor line with myeloid potential [20]; and iii) 70Z/3 $CD5^{+ve}$ B cells [21]; iv) early pre-B cells expressing transduced *fms* genes [22]; and v) Eμ-myc transgenic B cells [23].

Similarity between monocytes and B lymphocytes was also seen in early studies in which we observed that the patterns of cellular phosphoproteins in human myeloid cell lines that were restricted to monocyte differentiation and pre-B cell lines were similar [24, 25]. Functional similarity between monocytes and B cells is emphasised by the finding that both express MHC Class II molecules and both can present MHC-ClassII-associated antigenic peptides to T cells – in the case of B cells, antigen to which their surface immunoglobulin is directed [26]. Moreover, some B lymphocytes from teleost fish and *Xenopus laevis* are phagocytic which led to the suggestion that B lymphocytes may have evolved from an ancestral phagocytic cell type [27].

Sequential Determination Models of Hematopoiesis

Above we have argued that the existence of neutrophil/monocyte and B lymphocyte/monocyte progenitors reflect close developmental relationships between each of these pairs of hematopoietic fates. In 1985, one of us extended this notion by comparing a variety of bi-potent cells as to the extent to which each has at its disposal a particular subset of the 28 possible pairings of the eight hematopoietic differentiation potentials that were then recognised. Many of these pairings of fates seemed not to occur, and observations of the various permitted pairings were then combined and used as a basis for proposing the Sequential Determination (SD) model for the development of blood cells. (See Figure 2 [28-30]). Consideration of the spectrum of lineage involvement in various myelodysplastic disorders provided further support for the sequential arrangement of lineage options, in that myelodysplastic disorders can be classified in terms of a progressively more extensive involvement of lineages along the sequence shown in Figure 2 [31].

The model shown in Fig. 2A is an updated version of that proposed in 1985 and depicts the postulated relatedness between eight lineages. It does not include dendritic cells, the placing of which is discussed later.

The SD model is based upon the idea that a cell-intrinsic developmental program runs both in stem cells and in oligopotent progenitors derived from them and that this drives cells along a continuous progression through the various differentiation options. Each fate becomes available and then latent in an orderly and predetermined manner. At any stage in this progression, a cell only has a binary choice, and the available two fates change progressively over time. A cell may become committed to a particular pathway of maturation, or it may progress to the next option. To mature, progenitors may have to engage an appropriate extra-cellular signal, but the role of this signal is permissive rather than instructive.

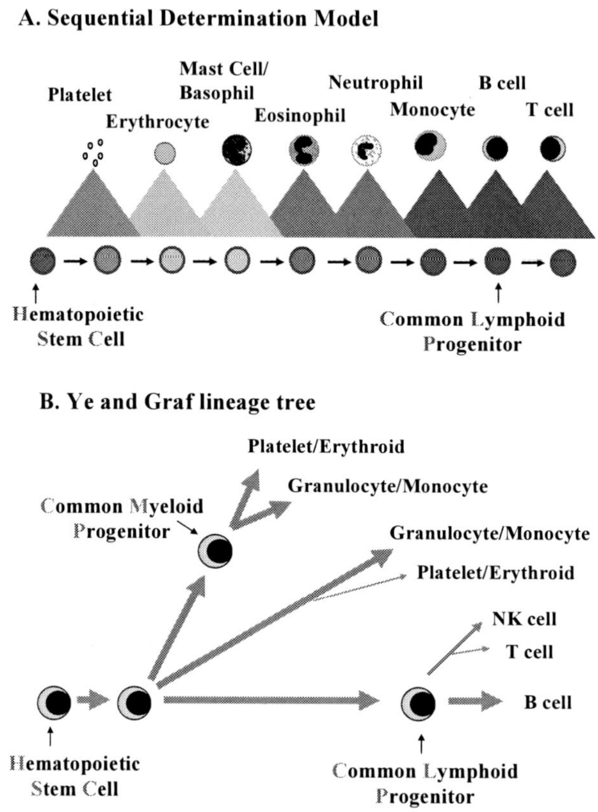

Figure 2. Sequential determination models of hematopoiesis. A. The sequential determination model proposes that there is a preferred sequence for the development of lineage potentials with an overlap between each pair of options (see the text: this is a slight variant of the model offered in [28]). B. Ye and Graf have presented a scheme for hematopoiesis within the bone marrow as a flow along branches of a lineage tree [32]. The thickness of the arrows represents the extent of flow along each route.

Ye and Graf have recently presented a hematopoietic lineage tree for the developmental of HSCs within the bone marrow [32]. They also subscribe to the notion of a preferred sequence of options, and some similarity exists between their organization of pathways and that depicted in the SD model. Based on the analysis of loss-of-function mutations in different transcription factors required for the specification of different cell fates, Singh has proposed a scheme that is a variant of the SD model [33]. Lemishka's group has also proposed a sequential map, this time based on gene expression analysis of different progenitor cells [34].

Other recent lineage maps for the blood cell system, including the disparate maps shown in Figure 3, also reject a strict dichotomy between myeloid and lymphoid fates. Rather, they envisage that individual progenitors are able to run one of a number of different intracellular programs and that there are multiple developmental routes to myeloid cells. Such schemes envisaging myeloid progenitor cells arising from two or more different points of origin leads to an interesting question. Should the end cells coming from different branches of the lineage tree be considered as belonging to different cell lineages?

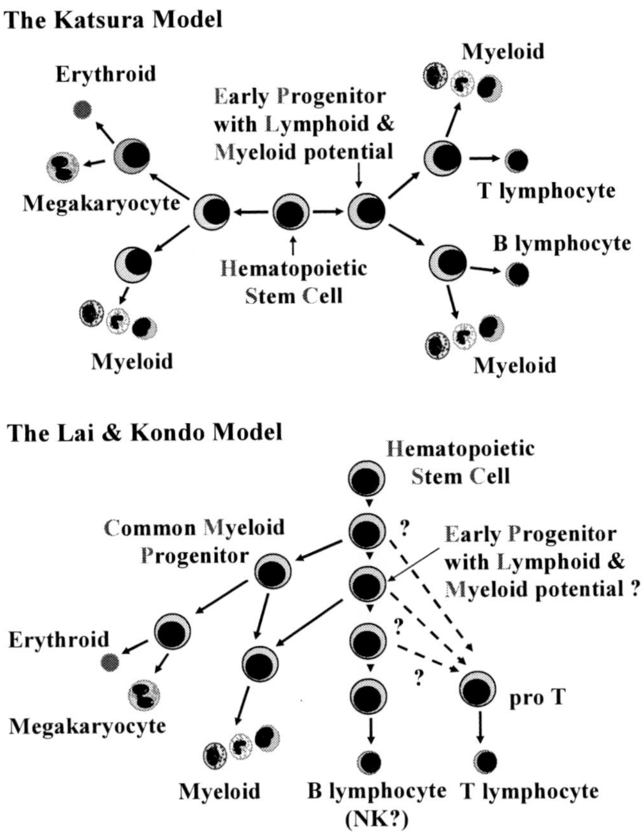

Figure 3. Branching models of hematopoiesis.

In the Katsura model [35] HSCs veer towards the erythroid potential whilst retaining the potentials for myeloid development and veer towards lymphocytes whilst also retaining the potentials for myeloid development. In the Lai and Kondo model [36] the initial branch point is towards erythroid/megakaryocyte development followed by the prospects of myeloid and subsequently B lymphoid development. In both models there is more than one branch point for myeloid development.

Katsura's model arose from the use of a clonal assay capable of supporting differentiation along the T lymphocyte, B lymphocyte, natural killer cell and myeloid pathways. Although bi-potent (at least) myeloid/T lymphocyte and myeloid/B lymphocyte cells were routinely observed, there were no bi-potent cells with only T and B lymphocyte potential [35]. Also, the fact that only some of the conceivable pairings of B, T and myeloid potentials were observed in the multi-lineage progenitor assay argues against any model involving random segregation of lineage potentials. Katsura [35] envisages a common myelolymphoid progenitor (CMLP) that, after a bifurcation, generates separate bi-potent myeloid/T lymphocyte and myeloid/B lymphocyte precursors (Fig. 3A). T lymphocytes and B lymphocytes are both seen as being more closely related to myeloid cells than to each other [35]. The pathway that gives rise to erythrocytes and megakaryocytes is considered as arising only after a closing down of the potentials for T and B lymphocyte differentiation. This model resembles the SD model by placing erythrocytes/megakaryocytes and T/B lymphocytes at the extreme poles of the spectrum of differentiation options, but it envisages multiple developmental programmes within stem/progenitor cells, with each involving stepwise lineage restriction.

The Lai and Kondo model (Fig. 3) envisages progenitor cells that can give rise to lymphocytes and neutrophils/macrophages rather than a simple myeloid/lymphoid dichotomy [36]. After examining three different sub-sets of cells that underwent lymphoid differentiation *in vivo* and that differed in their abilities to give rise to erythroid and myeloid cells, Lai and Kondo concluded that loss of erythroid differentiation potential is an early step in lineage specification, and is followed by loss of both neutrophil and monocyte potentials – and also that silencing of myeloid-associated genes might be a prerequisite for lymphoid lineage commitment. They concluded that the hierarchy of observed loss of potentials "supported the sequential differentiation model proposed by Brown et al".

Novel Lymphoid-Myeloid Progenitors Negate a Strict Myeloid/Lymphoid Dichotomy

When initially described, the cell population designated as common lymphoid progenitors (CLPs) could not be shown to possess myeloid potential. However, the apparent absence of myeloid potential in CLP has been challenged by Adolfsson and co-workers who identified a $CD34^{+ve}Flt3^{high}$ lympho/myeloid progenitor cell that lacks both erythroid and megakaryocyte differentiation potentials [37]. These cells have i) down-regulated transcription of megakaryocyte- and erythroid-related genes, ii) primed granulocyte/monocyte genes and iii) up-regulated genes common to both B and T lymphocytes.

Cells termed Early Progenitors with Lymphoid and Myeloid potential (EPLM) are present in the bone marrow of normal mice [38]. EPLM are $B220^{+ve}$, $CD117^{low}$, $CD93^{+ve}$, $CD19^{-ve}$, $CD3^{-ve}$ and $NK1.1^{-ve}$ and resemble pre-B cells that are found in Pax-5 (B cell transcription factor) knock-out mice (reviewed in [39]). Following *in vivo* transfer, EPLM give rise to mature T and B lymphocytes that can collaborate in mounting an antibody response. EPLM can differentiate *in vitro* into T lymphocytes, a process that requires a persistent Notch signal and the presence of IL-7. In the absence of Notch and presence of IL-7, B lymphocytes are generated. Development of natural killer cells requires a transient Notch signal and the presence of IL-2 or probably IL-15 whereas, dendritic cell or macrophage development is Notch-independent but requires macrophage colony-stimulating factor (M-CSF). Despite the considerable T lineage potential of EPLM, EPLM-like cells are not present in the thymus, and they appear to represent the last stage in a predominantly B lymphocyte developmental pathway at which T lymphopoiesis can be rescued by a Notch signal [39]. Cells with similar properties to EPLM have been described in man [40].

Bifurcation of Lineage Potentials

Bifurcation of families of fates, such as myeloid versus lymphoid, or of two individual fates is a feature of all of the ordered maps. How might this occur? Asymmetrical segregation of regulators of cell fate during cell division is important for lineage diversification in organisms such as *Caenorhabditis elegans* and *Drosophila melanogaster*. This process is clearly important for the diversification of vertebrate lymphocytes ([41], reviewed in [42]) and retinal precursor cells. Rat retinal precursor cells may divide with the cleavage plane parallel to or perpendicular to that of the neural epithelium to which they are attached. Both daughter cells from parallel divisions become photoreceptors, whereas one daughter cell from a perpendicular division becomes a rod and the other either an interneuron or a Muller cell [43]. The fate-determining protein Numb is concentrated at the apical pole before division and is preferentially inherited by the apical daughter during perpendicular divisions.

The extent to which asymmetrical cell division influences the adoption of fates by HSCs is, as yet, unclear. Cultured $CD34^{+ve}$ HSCs can polarize, as revealed by uneven distribution of the surface molecules CD133 [44], L-selectin, the transferrin receptor, CD53 and LAMP3 [45]. HSCs also associate homotypically, through CD34, to form bi-cellular or multi-cellular aggregates [46] and these cells co-polarize CD34 and Notch-1 [46]. Association and adhesion of HSCs at multiple stages of their development with mesenchymal stromal cells may also provide a polarizing signal. HSCs also undergo asymmetrical cell division [47], and its frequency declines as differentiation potentials become restricted [48, 49], and may be increased in polarized cells.

In 1975 Cairns showed that DNA can be semi-conservatively replicated such that the older of the two DNA strands is retained by one daughter and the newly synthesized strand by the other [50, 51]. He suggested that the older strand might be retained by the more stem-like of the daughter cells, with the newly synthesized cell staying with the daughter cell that goes on to further multiply and he termed the rarely replicated strand the ‘immortal strand’. This segregation appears to be important for protection of the genetic integrity of some types of stem cells with the newly synthesized strand that carries any replication errors going into those cells that multiply and undergo terminal maturation. Despite earlier suggestions to the contrary, a recent study of the prolonged retention of 5-bromo-2-deoxyuridine label in DNA indicates that bone marrow HSC which divide only about once a week, do not selectively retain their parental DNA strands [52].

Evidence for the Particular Sequence of Relationships Depicted in the SD Model

The SD model envisages that the megakaryocyte pathway branches off first, closely followed by the erythroid option. The close association between the megakaryocyte and erythroid pathways is implicit in alternative fates of a bi-potent progenitor in an update of the lymphoid/myeloid dichotomy model [11], and during primitive (in contrast to definitive) hematopoiesis there is transient production only of erythrocytes and megakaryocytes. The SD model envisages that as the erythroid option becomes available, megakaryopoiesis becomes latent. c-Myb is essential for definitive hematopoiesis [53, 54] and a study in mice of an enforced decrease in c-Myb expression has revealed that c-Myb is essential for the choice between erythroid and megakaryocyte potentials. In keeping with the SD model, high c-Myb both promotes erythropoiesis and suppresses megakaryopoiesis [55, 56].

The Rho GTPase Cdc42 is a regulator of cell cycle progression of HSCs and appears to play an important role in the transition from the erythroid to myeloid potentials. When Cdc42 was deleted (cdc42-floxed) in mouse bone marrow, the deficient mice developed fatal myeloproliferative disorders with neutrophilia, and erythroid differentiation was suppressed (as revealed by anemia and reduced numbers of BFU-E and CFU-E in bone marrow). Commensurate with increased myelopoiesis and decreased erythropoiesis, myeloid progenitors (CFU-GM) from knockout mice showed up-regulation of pro-myeloid genes (PU.1, C/EBP1a and Gfi-1), whilst megakaryocyte-erythoid progenitors showed down-regulation of the pro-erythroid gene GATA-2. In summary, active Cdc42 tips the balance from erythropoiesis towards myelopoiesis [57].

Basophils/mast cells, eosinophils, neutrophils and monocytes may be viewed as a family of closely related myeloid cells. The order of relatedness of these cells is not entirely clear, but is suggested by several observations. First, in mice carrying a GATA-1^{low} mutation [58], expansion of a megakaryocyte/erythrocyte/basophil progenitor indicates that the potential for basophil differentiation is closely related to the upstream potentials for megakaryocyte/erythroid differentiation as outlined in the SD model. Second, a close relationship between the potentials for eosinophil and neutrophil differentiation is revealed by the following recent studies. High protein kinase B (PKB/c-akt) promotes neutrophil and monocyte development, whereas a reduction in PKB activity is required for optimal eosinophil differentiation. PKB phosphorylates CCAAT enhancer-binding protein α (C/EBPα) and expression in hematopoietic progenitors of a form of C/EBPα that is non-phosphorylatable at an inhibitory site inhibits eosinophil differentiation and induces neutrophil development. [59]. Additionally, HT93 cells show eosinophil-neutrophil bipotency, and YJ cells eosinophil-neutrophil-monocyte tripotency [60]. Third, there is a sequence of transition from the potential for neutrophil differention to that for monocyte differentiation (see later).

Whether progenitors choose to follow each of the above four myeloid pathways is influenced by the levels of and co-ordinated expression of C/EBPα and GATA-2. Granulocyte/monocyte progenitors express C/EBPα. Elevated GATA-2 expression redirects granulocyte/monocyte progenitors to the eosinophil lineage, but if C/EBPα expression is suppressed prior to GATA-2 expression cells commit to the basophil/mast cell lineage [61]. Intriguingly, GATA-2 expression is high in S phase and low in G1 and M phases due to cell cycle-specific proteasome-dependent degradation [62]. This timing of expression of GATA-2 may be important to ordering lineage specification which may vary as cells progress through different phases of the cell-cycle. Commensurate with this notion is that hotspots for commitment to the megakaryocyte and granulocyte differentiation pathway occur in early S-phase and in mid S-phase respectively, coincident with accumulation of the transcription factor FOG-1 [63].

As mentioned above, early studies of hematopoiesis described a colony-forming unit that gave rise to neutrophils and monocytes. Such information does not infer a strict bifurcation of these two lineages. Instead, it has been suggested that progenitors are first able to undergo neutrophil differentiation, then move to a state in which both potentials are available, and finally into a monocyte-restricted state [64]. The human promyeloid cell line HL60 undergoes both neutrophil and monocyte differentiation, and analysis of variant lines with differing capacities for neutrophil and monocyte differentiation concurred with this notion [28].

Findings that support a transition of HSC from the potential for differentiation to monocytes to that for B lymphocytes have been considered above. B lymphocyte/monocyte progenitors clearly exist, and their recognition ended years of controversy about the closeness of the relationship between B lymphocytes and monocytes. Moreover, most investigators would agree that neutrophil/monocyte and monocyte/B lymphocyte progenitors exist whereas cells that can differentiate only into neutrophils and B lymphocytes, but not to monocytes, do not.

Placing the options for B and T lymphocyte development at the end of a continuous sequence of lineage choices has been uncomfortable to some investigators. For the past 20 years, immunologists considered that lymphocytes – by virtue of having to rearrange genes to make sophisticated antigen-specific receptors – were 'special' and would not follow the mundane rules of hematopoietic development applicable to other lineages. This contributed to perpetuation of the notion that lymphocyte and myeloid cell development diverge early on. But in reality lymphocyte development – especially prior to the acquisition of B or T cell receptors – is clearly analogous to that of other lineages.

Commitment to the B cell lineage in the bone marrow is shown as a relatively late event in the lineage tree model proposed by Ye and Graf [32]. Remarkably, even after reaching an advanced stage of their differentiation B lymphocytes retain the potential to de-differentiate and then re-differentiate to T lymphocytes. After Pax5 had been conditionally deleted in mice, their peripheral lymphoid mature B lymphocytes de-differentiated into cells that could rescue T lymphopoiesis in the thymus of T-lymphocyte-deficient mice, with the resulting T lymphocytes retaining the B-lymphoid pattern of re-arranged heavy- and light-chain immunoglobulin genes.

The final option in the SD model is that for T lymphocyte development. As originally proposed by Katsura [35], commitment to the T lymphocyte pathway is developmentally, a "late" event. Indeed, recent examination of the lineage potentials of immature progenitors that have recently migrated to the thymus (thymus-settling progenitors, TSP) confirms this notion. TSP ($CCR9^{+}/CD135^{+ve}/CD117^{high}/CD44^{+ve}/CD25^{-ve}$) [65] lose CCR9 and CD135 to become early thymocyte progenitors (ETP, [66]). ETP acquire CD25 to become DN2 cells ($CD135^{-ve}/CD117^{high}/CD44^{+ve}/CD24^{low}/CD25^{+ve}$) which become DN3 cells involving gradual loss of CD117 and CD44 and expression of cytoplasmic CD3 [67]. B lineage potential is lost at the first TSP>ETP transition [68]. Natural killer cell and myeloid/dendritic cell potentials are lost at the DN2>DN3 transition [69]. DN3 cells are fully T lymphocyte-committed. They contain abundant V-DJβ rearrangements, but such rearrangements are not indicative of T cell commitment since they are present in both natural killer and myeloid cells generated from DN2 cells [69].

Finally, the order of lineages predicted in the SD model is in keeping with ontogeny repeating phylogeny. The coagulation function and requirement for platelets to be functional during angiogenesis might be viewed as a primitive defense mechanism that evolved early for restricting the dispersal of pathogens within an organism. Other primordial defense mechanisms include non-specific recognition of pathogens by macrophages and phagocytosis (which is also needed to clear apoptotic cells during morphogenesis), with lymphocyte-mediated specific and adaptive immunity evolving later [70, 71]. The presence of innate and non-specific immunity and a lack of adaptive and lymphoid-mediate immunity are seen clearly, for example, in *Drosophila melanogaster* that has coagulation, crystal cells (neutrophil-like) and plasmatocytes (macrophage-like). If these arguments are valid then as organisms evolved and cells with immune function became more diverse and sophisticated, new lineage options have been added on, one by one, to the end of pre-existing sequences of progenitor cell diversification.

Closing Down Unwanted Pathways and Developmental Plasticity

Lineage options becoming latent as HSC cycle through them is an important feature of the SD model and lineage commitment is clearly a combination of gene activation and repression. This is emphasized by studies of the transcription factor Pax-5 which plays a role in B cell development by inducing B lineage genes and by repressing genes that are important to the generation of non-B cell lineages. Additionally and as outlined below, studies of Pax-5 gene deletion and re-introduction have highlighted the developmental plasticity of lineage committed cells in that multiple lineage options are recoverable at quite a late stage of lineage commitment (in pro-B/pre-B cells). This is not too surprising as, at the extreme, transplanting nuclei from fully differentiated lymphocytes into ES cells can generate whole mice ([72], for review see [73]).

Progenitor B cells in Pax-5-deficient mice have a block at the pro-B/pre-BI cell stage – these cells have rearranged D-J at their immunoglobulin heavy chain loci [74]. Progenitor B cells in the marrow of Pax-5-deficient mice and in wild-type mice share many features; both express λ5, VpreB, CD79b, TdT, and RAG-1 and 2 [75, 76]. In contrast to wild-type progenitors, Pax-5-deficient progenitors express B cell-lineage genes at either a very low level (CD79a) or not at all (CD19) [76] and genes required for the generation of T lymphocytes and myeloid cells (Notch1, the receptor for M-CSF (M-CSFR) and FLT3) [77]. Commensurate with this is that Pax-5 deficient progenitors differentiate both *in vivo* and *in vitro* into T lymphocytes, natural killer cells and various myeloid cells. By contrast, progenitors from wild-type mice are committed to B cell development. Re-introduction of the Pax-5 gene into Pax-5-deficient progenitors results in a loss of multi-lineage potential (mirrored by down-regulation of Notch1, M-CSFR and FLT3) and a gain of B cell developmental potential (mirrored by expression of CD19 and up-regulation of CD79a) [78]. In keeping with all of this, inactivation of the Pax-5 gene in Pax-5-expressing pre-B cells generates a cell that can differentiate into T lymphocytes and myeloid cells that is similar to Pax-5-deficient pre-B cells and to the EPLM of wild-type mice (see above).

That the duration of signal that a cell receives is important for closing down or provoking differentiation options has been revealed by studies of the Notch signaling pathway. Notch was first found in *Drosophila melanogaster* and is the prototype of a family of transmembrane receptors whose cognate ligands are proteins of the Jagged and Delta/Serrate families. Ligand binding to Notch provokes proteolytic release of an intracellular domain of Notch which associates with other cytoplasmic components, enters the nucleus and de-represses the transcription of certain genes. The differentiation options of TSP/ETP progenitor thymocytes are affected by Notch signaling. Continued Notch signaling is required in order to fully commit these cells to the T lymphocyte lineage - DN1 thymocytes may have already received a Notch signal prior to thymus entry, and they continue to receive this signal within the thymus. Notch signalling also inhibits the B lymphocyte and natural killer cell options of TSP/ETP [79]. However, rather than inhibiting natural killer cell development, we and others have shown that a transient Notch signal is required to initiate natural killer cell development [80, 81].

Are Lineage Options the Ends of Branches on Lineage Trees or Are They Repetitively Cycled Within HSCs?

Presentation of lineage commitment as flows along the branches of a tree has largely arisen from studies of lower organisms whereby the importance of binary decisions and of intracellular developmental programs have been clearly established. A succession of binary decisions that appears to be intrinsically programmed underpins the entire development of *Caenorhabditis elegans* [82, 83]. *Caenorhabditis elegans* larval development is depicted as a tree [84], with the sequential activation of necessary genes modelled as a unidirectional cascade analogous to the toppling of dominoes [85]. However, *Caenorhabditis elegans* is primitive and one might argue an evolutionary backwater/dead end. Might vertebrates have advanced a unidirectional cascade process of lineage commitment?

Figure 4 offers an updated variant of the SD model depicted as a possible outline of how a pluripotent HSC and the various oligopotent progenitors that it can generate may repetitively cycle through the available lineage options. The main circle depicts the behaviour of a cell that retains all hematopoietic lineage potentials, and the concentric arcs represent the various combinations of potentials that have been seen in the various intermediate progenitor populations discussed above. The possibility that HSCs and progenitors might repeatedly 'look' at the sequences of options available to them would mean that these cells are used sparingly and in a non-wasteful manner.

As discussed previously, in some models there is more than one branch point for myeloid development. For example, in the model proposed by Katsura [35], there is a branch point for myeloid versus megakaryocyte/erythroid development and one for myeloid versus B lymphocyte development. These apparent different origins of myeloid cells can be depicted in the cyclical specification model of hematopoietic lineages as follows. The myeloid versus megakaryocyte/erythroid branching can be represented as an arc encompassing the options megakaryocyte < > erythroid < > myeloid (neutrophil/eosinophil/basophil), and the myeloid versus B lymphocyte branching as an arc encompassing myeloid < > B lymphocyte. So, although associations of potentials in particular progenitor cells might be viewed as more than one origin for myeloid cells, in regard to the inner sequence in which potentials arise in the cyclical model there are essentially single developmental points as to veering towards myeloid cells.

Dendritic cells have also appeared to have two separate - lymphoid and monocyte - origins. However, if dendritic cells are placed as an option between those for monocyte and B lymphocyte development (Fig. 4), then the above two origins of dendritic cells may then appear within an arc encompassing monocyte < > dendritic cell < > B lymphocyte rather than two origins for dendritic cells. In keeping with this notion, it was recently shown that the surface phenotypes and gene transcription profiles of dendritic cell subsets derived from purified common myeloid progenitors and from purified common lymphoid progenitors are indistinguishable, indicating that the developmental program leading to dendritic cells operates independently of conventional myeloid and lymphoid pathways [87]. This poses the question as to whether the bipotent monocyte/B lymphocyte progenitor that has been identified in the adult mouse bone marrow and foetal liver might have dendritic cell potential.

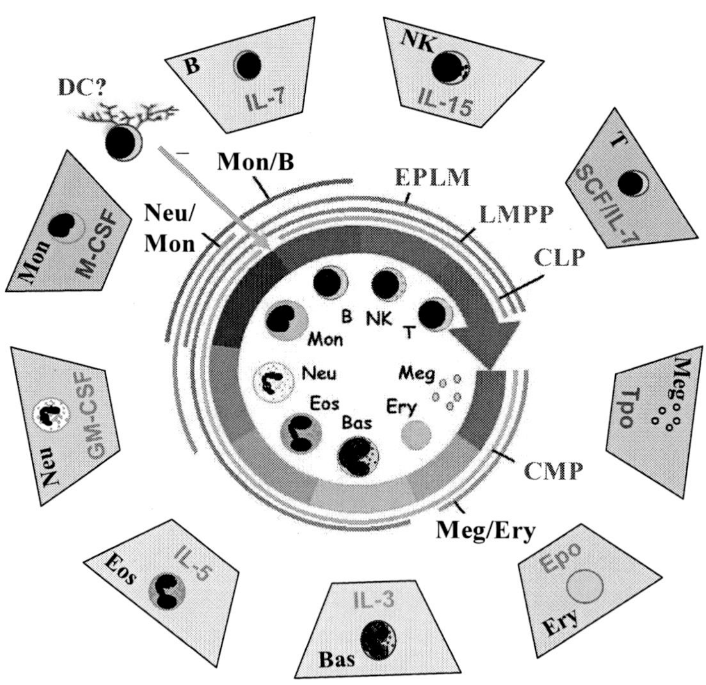

Figure 4. A simple cyclical summary of the sequential and progressive specification of hematopoietic lineages.

This model is a variant of that presented by Brown et al. [86] whereby there is a suggestion of where dendritic cell potential is placed in the SD spectrum of hematopoietic lineage potentials. The continuous main cycle indicates the relatedness between different hematopoietic lineages in HSCs, and the external arcs represent the progressively reducing subsets of lineage options that have been observed in various intermediate progenitor populations (for details, see the text). Meg, megakaryocyte/platelet; Ery, erythroid; Bas, basophil/mast cell; Eos, Eosinophil; Neu, neutrophil; Mon, monocyte/macrophage; B, B lymphocyte; NK, natural killer cell; T, T lymphocyte; DC, dendritic cell; EPLM, Early Progenitor with Lymphoid and Myeloid potential; LMPP, Lymphoid-primed Multi-Potent Progenitor; CLP, Common Lymphoid Progenitor; CMP, Common Myeloid Progenitor. Around the outside are some of the growth factors that are permissive for differentiation/survival of each of the cell types. Tpo, thrombopoietin; Epo, erythropoietin; GM-CSF, granulocyte/monocyte colony-stimulating factor; M-CSF, monocyte colony-stimulating factor; SCF, stem cell factor; IL, interleukin.

Hard-Wiring Hematopoiesis

To date, perhaps the most striking issue in examining our understanding of hematopoiesis is that we are faced with multiple lineage maps. This is not too surprising as most studies of hematopoiesis have examined proliferating HSC/progenitors, and the mature cell types yielded after a considerable number of cell divisions. Such experiments that equate the final fates of multi-generational offspring with the lineage options of the initial cells can never define the stages of lineage determination unambiguously. Essentially there is a need to track the fates of individual cells as achieved by time-lapse video recording of rat retinal precursor cells expressing green fluorescent protein [43]. Tracking what happens *in vivo* is crucial, and a recently-developed technique in zebra fish allows such by laser activation of a fluorescent tracer in individual progenitors [88].

Consideration of the cyclical model of specification of fates offers a simple explanation as to why there has been a plethora of lineage maps for hematopoiesis. In investigating the various sorts of progenitor cells in bone marrow it must be possible to 'pick-up' on any number of different 'arcs' that link together a continuum of any number of contiguous fates. As each 'arc' of varying extent and nature, as to spectrum of options covered, is identified in a particular study and thought of in terms of representing a branch of a tree, then many different branches might individually come to light and provoke the drawing of many different lineage maps.

However, the many maps that have appeared for HSC development in quite recent reviews favour a cell-intrinsic program driving an order to the availability of lineage options. As yet, it is difficult to envisage how changes in gene accessibility and/or expression, the latter by virtue of changes in transcription factor concentrations, might be 'hard-wired' to restrict HSCs to an ordered process of development. Perhaps the clearest evidence for intrinsic programming of progenitor cell development is from studies of neuronal cells (for review see [89, 90]). Astrocytes, ependymal cells and oligodendrocytes, the three types of glial cells in rat brain, appear sequentially at embryonic (E) days 15-16, E17-E18, and E21-22 (around birth). This timing of appearance of each of the three cell types is recapitulated when dissociated E10 brain cells are cultured *in vitro*, indicating that an intracellular mechanism hard-wires both the sequence and timing of glial cell development. A single retinal precursor gives rise to seven cell types in the order: ganglion cells, cones, horizontal cells, rods and amacrines, bipolar cells and Muller cells [91-93]. Experiments using dissociated cell cultures of rat retina concluded that retinal precursor cells "become specified before E16-E17 and then step through an intrinsic developmental programme" [94].

Concluding Remarks

We favour a simple ordered process of lineage commitment of HSCs, as presented in the original and in updated versions of the SD model. This model has stood the test of time, and other recent models focus on windows opening and closing in an orderly manner. Moreover, the emergence of information on fluctuations in and polarization of transcription factor concentrations, on asymmetric cell division and on the role of Notch in binary cell fate choices has illuminated how stepwise binary development of HSC might occur. Hematopoiesis may well be considered as a paradise for an experimental biologist, but we are still some way from fully understanding how hematopoiesis really works. This is of substantial practical importance in regard to efforts that are being made to grow stem cells that retain as much pluripotency as possible.

Acknowledgements

Antonius G. Rolink is holder of the chair in Immunology endowed by R. Hoffman-La Roche Ltd., Basel. This work was supported by grants from the Swiss National Science Foundation to A.G.R.. Rod Ceredig thanks INSERM for continued support. We thank Naomi Oppenheim for help with figure 4.

References

[1] Metcalf, D., Colony-stimulating factors. *Cell* 1990. 61: 756-757.

[2] Holtzer, H., Comments. *Differentiation* 1979. 14: 33-34.

[3] Cross, M. A. and Enver, T., The lineage commitment of haemopoietic progenitor cells. *Curr Opin Genet Dev* 1997. 7: 609-613.

[4] Enver, T. and Greaves, M., Loops, lineage, and leukemia. *Cell* 1998. 94: 9-12.

[5] McAdams, H. H. and Arkin, A., Stochastic mechanisms in gene expression. *Proc Natl Acad Sci U S A* 1997. 94: 814-819.

[6] Hu, M., Krause, D., Greaves, M., Sharkis, S., Dexter, M., Heyworth, C. and Enver, T., Multilineage gene expression precedes commitment in the hemopoietic system. *Genes Dev* 1997. 11: 774-785.

[7] Suda, T., Suda, J. and Ogawa, M., Disparate differentiation in mouse hemopoietic colonies derived from paired progenitors. *Proc Natl Acad Sci U S A* 1984. 81: 2520-2524.

[8] Morrison, S. J., Shah, N. M. and Anderson, D. J., Regulatory mechanisms in stem cell biology. *Cell* 1997. 88: 287-298.

[9] Bradley, T. R. and Metcalf, D., The growth of mouse bone marrow cells in vitro. *Aust J Exp Biol Med Sci* 1966. 44: 287-299.

[10] Kondo, M., Scherer, D. C., King, A. G., Manz, M. G. and Weissman, I. L., Lymphocyte development from hematopoietic stem cells. *Curr Opin Genet Dev* 2001. 11: 520-526.

[11] Weissman, I. L., Anderson, D. J. and Gage, F., Stem and progenitor cells: origins, phenotypes, lineage commitments, and transdifferentiations. *Annu Rev Cell Dev Biol* 2001. 17: 387-403.

[12] Akashi, K., Traver, D., Miyamoto, T. and Weissman, I. L., A clonogenic common myeloid progenitor that gives rise to all myeloid lineages. *Nature* 2000. 404: 193-197.

[13] Kondo, M., Weissman, I. L. and Akashi, K., Identification of clonogenic common lymphoid progenitors in mouse bone marrow. *Cell* 1997. 91: 661-672.

[14] Galy, A., Travis, M., Cen, D. and Chen, B., Human T, B, natural killer, and dendritic cells arise from a common bone marrow progenitor cell subset. *Immunity* 1995. 3: 459-473.

[15] Cumano, A., Paige, C. J., Iscove, N. N. and Brady, G., Bipotential precursors of B cells and macrophages in murine fetal liver. *Nature* 1992. 356: 612-615.

[16] Montecino-Rodriguez, E., Leathers, H. and Dorshkind, K., Bipotential B-macrophage progenitors are present in adult bone marrow. *Nat Immunol* 2001. 2: 83-88.

[17] Borrello, M. A., Palis, J. and Phipps, R. P., The relationship of CD5+ B lymphocytes to macrophages: insights from normal biphenotypic B/macrophage cells. *Int Rev Immunol* 2001. 20: 137-155.

[18] Fisher, A. G., Burdet, C., Bunce, C., Merkenschlager, M. and Ceredig, R., Lymphoproliferative disorders in IL-7 transgenic mice: expansion of immature B cells which retain macrophage potential. *Int Immunol* 1995. 7: 415-423.

[19] Boyd, A. W. and Schrader, J. W., Derivation of macrophage-like lines from the pre-B lymphoma ABLS 8.1 using 5-azacytidine. *Nature* 1982. 297: 691-693.

[20] Davidson, W. F., Pierce, J. H., Rudikoff, S. and Morse, H. C., 3rd, Relationships between B cell and myeloid differentiation. Studies with a B lymphocyte progenitor line, HAFTL-1. *J Exp Med* 1988. 168: 389-407.

[21] Tanaka, T., Wu, G. E. and Paige, C. J., Characterization of the B cell-macrophage lineage transition in 70Z/3 cells. *Eur J Immunol* 1994. 24: 1544-1548.

[22] Borzillo, G. V., Ashmun, R. A. and Sherr, C. J., Macrophage lineage switching of murine early pre-B lymphoid cells expressing transduced fms genes. *Mol Cell Biol* 1990. 10: 2703-2714.

[23] Klinken, S. P., Alexander, W. S. and Adams, J. M., Hemopoietic lineage switch: v-raf oncogene converts Emu-myc transgenic B cells into macrophages. *Cell* 1988. 53: 857-867.

[24] Bunce, C. M., Lord, J. M., Wong, A. K. and Brown, G., Near neighbour analysis of variant cell lines derived from the promyeloid cell line HL60. *Br J Cancer* 1988. 57: 559-563.

[25] Wong, A. K., Bunce, C. M., Lord, J. M., Salt, J. and Brown, G., Evidence that precursor cells of monocytes and B-lymphocytes are closely related. *Exp Hematol* 1989. 17: 968-973.

[26] Lanzavecchia, A., Antigen-specific interaction between T and B cells. *Nature* 1985. 314: 537-539.

[27] Li, J., Barreda, D. R., Zhang, Y. A., Boshra, H., Gelman, A. E., Lapatra, S., Tort, L. and Sunyer, J. O., B lymphocytes from early vertebrates have potent phagocytic and microbicidal abilities. *Nat Immunol* 2006. 7: 1116-1124.

[28] Brown, G., Bunce, C. M. and Guy, G. R., Sequential determination of lineage potentials during haemopoiesis. *Br J Cancer* 1985. 52: 681-686.

[29] Brown, G., Bunce, C. M., Howie, A. J. and Lord, J. M., Stochastic or ordered lineage commitment during hemopoiesis? *Leukemia* 1987. 1: 150-153.

[30] Brown, G., Buncc, C. M., Lord, J. M. and McConnell, F. M., The development of cell lineages: a sequential model. *Differentiation* 1988. 39: 83-89.

[31] Rose, P. E., Howie, A. J., Galvin, G. P., Bunce, C. M. and Brown, G., Lineage determination during haemopoiesis and classification of myelodysplastic disorders. *Br J Haematol* 1986. 64: 626-628.

[32] Ye, M. and Graf, T., Early decisions in lymphoid development. *Curr Opin Immunol* 2007. 19: 123-128.

[33] Singh, H., Gene targeting reveals a hierarchy of transcription factors regulating specification of lymphoid cell fates. *Curr Opin Immunol* 1996. 8: 160-165.

[34] Phillips, R. L., Ernst, R. E., Brunk, B., Ivanova, N., Mahan, M. A., Deanehan, J. K., Moore, K. A., Overton, G. C. and Lemischka, I. R., The genetic program of hematopoietic stem cells. *Science* 2000. 288: 1635-1640.

[35] Katsura, Y., Redefinition of lymphoid progenitors. *Nat Rev Immunol* 2002. 2: 127-132.

[36] Lai, A. Y. and Kondo, M., Asymmetrical lymphoid and myeloid lineage commitment in multipotent hematopoietic progenitors. *J Exp Med* 2006. 203: 1867-1873.

[37] Adolfsson, J., Mansson, R., Buza-Vidas, N., Hultquist, A., Liuba, K., Jensen, C. T., Bryder, D., Yang, L., Borge, O. J., Thoren, L. A., Anderson, K., Sitnicka, E., Sasaki, Y., Sigvardsson, M. and Jacobsen, S. E., Identification of Flt3+ lympho-myeloid stem cells lacking erythro-megakaryocytic potential a revised road map for adult blood lineage commitment. *Cell* 2005. 121: 295-306.

[38] Balciunaite, G., Ceredig, R., Massa, S. and Rolink, A. G., A B220+ CD117+ CD19- hematopoietic progenitor with potent lymphoid and myeloid developmental potential. *Eur J Immunol* 2005. 35: 2019-2030.

[39] Rolink, A. G., Massa, S., Balciunaite, G. and Ceredig, R., Early lymphocyte development in bone marrow and thymus. *Swiss Med Wkly* 2006. 136: 679-683.

[40] Reynaud, D., Lefort, N., Manie, E., Coulombel, L. and Levy, Y., In vitro identification of human pro-B cells that give rise to macrophages, natural killer cells, and T cells. *Blood* 2003. 101: 4313-4321.

[41] Chang, J. T., Palanivel, V. R., Kinjyo, I., Schambach, F., Intlekofer, A. M., Banerjee, A., Longworth, S. A., Vinup, K. E., Mrass, P., Oliaro, J., Killeen, N., Orange, J. S., Russell, S. M., Weninger, W. and Reiner, S. L., Asymmetric T Lymphocyte Division in the Initiation of Adaptive Immune Responses. *Science* 2007. 315: 1687-1691.

[42] Littman, D. R. and Singh, H., Immunology. Asymmetry and immune memory. *Science* 2007. 315: 1673-1674.

[43] Cayouette, M. and Raff, M., The orientation of cell division influences cell-fate choice in the developing mammalian retina. *Development* 2003. 130: 2329-2339.

[44] Giebel, B., Corbeil, D., Beckmann, J., Hohn, J., Freund, D., Giesen, K., Fischer, J., Kogler, G. and Wernet, P., Segregation of lipid raft markers including CD133 in polarized human hematopoietic stem and progenitor cells. *Blood* 2004. 104: 2332-2338.

[45] Beckmann, J., Scheitza, S., Wernet, P., Fischer, J. C. and Giebel, B., Asymmetric cell division within the human hematopoietic stem and progenitor cell compartment: Identification of asymmetrically segregating proteins. *Blood* 2007. 109: 5494-5501.

[46] Bullock, T. E., Wen, B., Marley, S. B. and Gordon, M. Y., Potential Of Cd34 In The Regulation Of Symmetrical And Asymmetrical Divisions By Hematopoietic Progenitor Cells. *Stem Cells* 2007. 25: 844-851.

[47] Faubert, A., Lessard, J. and Sauvageau, G., Are genetic determinants of asymmetric stem cell division active in hematopoietic stem cells? *Oncogene* 2004. 23: 7247-7255.

[48] Giebel, B., Zhang, T., Beckmann, J., Spanholtz, J., Wernet, P., Ho, A. D. and Punzel, M., Primitive human hematopoietic cells give rise to differentially specified daughter cells upon their initial cell division. *Blood* 2006. 107: 2146-2152.

[49] Takano, H., Ema, H., Sudo, K. and Nakauchi, H., Asymmetric division and lineage commitment at the level of hematopoietic stem cells: inference from differentiation in daughter cell and granddaughter cell pairs. *J Exp Med* 2004. 199: 295-302.

[50] Cairns, J., Mutation selection and the natural history of cancer. *Nature* 1975. 255: 197-200.

[51] Cairns, J., Cancer and the immortal strand hypothesis. *Genetics* 2006. 174: 1069-1072.

[52] Kiel, M. J., He, S., Ashkenazi, R., Gentry, S. N., Teta, M., Kushner, J. A., Jackson, T. L. and Morrison, S. J., Haematopoietic stem cells do not asymmetrically segregate chromosomes or retain BrdU. *Nature* 2007. 449: 238-242.

[53] Emambokus, N., Vegiopoulos, A., Harman, B., Jenkinson, E., Anderson, G. and Frampton, J., Progression through key stages of haemopoiesis is dependent on distinct threshold levels of c-Myb. *Embo J* 2003. 22: 4478-4488.

[54] Sakamoto, H., Dai, G., Tsujino, K., Hashimoto, K., Huang, X., Fujimoto, T., Mucenski, M., Frampton, J. and Ogawa, M., Proper levels of c-Myb are discretely defined at distinct steps of hematopoietic cell development. *Blood* 2006. 108: 896-903.

[55] Mukai, H. Y., Motohashi, H., Ohneda, O., Suzuki, N., Nagano, M. and Yamamoto, M., Transgene insertion in proximity to the c-myb gene disrupts erythroid-megakaryocytic lineage bifurcation. *Mol Cell Biol* 2006. 26: 7953-7965.

[56] Vegiopoulos, A., Garcia, P., Emambokus, N. and Frampton, J., Coordination of erythropoiesis by the transcription factor c-Myb. *Blood* 2006. 107: 4703-4710.

[57] Yang, L., Wang, L., Kalfa, T., Cancelas, J. A., Shang, X., Pushkaran, S., Mo, J., Williams, D. A. and Zheng, Y., Cdc42 critically regulates the balance between myelopoiesis and erythropoiesis. *Blood* 2007. 110: 3853-3861.

[58] Ghinassi, B., Sanchez, M., Martelli, F., Amabile, G., Vannucchi, A. M., Migliaccio, G., Orkin, S. H. and Migliaccio, A. R., The hypomorphic Gata1low mutation alters the

proliferation/differentiation potential of the common megakaryocytic-erythroid progenitor. *Blood* 2007. 109: 1460-1471.

[59] Buitenhuis, M., Verhagen, L. P., van Deutekom, H. W., Castor, A., Verploegen, S., Koenderman, L., Jacobsen, S. E. and Coffer, P. J., Protein kinase B (C-AKT) regulates hematopoietic lineage choice decisions during myelopoiesis. *Blood* 2008. 111: 112-121.

[60] Yamaguchi, Y., Nishio, H., Kasahara, T., Ackerman, S. J., Koyanagi, H. and Suda, T., Models of lineage switching in hematopoietic development: a new myeloid-committed eosinophil cell line (YJ) demonstrates trilineage potential. *Leukemia* 1998. 12: 1430-1439.

[61] Iwasaki, H., Mizuno, S., Arinobu, Y., Ozawa, H., Mori, Y., Shigematsu, H., Takatsu, K., Tenen, D. G. and Akashi, K., The order of expression of transcription factors directs hierarchical specification of hematopoietic lineages. *Genes Dev* 2006. 20: 3010-3021.

[62] Koga, S., Yamaguchi, N., Abe, T., Minegishi, M., Tsuchiya, S., Yamamoto, M. and Minegishi, M., Cell cycle-dependent oscillation of GATA2 expression in hematopoietic cells. *Blood* 2007. 109, 4200-4208

[63] Colvin, G. A., Dooner, M. S., Dooner, G. J., Sanchez-Guijo, F. M., Demers, D. A., Abedi, M., Ramanathan, M., Chung, S., Pascual, S. and Quesenberry, P. J., Stem cell continuum: directed differentiation hotspots. *Exp Hematol* 2007. 35: 96-107.

[64] Dexter, T. M., Garland, J., Scott, D., Scolnick, E. and Metcalf, D., Growth of factor-dependent hemopoietic precursor cell lines. *J Exp Med* 1980. 152: 1036-1047.

[65] Zediak, V. P., Maillard, I. and Bhandoola, A., Closer to the source: notch and the nature of thymus-settling cells. *Immunity* 2005. 23: 245-248.

[66] Heinzel, K., Benz, C., Martins, V. C., Haidl, I. D. and Bleul, C. C., Bone marrow-derived hemopoietic precursors commit to the T cell lineage only after arrival in the thymic microenvironment. *J Immunol* 2007. 178: 858-868.

[67] Wilson, A., Capone, M. and MacDonald, H. R., Unexpectedly late expression of intracellular CD3epsilon and TCR gammadelta proteins during adult thymus development. *Int Immunol* 1999. 11: 1641-1650.

[68] Ceredig, R., Bosco, N. and Rolink, A. G., The B lineage potential of thymus settling progenitors is critically dependent on mouse age. *Eur J Immunol* 2007. 37: 830-837.

[69] Balciunaite, G., Ceredig, R. and Rolink, A. G., The earliest subpopulation of mouse thymocytes contains potent T, significant macrophage, and natural killer cell but no B-lymphocyte potential. *Blood* 2005. 105: 1930-1936.

[70] Fisher, H., *Macrophages*. Springer, Berlin, Heidelberg, New York.: 1976.

[71] Manning, M. and Turner, R., *Comparative Immunobiology*. Blackie, Glasgow, London.: 1976.

[72] Hochedlinger, K. and Jaenisch, R., Monoclonal mice generated by nuclear transfer from mature B and T donor cells. *Nature* 2002. 415: 1035-1038.

[73] Laiosa, C. V., Stadtfeld, M. and Graf, T., Determinants of lymphoid-myeloid lineage diversification. *Annu Rev Immunol* 2006. 24: 705-738.

[74] Urbanek, P., Wang, Z. Q., Fetka, I., Wagner, E. F. and Busslinger, M., Complete block of early B cell differentiation and altered patterning of the posterior midbrain in mice lacking Pax5/BSAP. *Cell* 1994. 79: 901-912.

[75] Nutt, S. L., Heavey, B., Rolink, A. G. and Busslinger, M., Commitment to the B-lymphoid lineage depends on the transcription factor Pax5. *Nature* 1999. 401: 556-562.

[76] Nutt, S. L., Morrison, A. M., Dorfler, P., Rolink, A. and Busslinger, M., Identification of BSAP (Pax-5) target genes in early B-cell development by loss- and gain-of-function experiments. *Embo J* 1998. 17: 2319-2333.

[77] Holmes, M. L., Carotta, S., Corcoran, L. M. and Nutt, S. L., Repression of Flt3 by Pax5 is crucial for B-cell lineage commitment. *Genes Dev* 2006. 20: 933-938.

[78] Mikkola, I., Heavey, B., Horcher, M. and Busslinger, M., Reversion of B cell commitment upon loss of Pax5 expression. *Science* 2002. 297: 110-113.

[79] Lehar, S. M., Dooley, J., Farr, A. G. and Bevan, M. J., Notch ligands Delta 1 and Jagged1 transmit distinct signals to T-cell precursors. *Blood* 2005. 105: 1440-1447.

[80] Carotta, S., Brady, J., Wu, L. and Nutt, S. L., Transient Notch signaling induces NK cell potential in Pax5-deficient pro-B cells. *Eur J Immunol* 2006. 36: 3294-3304.

[81] Rolink, A. G., Balciunaite, G., Demoliere, C. and Ceredig, R., The potential involvement of Notch signaling in NK cell development. *Immunol Lett* 2006. 107: 50-57.

[82] Deppe, U., Schierenberg, E., Cole, T., Krieg, C., Schmitt, D., Yoder, B. and von Ehrenstein, G., Cell lineages of the embryo of the nematode Caenorhabditis elegans. *Proc Natl Acad Sci U S A* 1978. 75: 376-380.

[83] Sulston, J. E. and Horvitz, H. R., Post-embryonic cell lineages of the nematode, Caenorhabditis elegans. *Dev Biol* 1977. 56: 110-156.

[84] Sulston, J. E., Schierenberg, E., White, J. G. and Thomson, J. N., The embryonic cell lineage of the nematode Caenorhabditis elegans. *Dev Biol* 1983. 100: 64-119.

[85] Bailey, D. W., Genetic programming of development: a model. *Differentiation* 1986. 33: 89-100.

[86] Brown, G., Hughes, P. J., Michell, R. H., Rolink, A. G. and Ceredig, R., The sequential determination model of hematopoiesis. *Trends Immunol* 2007. 28: 442-448.

[87] Ishikawa, F., Niiro, H., Iino, T., Yoshida, S., Saito, N., Onohara, S., Miyamoto, T., Minagawa, H., Fujii, S. I., Shultz, L. D., Harada, M. and Akashi, K., The developmental program of human dendritic cells is operated independently of conventional myeloid and lymphoid pathways. *Blood* 2007. 110: 3591-3560.

[88] Murayama, E., Kissa, K., Zapata, A., Mordelet, E., Briolat, V., Lin, H. F., Handin, R. I. and Herbomel, P., Tracing hematopoietic precursor migration to successive hematopoietic organs during zebrafish development. *Immunity* 2006. 25: 963-975.

[89] Qian, X., Shen, Q., Goderie, S. K., He, W., Capela, A., Davis, A. A. and Temple, S., Timing of CNS cell generation: a programmed sequence of neuron and glial cell production from isolated murine cortical stem cells. *Neuron* 2000. 28: 69-80.

[90] Raff, M., The mystery of intracellular developmental programmes and timers. *Biochem Soc Trans* 2006. 34: 663-670.

[91] Holt, C. E., Bertsch, T. W., Ellis, H. M. and Harris, W. A., Cellular determination in the Xenopus retina is independent of lineage and birth date. *Neuron* 1988. 1: 15-26.

[92] Turner, D. L. and Cepko, C. L., A common progenitor for neurons and glia persists in rat retina late in development. *Nature* 1987. 328: 131-136.
[93] Wetts, R. and Fraser, S. E., Multipotent precursors can give rise to all major cell types of the frog retina. *Science* 1988. 239: 1142-1145.
[94] Cayouette, M., Barres, B. A. and Raff, M., Importance of intrinsic mechanisms in cell fate decisions in the developing rat retina. *Neuron* 2003. 40: 897-904.

In: Stem Cell Applications in Disease and Health
Editors: W. Burnsides and R. Ellsley
ISBN: 978-1-60456-255-2

Chapter XIII

Incomplete Differentiation of Embryonic Stem Cell: A Risk of Teratoma Formation in Regenerative Medicine

Eva Pericuesta, Miguel A. Ramírez, Raúl Fernández-González and Alfonso Gutiérrez-Adán
Department of Animal Reproduction, INIA, Madrid, Spain

Abstract

The use of embryonic stem (ES) cells in regenerative medicine is one of the most important promises of this decade. The knowledge about culture medium, isolation protocols and pluripotency-related genes is growing, but their therapeutic use is still restricted by a risk of teratoma formation. To test the hypothesis that some undifferentiated cells remain after the differentiation culture process, we have electroporated mouse ES cells with a construct that comprise the mTert promoter coupled to green fluorescent protein (GFP). When ES cell colonies are in culture, they express the green fluorescent protein as a consequence of their telomerase activity. After LIF removal a natural differentiation process into embryoid bodies (EB) occurs, so the GFP signal gradually diminishes due to this differentiation. However, after two weeks of incubation, we could observe small groups of fluorescent cells remaining in the differentiated monolayer. After isolation and expansion of these groups under ES cell culture conditions, ES cell morphology was recovered. Molecular analysis showed that after several passages in ES medium, pluripotency quality characteristics of these cells are completely recovered. According to our results, specific screening of differentiated cells before their therapeutic use should be done to avoid the risk of teratoma formation stimulated by an adequate environment.

Introduction

An embryonic stem (ES) cell is defined as a cell derived from the inner cell mass (ICM) of a blastocyst that has properties of self renewal, indefinite divisions and can differentiate into most adult cell types. They have attracted intense interest because of their potential therapeutic applications (O'Shea 2004). These characteristics have made them become a promising tool to tackle diseases such as Parkinson's, diabetes, chronic heart disease, end-stage kidney disease, liver failure, and cancer (Kirschstein et al., 2001) or in screening potential toxins and drug testing. Thus, if their differentiation could be controlled, cultured ES cells would be an excellent source of cells for transplantation to replace cells lost through disease or injury or cells lacking hormones or genes. Although the therapeutic potential of ES-derived cells is huge, several problems such as teratoma formation have yet to be resolved (Drukker et al. 2002). Each differentiated stem-cell transplant may contain cells that had failed to differentiate, and which remained undifferentiated. These cells keep dividing, and can turn into tumours (Roy et al. 2006). It is well established that pluripotent cells, in general, have a propensity to be more tumorigenic than others (Brickman and Burdon 2002). When ES cells are injected into an ectopic site in immuno-compromised mice they will form a benign tumour (teratoma), comprised of differentiated tissue originating from all three embryonic germ layers. This chapter will focus on the identification of undifferentiated cells after ES cell differentiation, and the particular safety issues surrounding the clinical use of ES cells.

Directed Differentiation of Mice ES Cells

Many laboratories have focused on the differentiation of ES cells, and the literature in this area is extensive (Keller 2005). The main pathway through which ES-derived cells become differentiated cells involves the formation of embryoid bodies "EB", that consist of a disorganized array of differentiated or partially differentiated cell types that are derived from the three primary germ layers of the embryo—endoderm, mesoderm, and ectoderm— and have been described as the developmental equivalent of the egg cylinder-stage mouse embryo, with an outer endodermal layer and a core of differentiating cells (O'Shea 2004). EBs are not part of a normal development and they can be observed only in *in vitro* conditions. Mouse EB form after several days of growth of ES cells in the absence of leukaemia inhibitory factor (LIF). When growth factors are added to the medium for ES cells differentiation, an incomplete differentiation is commonly observed in the EB because the outer endoderm of the EB is exposed to added growth factors and inner cells may be exposed to factors produced by differentiated cells in the EB, like mesoderm cells that produce LIF (O'Shea 2004). Although during differentiation process the pluripotency of the EB is lost and a mixture of multiple cell types is generated, the possibility that some multipotent cells remain causing teratoma exists. It has been reported (Hilario et al., 2001) that in monolayer culture derived from teratocarcinoma EB there are four stable cells which show different morphology of undifferentiated or stem cells. The cell ratio of these four populations does not vary through the culture time (Hilario et al. 2001). This suggests that intrinsic mechanisms of regulation may exist which would strike a balance between differentiation and proliferation as happens with the negative feedback loop of Nanog, Oct4 and FoxD3, that controls stem cell pluripotency and self-renewal (Pan et al. 2006).

Can We Recover ES-Like Cells from Monolayers of Differentiated EBs?

Several laboratories have demonstrated that *in vitro* human ES cells are pluripotent and they can generate cell types derived from the three embryonic germ layers; however, the experimental transfer of ES cells has been related to an unusually high risk of genetic changes and teratoma formation. The study by Ensenat-Waser (Ensenat-Waser et al. 2006) presented the possibility that residual undifferentiated cells can survive after standard *in vitro* procedures. They created a model with Oct4 promoter, which is normally expressed in undifferentiated ES cells but down-regulated on differentiated cells, coupled with the green fluorescent protein. This model allowed them to analyze the differentiation process of a R1 transfected line during the formation of EBs. After 14 days without LIF, residual fluorescent cells still displayed a significant Oct4 expression. This expression was maintained after the expansion of these cells in absence of LIF, showing an intrinsic resistance to differentiate. When these cells were subcutaneously injected into immunodeficient mice, they induced teratomas, suggesting that they possess a limited developmental potency *in vitro* that can be reactivated *in vivo*.

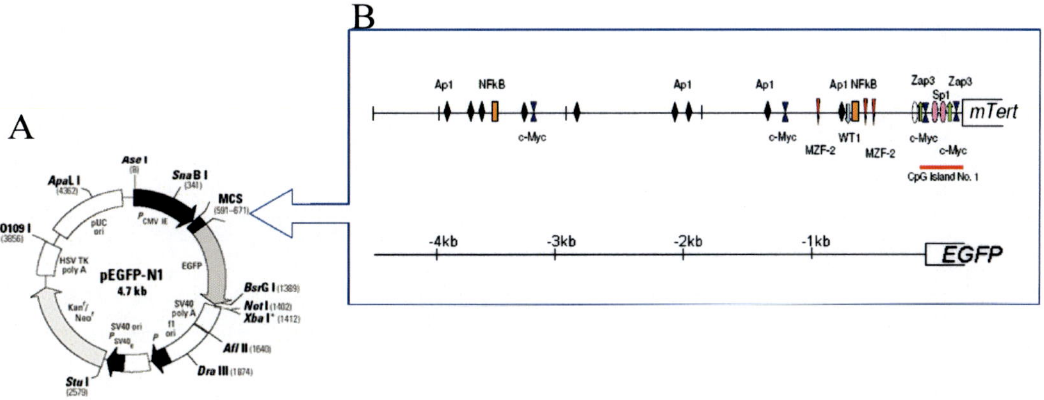

Figure 1. mTert promoter coupled with Green Fluorescent Protein (GFP). (A). Promotor region of mouse Tert was cloned in a plasmid with the Green Fluorescent Protein. (B). Promoter and regulators elements of mouse Tert.

We have development a model using the promoter of telomerase reverse transcriptase (mTert) regulating the expression of GFP in ES cells to test if the use of differentiated cells in regenerative medicine has any risk of teratoma formation. Telomerase is an enzyme complex that maintains the ends of all eukaryotic linear chromosomes. Murine ES cells have been shown to exhibit high telomerase activity, which is strictly regulated during cell differentiation (Albanell et al. 1996; Armstrong et al. 2000). The murine reverse transcriptase unit (mTert) of telomerase is the catalytic subunit of the telomerase complex and is closely linked to telomerase activity (Armstrong et al. 2000; Pericuesta et al. 2006). In a previous work, we reported that ES cell telomerase expression levels induced by the mTert promoter diminish as differentiation progresses (Pericuesta et al. 2006). Using this model, we have also analyzed the presence of pluripotent cells remain undifferentiated in a monolayer derived from differentiated EB (Ramirez et al. 2007). To do so, we transformed R1 ES cell line with a 4.5 Kb fragment of the mTert promoter coupled to pEGFP (Figure 1). Transformed ES cells (R1-mTert-GFP) were able to mimic the expected mTert expression, which was associated to green fluorescence. ES cell telomerase expression levels, induced by the mTert promoter, are higher when EBs are in formation and diminish during the differentiation process. After 2 weeks of differentiation process, small groups of morphologically differentiated and fluorescent cells remained in the differentiated monolayer (Figure 2). To analyze pluripotent characteristics of these green clumps, we isolated and cultured those under stem cell conditioned medium. In just two passages ES-like cells recovered stem cell morphology. We called them ES-EB cells as they were ES-like cells recovered from a differentiated monolayer derived from EBs. Nevertheless, when we picked morphologically differentiated cells that did not express GFP from the differentiated monolayer, after one month of culture in stem cells conditioned media; they did not recover the stem cell morphology.

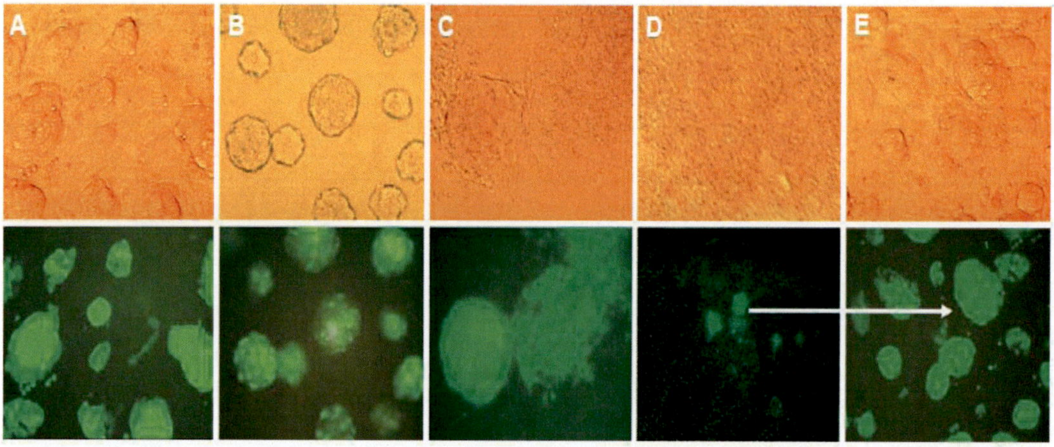

Figure 2. Sequence of micrographs showing embryonic stem (ES) cells transformed with mTert-GFP during differentiation induced by leukemia inhibitory factor (LIF) removal. (Upper photographs were taken under light microscopy at 200x magnification and lower photographs under epifluorescence). (A) ES cell expressing green fluorescent protein (GFP); (B) four days after LIF removal, ES cells plated in non-adherent dishes form embryoid bodies (EB); (C) after 6 days in the absence of LIF, EBs plated in gelatine coated tissue dishes start to differentiate into different cell types showing diminished fluorescence intensity; (D) after 12 days in the absence of LIF, EBs differentiate into monolayer cells and loose fluorescence; in some wells, groups of GFP expressing cells remain among the differentiated cell colonies; (E) these cells expressing GFP under the mTert promoter can be picked up, cultured and expanded under ES cell culture conditions, generating new ES cell-like colonies in the presence of LIF.

Pluripotent Characteristics of Undifferentiated Cells Recovered from Monolayers of Differentiated EBs

To analyze pluripotent characteristics of the isolated ES-EB cells, we tested the expression of pluripotency and differentiation markers by RT-PCR and by immunochemistry as well as the capacity of these ES-EB cells for chimera formation. The pluripotency markers analyzed by RT-PCR were Nanog, Oct3/4, Rex1, GENESIS (Foxd3), FGFR-4, TERF1, Cx43, GLUT1, BCRP1. All these markers are expressed in early embryo development stage and in ES cells. The differentiation markers analyzed, GATA4, GATA2, AFP, Msx-1, Brachyury, Myf5, Keratine 15, HNF3-beta, Nestin, Vimentine, β3 Tubuline, make up different stages of differentiation of the three germ layers. Table 1 shows mRNA expression of pluripotency markers. As pluripotency control we chose the original stem cell line R1 and the transfected R1 line (R1-mTert-GFP). These two ES cell lines express all the pluripotency markers as shown in Table 1. We analyzed 8 of the 23 ES-EB lines obtained from the differentiated monolayer and named them from ES-EB1 to ES-EB8 line. ES-EB8 line maintains all the pluripotency markers as well as the controls. ES-EB7 line maintains all the pluripotency markers except Abcg2. The others ES-EB analyzed lines lost some pluripotency markers. Interestingly Nanog, Ter1 and Foxd3 were detected in almost all the ES-EB lines (Table1). To confirm the pluripotent characteristics of the ES-EB cells, we selected ES-EB2 line at early passage to analyze the pluripotency markers Nanog and SSEA-1 by immunochemistry. Results showed a homogeneous expression of these two markers (Figure 3).

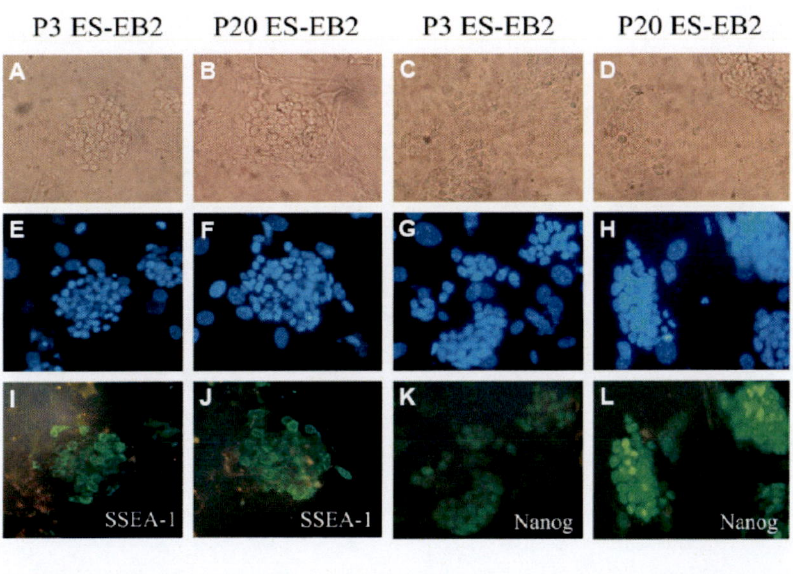

Figure 3. Expression of Nanog and SSEA-1 in an ES-EB line at passages 3 and 20 (A,B,C,D) Phase contrast microphotographs. (E,F,G,H) Nuclear stain (DAPI). Membrane SSEA-1 immunofluorescence **(I,J)** and nuclear Nanog immunofluorescence **(K,L)** are detected in the ES-EB line at early and late passage (200x magnification).

Table 1. mRNA expression of pluripotency markers.

	R1	R1-mTert-GFP	ES-EB1		ES-EB2		ES-EB3		ES-EB4		ES-EB5	ES-EB6	ES-EB7	ES-EB8
			p2	P7	p2	P10	p2	P7	p2	P10	p3	p3	p3	p3
Nanog	+	+	+	+	+	+	+	+	+	+	+	+	+	+
Oct3/4	+	+	-	+	-	+	-	+	-	+	-	-	+	+
Rex1	+	+	-	+	-	+	-	+	-	+	-	-	+	+
Foxd3	+	+	-	+	+	+	+	+	+	+	+	+	+	+
Fgfr4	+	+	-	+	-	+	-	+	+	+	+	+	+	+
Terf1	+	+	+	+	+	+	+	+	+	+	+	+	+	+
Cx43	+	+	-	+	-	+	+	+	+	+	+	+	+	+
Glut1	+	+	-	+	-	+	+	+	-	+	+	+	+	+
Abcg2	+	+	-	-	-	-	-	-	-	-	-	-	-	+

Table 2. mRNA expression of markers of differentiated phenotypes

	R1	R1-mTert-GFP	ES-EB1		ES-EB2		ES-EB3		ES-EB4		ES-EB5	ES-EB6	ES-EB7	ES-EB8
			p2	P7	p2	P10	p2	P7	p2	P10	p3	p3	p3	p3
Gata-4	-	-	+	-	+	-	-	-	-	-	-	-	-	-
Gata-2	+	+	+	+	+	+	+	+	+	+	+	+	+	+
Afp	-	-	+	-	-	-	-	-	-	-	-	-	-	-
Msx1	-	-	+	-	+	-	-	-	-	-	+	-	-	+
T	-	-	+	-	-	-	-	-	-	-	-	-	-	-
Myf5	-	-	+	-	-	-	+	-	+	-	-	+	+	-
Krt15	-	-	+	-	+	-	+	-	+	-	+	-	-	-
Foxa2	-	-	+	+	+	-	+	-	+	-	+	+	+	-
Nes	-	-	+	-	+	-	-	-	-	-	-	-	-	-
Vim	-	-	+	-	+	-	-	-	-	-	-	-	-	-
Tubb3	-	-	+	-	+	-	+	-	+	-	-	+	-	-
Gapdh	+	+	+	+	+	+	+	+	+	+	+	+	+	+

Table 2 shows the mRNA expression markers of differentiated phenotypes of the analyzed cell lines. Controls R1 and R1-mTert-GFP only express the differentiation marker GATA2, which is expressed by the inactivated fibroblasts used for murine stem cell culture. Lines ES-EB8 and ES-EB7, which express almost all the pluripotency markers, express only three of the differentiation markers. The other ES-EB analyzed lines have different mRNA expression of these markers, indicating different stages of differentiation.

Attending to the fact that only Nanog, Foxd3 and Tert1 were detected in all ES-EB cells, we propose these three markers to identify the inadvertent presence of pluripotent cell stages among differentiated cells. Nanog, a homeobox transcription factor, is a pluripotent cell specific gene that plays a crucial role in maintaining the undifferentiated state of early postimplantation embryos and ES cells. Nanog mRNA is present in pluripotent mouse and human cell lines and absent from differentiated cells (Chambers et al. 2003). Two important properties conferred by Nanog have been described: a) a fundamental role in pluripotency of both ICM and ES cells and b) an ability to maintain ES self-renewal in the absence of LIF, or in absence of Oct3/4 and Stat3 expression (Chambers et al. 2003; Mitsui et al. 2003). Our results are in agreement with the importance of Nanog as a master transcriptional organizer and suggest that even in Oct4 absence, Tert, Foxd3 and Nanog may control different but partly overlapping pools of target genes. Foxd3 is a fork-head transcription factor highly expressed in mouse ES cells. In agreement with our results, it has been recently published that Oct4, Nanog and Foxd3 form a loop anchors, an interdependent network of transcription factors that regulate stem cell pluripotency (Pan et al. 2006). The authors reported that Nanog and Foxd3 were able to activate Oct4 promoter. Ectopic expression of Ter1 in mouse ES cells does not affect differentiation but confers resistance to differentiation- and stress-induced p53-dependent apoptosis (Lee et al. 2005). These recent data suggest that the decreased telomerase activity observed during differentiation is a consequence of the differentiation process, during which the pluripotency of the ES cell is lost. It has been also demonstrated that over-expression of telomerase confers a growth advantage, stress resistance and enhanced differentiation of ES cells towards the hematopoietic linage (Armstrong et al. 2005).

Appropriate Culture Recuperates Full Pluripotent of Undifferentiated Cells Recovered from Monolayers of Differentiated EBs

To determine if ES-EB lines were able to recover all the pluripotency characteristics after an adequate culture in ES cell media, we analyzed the capacity for chimera formation at different cell passages. We selected three of the ES-EB lines at early passages: ES-EB1 line that had lost most of the pluripotency markers, ES-EB2 that had lost the majority of pluripotency markers but kept Nanog, Terfl, Foxd3, and ES-EB3 line that expressed the majority of pluripotency markers except Oct4, Rex1, Fgrt4. We injected and aggregated more than 200 blastocysts and eight cells stage embryos and we only obtained one chimera from ES-EB2 line with less than 25% coat colour chimerism. These results showed the limited capacity for chimera formation of these original ES-like cells. However, after 7-9 passages,

ES-EB1-4 lines, which had lost the expression of most pluripotency markers, recovered it (Table 1). The only two markers that were lost in the four ES-EB lines, Oct3/4 and Rex1, were recovered in all the lines. The expression of differentiation markers after culture in ES cells conditioned media was over, showing more pluripotent state of these ES-EB cells (Table 2). By immunochemistry we confirmed that ES-EB2 line after thirteen passages in ES cells conditioned medium maintained a homogeneous expression of the pluripotent markers SSEA-1 and Nanog (Fig. 3). To analyze if the capacity for chimera formation of the ES-EB lines had increased after culturing in ES cells conditions, we selected ES-EB2 line, which had lost the majority of pluripotency markers but kept Nanog, Terfl and FoxD3 before culturing in conditioned ES cells media, and ES-EB3 line that expressed the majority of pluripotency markers except Oct4, Rex1, Fgrt4. After 9 passages, these two lines recovered all the pluripotency markers. We obtained 8 chimeras out of the 21 born alive from 212 injected and aggregated embryos with ES-EB2 line, and 5 chimeras out of the 13 born alive from 153 blastocysts injected or aggregated with the ES-EB3 line. The level of chimerism observed was similar to that obtained with the original R1-mTert-GFP ES line, ranging from 15-50% based on coat colour contribution. These results indicate that the ES-EB cells had recovered the pluripotential nature of the original ES cell line.

In the adult, stem cells persist in some tissues such as hematopoietic, gastrointestinal, testicular tissue and the integumentary system and carry out lifelong processes of regeneration and renewal. A recent study has reported that skin-derived adult stem cells are capable of recalling the program for differentiating into an oocyte (Dyce et al. 2006). It has been suggested that the differentiation of adult stem cells is incomplete and lacks certain cues necessary to acquire a truly functional status (Belema Bedada et al. 2005). Such that the multipotency potential observed for spermatogonial stem cells or skin cells may actually be imposed by the culture conditions rather than reflect the natural behaviour of a stem cell in a given niche. Just as stem cells depend on their environment to maintain their properties, the culture environment may also modulate the stem cell properties with some degree of multipotency (Ramirez et al. 2006). In our experiments, due to specific culture conditions, ES-EB cells acquired the remarkable potential to turn on a large variety of markers specific for ES cells. Even cell lines that had lost the expression of Oct-4 were able to recover it and then recover also the potential to form chimeras. We propose that during the differentiation protocol the mTert-GFP positive cells that we selected after two weeks of differentiation still retain some pluripotency potential that is stimulated by an adequate ES cells culture conditions. These findings and the risk of teratoma formation suggest that more specific analysis of markers are required to confirm the loss of multipotent characteristics of the cells before being used for regenerative therapy. We must consider that an adequate environment may reprogram cells with memory for pluripotency that could cause a tumour.

The mTert-GFP model has allowed us to identify cells with a level of multipotency or of memory factors (or epi-memory factors) enough for them to recover their pluripotency. We do not know if some differentiated cells derived from human ES cells lines retain this mouse ES cells ability to regain pluripotency. However, to transplant even a small number of incompletely differentiated cells represents a serious risk of including cells with tumour-forming properties into patients. We propose the analysis of the markers Ter1 (survival enzyme marker), Nanog and Foxd3 (pluripotency control markers) in order to define ES cell identity. These results suggest that any adult or foetal stem cell that has a multipotent memory may be reprogrammed by culturing *in vitro* to acquire pluripotency characteristics similar to those of ES cells. Cells with a genetic or epigenetic memory of a multipotent development stage can be reprogrammed to confer an ES-like pluripotency state (with a capacity for chimera production). Moreover, it is important to verify the presence of specific pluripotency markers in the homogeneous monolayer of differentiated cells used for cell therapy purposes, in order to avoid the possible transfer of a significant proportion of undifferentiated cells with tumourogenic properties.

Acknowledgments

This work was supported by the Grant AGL2006-04799 from the Spanish Ministry of Science and Technology. Eva Pericuesta was supported by a FPI grant from the Spanish Ministry of Education and Science.

References

Albanell J, Han W, Mellado B, Gunawardane R, Scher HI, Dmitrovsky E, Moore MA. 1996. Telomerase activity is repressed during differentiation of maturation-sensitive but not resistant human tumor cell lines. *Cancer research* 56(7):1503-1508.

Armstrong L, Lako M, Lincoln J, Cairns PM, Hole N. 2000. mTert expression correlates with telomerase activity during the differentiation of murine embryonic stem cells. *Mechanisms of development* 97(1-2):109-116.

Armstrong L, Saretzki G, Peters H, Wappler I, Evans J, Hole N, von Zglinicki T, Lako M. 2005. Overexpression of telomerase confers growth advantage, stress resistance, and enhanced differentiation of ESCs toward the hematopoietic lineage. *Stem cells* (Dayton, Ohio) 23(4):516-529.

Belema Bedada F, Technau A, Ebelt H, Schulze M, Braun T. 2005. Activation of myogenic differentiation pathways in adult bone marrow-derived stem cells. *Molecular and cellular biology* 25(21):9509-9519.

Brickman JM, Burdon TG. 2002. Pluripotency and tumorigenicity. Nature genetics 32(4):557-558.

Chambers I, Colby D, Robertson M, Nichols J, Lee S, Tweedie S, Smith A. 2003. Functional expression cloning of Nanog, a pluripotency sustaining factor in embryonic stem cells. *Cell* 113(5):643-655.

Drukker M, Katz G, Urbach A, Schuldiner M, Markel G, Itskovitz-Eldor J, Reubinoff B, Mandelboim O, Benvenisty N. 2002. Characterization of the expression of MHC proteins in human embryonic stem cells. *Proceedings of the National Academy of Sciences of the United States of America* 99(15):9864-9869.

Dyce PW, Wen L, Li J. 2006. In vitro germline potential of stem cells derived from fetal porcine skin. *Nature cell biology* 8(4):384-390.

Ensenat-Waser R, Santana A, Vicente-Salar N, Cigudosa JC, Roche E, Soria B, Reig JA. 2006. Isolation and characterization of residual undifferentiated mouse embryonic stem cells from embryoid body cultures by fluorescence tracking. *In vitro cellular & developmental biology* 42(5-6):115-123.

Hilario E, Alvarez A, Simon J, Garcia-Sanz M, Lacalle J, Arechaga J. 2001. Presence of four stem cell populations in monolayer cultures derived from teratocarcinoma embryoid bodies. *In vivo (Athens, Greece)* 15(3):217-226.

Keller G. 2005. Embryonic stem cell differentiation: emergence of a new era in biology and medicine. *Genes & development* 19(10):1129-1155.

Kirschstein, et al. (2001). "Stem Cells: Scientific Progress and Future Reasearch Directions. National Institutes of Health. Terese Winslow.

Lee MK, Hande MP, Sabapathy K. 2005. Ectopic mTERT expression in mouse embryonic stem cells does not affect differentiation but confers resistance to differentiation- and stress-induced p53-dependent apoptosis. *Journal of cell science* 118(Pt 4):819-829.

Mitsui K, Tokuzawa Y, Itoh H, Segawa K, Murakami M, Takahashi K, Maruyama M, Maeda M, Yamanaka S. 2003. The homeoprotein Nanog is required for maintenance of pluripotency in mouse epiblast and ES cells. *Cell* 113(5):631-642.

O'Shea KS. 2004. Self-renewal vs. differentiation of mouse embryonic stem cells. *Biology of reproduction* 71(6):1755-1765.

Pericuesta E, Ramirez MA, Villa-Diaz A, Relano-Gines A, Torres JM, Nieto M, Pintado B, Gutierrez-Adan A. 2006. The proximal promoter region of mTert is sufficient to regulate telomerase activity in ES cells and transgenic animals. *Reprod Biol Endocrinol* 4:5.

Ramirez MA, Pericuesta E, Fernandez-Gonzalez R, Moreira P, Pintado B, Gutierrez-Adan A. 2006. Transcriptional and post-transcriptional regulation of retrotransposons IAP and MuERV-L affect pluripotency of mice ES cells. *Reprod Biol Endocrinol* 4:55.

Ramirez MA, Pericuesta E, Fernandez-Gonzalez R, Pintado B, Gutierrez-Adan A. 2007. Inadvertent presence of pluripotent cells in monolayers derived from differentiated embryoid bodies. *The International journal of developmental biology* 51(5):397-407.

Roy NS, Cleren C, Singh SK, Yang L, Beal MF, Goldman SA. 2006. Functional engraftment of human ES cell-derived dopaminergic neurons enriched by coculture with telomerase-immortalized midbrain astrocytes. *Nature medicine* 12(11):1259-1268.

Index

B

C

E

F

G

H

I

J

K

L

M

N

O

P

Q

R

S

T

U

V

W

X

Y

Z